Experiments in Modern Electronics

Third Edition

W. Marshall Leach, Jr.
Thomas E. Brewer
Georgia Institute of Technology

KENDALL/HUNT PUBLISHING COMPANY
4050 Westmark Drive   Dubuque, Iowa 52002

Copyright © 1999, 2000, 2006 by W. Marshall Leach, Jr. and Thomas E. Brewer.

ISBN 13: 978-0-7575-3174-3
ISBN 10: 0-7575-3174-1

Kendall/Hunt Publishing Company has the exclusive rights to reproduce this work,
to prepare derivative works from this work, to publicly distribute this work,
to publicly perform this work and to publicly display this work.

All rights reserved. No part of this publication may be reproduced,
stored in a retrieval system, or transmitted, in any form or by any
means, electronic, mechanical, photocopying, recording, or otherwise,
without the prior written permission of the copyright owner.

Printed in the United States of America
10 9 8 7 6 5 4 3 2 1

Contents

1 A Single Stage Common-Emitter BJT Amplifier — 1
  1.1 Object — 1
  1.2 Theory — 1
    1.2.1 Voltage Gain — 2
    1.2.2 Biasing — 3
  1.3 Example Design — 5
  1.4 SPICE Simulation — 6
  1.5 Procedure — 7
    1.5.1 Breadboard Preparation — 7
    1.5.2 Transistor Leads — 9
    1.5.3 Circuit Assembly — 9
    1.5.4 Quiescent dc Bias Point — 9
    1.5.5 Small-Signal Amplifier — 10
    1.5.6 Frequency Response — 10
    1.5.7 Large-Signal Clipping Behavior — 10
    1.5.8 Spectral Analysis — 11
  1.6 Laboratory Report — 11
    1.6.1 Bias Values — 11
    1.6.2 Small-Signal Gain — 11
    1.6.3 Frequency Response — 11
    1.6.4 Large-Signal Clipping Behavior — 12
    1.6.5 Spectral Analysis — 12
  1.7 References — 12

2 Basic Op-Amp Circuits 1 — 13
  2.1 Object — 13
  2.2 Theory — 13
    2.2.1 Voltage and Current Notation — 13
    2.2.2 The Ideal Op-Amp — 13
    2.2.3 Voltage-Gain Transfer Function of Physical op amps — 14
    2.2.4 Non-Inverting Amplifier — 14
    2.2.5 Voltage-Gain Transfer Function of Non-Inverting Amplifier — 15
    2.2.6 Inverting Amplifier — 15
    2.2.7 Voltage-Gain Transfer Function of Inverting Amplifier — 16
    2.2.8 Inverting Amplifier with T Feedback Network — 17
    2.2.9 Voltage-Gain Transfer Function of Inverting Amplifier with T Feedback Network — 17
    2.2.10 Integrator — 18
    2.2.11 Practical Integrator — 19
    2.2.12 Differentiator — 20
    2.2.13 Practical Differentiator — 20

- 2.3 SPICE Op-Amp Macromodel Circuits 21
  - 2.3.1 Quasi-Ideal Macromodel 21
  - 2.3.2 Small-Signal AC Macromodel 22
- 2.4 Preliminary Derivations 22
- 2.5 Preliminary Calculations 23
  - 2.5.1 Non-Inverting Amplifier 23
  - 2.5.2 Inverting Amplifier 23
  - 2.5.3 Inverting Amplifier with T Feedback Network 23
  - 2.5.4 Integrator 23
  - 2.5.5 Differentiator 23
- 2.6 Preliminary SPICE Simulations 23
  - 2.6.1 Non-Inverting Amplifier 23
  - 2.6.2 Inverting Amplifier 24
  - 2.6.3 Inverting Amplifier with T Feedback Network 24
  - 2.6.4 Integrator 24
  - 2.6.5 Differentiator 24
- 2.7 Experimental Procedures 24
  - 2.7.1 Preparation 24
  - 2.7.2 Non-Inverting Amplifier 25
  - 2.7.3 Inverting Amplifier 26
  - 2.7.4 Inverting Amplifier with T Feedback Network 26
  - 2.7.5 Integrator 27
  - 2.7.6 Differentiator 28
- 2.8 Laboratory Report 29
- 2.9 References 29

3 Elementary Op-Amps 2 — 31
- 3.1 Object 31
- 3.2 Theory 31
  - 3.2.1 Peak Clipping 31
  - 3.2.2 Output Current Limiting 32
  - 3.2.3 Slew Rate Limiting 32
  - 3.2.4 DC Offset and Bias Currents 32
  - 3.2.5 Band-Pass Inverting Amplifier 33
  - 3.2.6 Non-inverting Amplifier with 100% DC Feedback 34
  - 3.2.7 Single-Power Supply Amplifiers 34
- 3.3 SPICE Macromodels 36
- 3.4 Preliminary Derivations 37
- 3.5 Preliminary Calculations 38
- 3.6 Preliminary SPICE Simulations 38
  - 3.6.1 Nonideal Properties of Op-Amp 38
  - 3.6.2 Inverting and Noninverting Amplifiers 39
- 3.7 Experimental Procedures 39
  - 3.7.1 Preparation 39
  - 3.7.2 Peak Clipping 39
  - 3.7.3 Current Limiting 39
  - 3.7.4 Slewing 40
  - 3.7.5 DC Offset and Bias Currents 40
  - 3.7.6 Inverting Amplifier 40
  - 3.7.7 Non-Inverting Amplifier 40
  - 3.7.8 Single Power Supply Amplifiers 41
- 3.8 Laboratory Report 41

	3.9 References	41

4 Differential and Instrumentation Amplifier — 43
- 4.1 Object … 43
- 4.2 Theory … 43
  - 4.2.1 Differential Amplifier … 43
  - 4.2.2 Instrumentation Amplifier … 44
  - 4.2.3 Op-Amp Curve Tracer … 45
- 4.3 Preliminary Derivations … 46
  - 4.3.1 Differential Amplifier … 46
  - 4.3.2 Instrumentation Amplifier … 46
- 4.4 Preliminary Calculations … 46
  - 4.4.1 Differential Amplifier … 46
  - 4.4.2 Instrumentation Amplifier … 47
- 4.5 Preliminary SPICE Simulations … 47
  - 4.5.1 Differential Amplifier … 47
  - 4.5.2 Instrumentation Amplifier … 47
- 4.6 Experimental Procedures … 47
  - 4.6.1 Preparation … 47
  - 4.6.2 Differential Amplifier … 48
  - 4.6.3 Instrumentation Amplifier … 49
  - 4.6.4 Curve Tracer—Differential Amplifier … 50
- 4.7 Laboratory Report … 50
- 4.8 References … 51

5 Non-Linear Op-Amp Circuits — 53
- 5.1 Object … 53
- 5.2 Theory … 53
  - 5.2.1 Half-Wave Precision Rectifier … 53
  - 5.2.2 Full-Wave Precision Rectifiers … 54
  - 5.2.3 Peak-Detector Circuit … 57
  - 5.2.4 Comparators … 57
- 5.3 Preliminary Derivations … 58
  - 5.3.1 Precision Half-Wave Rectifier … 58
  - 5.3.2 First Precision Full-Wave Rectifier … 58
  - 5.3.3 Second Precision Full-Wave Rectifier … 58
  - 5.3.4 Third Precision Full-Wave Rectifier … 60
  - 5.3.5 Fourth Precision Full-Wave Rectifier … 60
- 5.4 Preliminary Calculations … 60
  - 5.4.1 Peak-Hold Circuit … 60
  - 5.4.2 Half-Wave Rectifier … 60
  - 5.4.3 Full-Wave Rectifier … 60
- 5.5 Preliminary SPICE Simulations … 60
  - 5.5.1 Transient Analysis … 60
  - 5.5.2 DC Sweep … 60
- 5.6 Experimental Procedures … 61
  - 5.6.1 Preparation … 61
  - 5.6.2 Half-Wave Rectifier … 61
  - 5.6.3 Full-Wave Rectifier … 61
  - 5.6.4 Peak Hold Circuit … 62
  - 5.6.5 Bar-Graph Array … 62
  - 5.6.6 Level Indicator … 63

	5.7 Laboratory Report	63
	5.8 References	63

6 Active Filters — 65

- 6.1 Object . . . 65
- 6.2 Introduction . . . 65
- 6.3 Classes of Filter Functions . . . 65
- 6.4 Transfer Functions . . . 66
- 6.5 Frequency Transformations . . . 68
  - 6.5.1 Transformations of First-Order Functions . . . 68
  - 6.5.2 Transformations of Second-Order Functions . . . 68
- 6.6 Butterworth Transfer Functions . . . 69
  - 6.6.1 Even-Order Butterworth Filters . . . 71
  - 6.6.2 Odd-Order Butterworth Filters . . . 71
- 6.7 Chebyshev Filter Transfer Functions . . . 72
  - 6.7.1 The Chebyshev Approximation . . . 72
  - 6.7.2 The dB Ripple . . . 74
  - 6.7.3 The Cutoff Frequency . . . 74
  - 6.7.4 The Parameter h . . . 75
  - 6.7.5 Even-Order Chebyshev Filters . . . 75
  - 6.7.6 Odd-Order Chebyshev Filters . . . 76
- 6.8 Elliptic Filter Transfer Functions . . . 77
- 6.9 The Thompson Phase Approximation . . . 79
- 6.10 First-Order Filter Topologies . . . 80
  - 6.10.1 First-Order Low-Pass Filter . . . 80
  - 6.10.2 First-Order High-Pass Filter . . . 82
- 6.11 Second-Order Filter Topologies . . . 82
  - 6.11.1 Sallen-Key Low-Pass Filter . . . 82
  - 6.11.2 Infinite-Gain Multi-Feedback Low-Pass Filter . . . 84
  - 6.11.3 Sallen-Key High-Pass Filter . . . 85
  - 6.11.4 Infinite-Gain Multi-Feedback High-Pass Filter . . . 86
  - 6.11.5 Sallen-Key Band-Pass Filter . . . 86
  - 6.11.6 Infinite-Gain Multi-Feedback Band-Pass Filter . . . 87
  - 6.11.7 Second-Order Band-Pass Filter Bandwidth . . . 87
  - 6.11.8 A Biquad Filter . . . 87
  - 6.11.9 A Second Biquad Filter . . . 90
  - 6.11.10 State-Variable Filter . . . 90
- 6.12 Third-Order Sallen-Key Filter Circuits . . . 92
  - 6.12.1 Low-Pass Filters . . . 92
  - 6.12.2 High-Pass Filters . . . 93
- 6.13 Impedance Transfer Functions . . . 93
  - 6.13.1 RC Network . . . 93
  - 6.13.2 RL Network . . . 94
- 6.14 Voltage Divider Transfer Functions . . . 94
  - 6.14.1 RC Network . . . 94
  - 6.14.2 High-Pass RC Network . . . 95
  - 6.14.3 RL Network . . . 95
  - 6.14.4 High-Pass RL Network . . . 95
- 6.15 Preliminary Derivations . . . 96
- 6.16 Preliminary Calculations . . . 96
- 6.17 Preliminary SPICE Simulations . . . 97
- 6.18 Experimental Procedures . . . 97

		6.18.1 Preparation	97

                6.18.1 Preparation 97
                6.18.2 Low-Pass Filter 97
                6.18.3 High-Pass Filter 98
                6.18.4 Band-Pass Filter 98
                6.18.5 Band-Reject Filter 98
                6.18.6 State-Variable Filter 98
        6.19 Laboratory Report 98
        6.20 References 98

7 Characteristics of Active Devices   99
        7.1   Object 99
        7.2   The Bipolar Junction Transistor 99
                7.2.1   Large-Signal Model 99
                7.2.2   Active-Mode Equations 100
                7.2.3   Output Characteristics 100
                7.2.4   Transfer Characteristics 102
                7.2.5   SPICE Input Decks 103
        7.3   The Junction Field Effect Transistor 103
                7.3.1   Saturation-Mode Equations 103
                7.3.2   Triode- or Linear-Mode Equations 104
                7.3.3   Saturation-Linear Boundary 104
                7.3.4   Output Characteristics 104
                7.3.5   Transfer Characteristics 106
                7.3.6   SPICE Input Decks 106
                7.3.7   Measuring β, I_{DSS}, and V_{TO} 107
                7.3.8   Bias Considerations 107
        7.4   Metal Oxide Semiconductor Field Effect Transistor 108
                7.4.1   Circuit Symbols 108
                7.4.2   Large-Signal Device Equations 108
                7.4.3   Bulk-to-Source Bias Voltage 109
                7.4.4   Linear-Saturation Boundary 109
                7.4.5   Comparison with the JFET 110
                7.4.6   Output Characteristics 110
                7.4.7   Transfer Characteristics 111
                7.4.8   SPICE Input Decks 111
                7.4.9   P-Channel Device Equations 112
        7.5   Preliminary Derivations 112
                7.5.1   BJT 112
                7.5.2   MOSFET 112
        7.6   Preliminary Calculations 112
        7.7   Preliminary SPICE Simulations 113
                7.7.1   BJT 113
                7.7.2   JFET 113
                7.7.3   MOSFET 113
        7.8   Procedure 113
                7.8.1   Safety Precautions 114
                7.8.2   NPN BJT Output Characteristic 114
                7.8.3   NPN BJT Transfer Characteristic 116
                7.8.4   PNP BJT Output Characteristic 116
                7.8.5   PNP BJT Transfer Characteristic 117
                7.8.6   N Channel Enhancement Mode MOSFET 118
                7.8.7   P Channel Enhancement Mode MOSFET 120

         7.8.8   N Channel JFET 121
         7.8.9   Safety Precautions 123
         7.8.10 NPN BJT Output Characteristic 124
         7.8.11 NPN BJT Transfer Characteristic 125
         7.8.12 PNP BJT Output Characteristic 126
         7.8.13 PNP BJT Transfer Characteristic 127
         7.8.14 N Channel Enhancement Mode MOSFET 127
         7.8.15 P Channel Enhancement Mode MOSFET 129
         7.8.16 N Channel JFET 130
    7.9 Laboratory Report 132
         7.9.1   NPN BJT 132
         7.9.2   PNP BJT 133
         7.9.3   N Channel Enhancement Mode MOSFET 133
         7.9.4   P Channel Enhancement Mode MOSFET 133
         7.9.5   N Channel JFET 134
    7.10 References 134

8 Digital Electronic Circuits   137
    8.1 Object 137
    8.2 Theory 137
         8.2.1   Logic Variables and Voltage Levels 138
         8.2.2   Ideal inverter 138
         8.2.3   Physical Inverter 139
         8.2.4   BJT Inverter 143
         8.2.5   IC Digital Electronic Circuits 145
         8.2.6   Standard TTL Logic Family 148
         8.2.7   Modes of Operation of the BJT 148
         8.2.8   Standard 7404 TTL Inverter 152
         8.2.9   Standard TTL NAND Gate 155
         8.2.10 Low-Power Schottky TTL 155
         8.2.11 CMOS Inverter 159
         8.2.12 Ring Oscillator 161
    8.3 Preliminary Derivations 162
    8.4 Preliminary Calculations 162
    8.5 Preliminary SPICE Simulations 162
    8.6 Experimental Procedures 162
         8.6.1   Preparation 162
         8.6.2   BJT Inverter 162
         8.6.3   Standard TTL Inverter 163
         8.6.4   74LS TTL Inverter 164
         8.6.5   CMOS Inverter 166
    8.7 Laboratory Report 167
    8.8 References 167

9 Analog to Digital and Digital to Analog Conversion Systems   169
    9.1 Object 169
    9.2 Theory 169
         9.2.1   Introduction 169
         9.2.2   Analog-to-Digital Converter Fundamentals 169
         9.2.3   Digital-to-Analog Converters 172
         9.2.4   Analog-to-Digital Converters (A/D) 174
         9.2.5   Components of A/D Converters 181

		9.2.6 Bipolar DAC	184

- 9.3 Preliminary Derivations ... 185
- 9.4 Preliminary Calculations ... 185
- 9.5 Preliminary SPICE Simulations ... 186
- 9.6 Experimental Procedures ... 186
  - 9.6.1 Power Supply ... 186
  - 9.6.2 Breadboard ... 186
  - 9.6.3 Clock ... 187
  - 9.6.4 Four Bit Binary Counter ... 187
  - 9.6.5 Four Bit D/A Converter ... 188
  - 9.6.6 Four Bit A/D System ... 189
  - 9.6.7 Eight Bit Binary Counter ... 192
  - 9.6.8 Eight Bit D/A Converter ... 193
  - 9.6.9 Eight Bit ADC System ... 194
- 9.7 Laboratory Report ... 194
- 9.8 References ... 195

10 The Bipolar Junction Transistor — 197

- 10.1 Object ... 197
- 10.2 Theory ... 197
  - 10.2.1 Notation ... 197
  - 10.2.2 Device Equations ... 197
  - 10.2.3 Transfer and Output Characteristics ... 198
  - 10.2.4 Hybrid-π Model ... 200
  - 10.2.5 T Model ... 201
  - 10.2.6 Simplified T Model ... 201
  - 10.2.7 Norton Collector Circuit ... 202
  - 10.2.8 Thévenin Emitter Circuit ... 204
  - 10.2.9 Thévenin Base Circuit ... 205
  - 10.2.10 Summary of Models ... 206
- 10.3 Small-Signal High-Frequency Models ... 206
- 10.4 Preliminary Derivations ... 208
- 10.5 Preliminary SPICE Simulations ... 208
- 10.6 Experimental Procedures ... 208
  - 10.6.1 Junction Resistance Measurement ... 208
  - 10.6.2 Preparation ... 209
  - 10.6.3 Threshold Measurement ... 209
  - 10.6.4 Junction Voltage Measurement ... 210
  - 10.6.5 Verge of Saturation Measurement ... 210
  - 10.6.6 Hard Saturation ... 210
  - 10.6.7 Early Voltage Measurement ... 211
  - 10.6.8 Saturation Current Measurement ... 211
  - 10.6.9 Current Gain Measurement ... 211
  - 10.6.10 PNP Transistor Measurement ... 211
  - 10.6.11 Curve Tracer ... 211
- 10.7 **Laboratory Report** ... 211
- 10.8 References ... 212

11 The Common-Emitter Amplifier — 213
- 11.1 Object — 213
- 11.2 Theory — 213
  - 11.2.1 Circuit Description — 214
  - 11.2.2 dc Bias Equation — 214
  - 11.2.3 Midband Small-Signal Voltage Gain — 215
  - 11.2.4 Approximate Voltage-Gain Expressions — 216
  - 11.2.5 Modeling the Effect of r_0 — 217
  - 11.2.6 Input Resistance — 217
  - 11.2.7 Output Resistance — 218
  - 11.2.8 Design Criteria — 218
  - 11.2.9 Biasing for Equal Voltage Drops — 218
  - 11.2.10 Biasing for Symmetrical Clipping — 218
  - 11.2.11 Lower Cutoff Frequency — 220
  - 11.2.12 Upper Cutoff Frequency — 221
- 11.3 Preliminary Derivations — 222
- 11.4 Preliminary Calculations — 222
- 11.5 Preliminary SPICE Simulations — 222
- 11.6 Experimental Procedures — 222
  - 11.6.1 Preparation — 223
  - 11.6.2 Bias Measurement — 223
  - 11.6.3 Gain Measurement — 223
  - 11.6.4 Clipping — 224
  - 11.6.5 Input Resistance Measurement — 224
  - 11.6.6 Output Resistance Measurement — 224
  - 11.6.7 Symmetric Clipping — 224
- 11.7 Laboratory Report — 224
- 11.8 References — 225

12 The Common-Base and Cascode Amplifiers — 227
- 12.1 Object — 227
- 12.2 Theory — 227
  - 12.2.1 The Common-Base Amplifier — 227
  - 12.2.2 The Cascode Amplifier — 231
- 12.3 Preliminary Derivations — 234
- 12.4 Preliminary Calculations — 234
- 12.5 Preliminary SPICE Simulations — 234
- 12.6 Experimental Procedures — 235
  - 12.6.1 Preparation — 235
  - 12.6.2 Bias Measurement — 235
  - 12.6.3 Gain Measurement — 235
  - 12.6.4 Clipping — 236
  - 12.6.5 Design Circuit — 236
  - 12.6.6 Cascode Amplifier — 236
- 12.7 Laboratory Report — 236
- 12.8 References — 236

13 The Common-Collector Amplifier — 239
- 13.1 Object . . . 239
- 13.2 Theory . . . 239
  - 13.2.1 The Common-Collector Amplifier . . . 240
  - 13.2.2 Cascade CE-CC Amplifier . . . 241
- 13.3 Preliminary Derivations . . . 244
- 13.4 Preliminary Calculations . . . 244
- 13.5 Preliminary SPICE Simulations . . . 244
- 13.6 Experimental Procedures . . . 245
  - 13.6.1 Preparation . . . 245
  - 13.6.2 Bias Measurement . . . 245
  - 13.6.3 Gain Measurement . . . 245
  - 13.6.4 Clipping . . . 246
  - 13.6.5 Input Resistance Measurement . . . 246
  - 13.6.6 Output Resistance Measurement . . . 246
  - 13.6.7 Design Circuits . . . 246
  - 13.6.8 Impedance Buffer . . . 246
  - 13.6.9 Common-Emitter Amplifier . . . 246
- 13.7 Laboratory Report . . . 247
- 13.8 References . . . 247

14 The Junction Field Effect Transistor — 249
- 14.1 Object . . . 249
- 14.2 Theory . . . 249
  - 14.2.1 Terminal Characteristics . . . 249
  - 14.2.2 Output Characteristics Curve Tracer . . . 253
  - 14.2.3 Transfer Characteristics Curve Tracer . . . 254
  - 14.2.4 Common Source Amplifier . . . 255
  - 14.2.5 Voltage Controlled Attenuator . . . 257
  - 14.2.6 Chopper Modulator . . . 259
- 14.3 Preliminary Derivations . . . 261
- 14.4 Preliminary Calculations . . . 261
- 14.5 Preliminary SPICE Simulations . . . 261
- 14.6 Experimental Procedures . . . 261
  - 14.6.1 Preparation . . . 261
  - 14.6.2 Output Characteristics Curve Tracer . . . 261
  - 14.6.3 Transfer Characteristics Curve Tracer . . . 262
  - 14.6.4 Common Source Amplifier Assembly . . . 262
  - 14.6.5 Common-Source Amplifier Gain Measurement . . . 262
  - 14.6.6 Common-Source Amplifier Clipping Measurement . . . 263
  - 14.6.7 Voltage Controlled Attenuator . . . 263
  - 14.6.8 Relaxation Oscillator . . . 263
  - 14.6.9 Chopper Modulator . . . 263
  - 14.6.10 Curve Tracer . . . 264
- 14.7 Laboratory Report . . . 264
- 14.8 References . . . 264

15 The BJT Differential Amplifier 265
- 15.1 Object . . . 265
- 15.2 Theory . . . 265
  - 15.2.1 Differential and Common-Mode Transconductances . . . 266
  - 15.2.2 Small-Signal Output Voltages . . . 266
  - 15.2.3 Differential and Common-Mode Voltage Gains . . . 267
  - 15.2.4 Common-Mode Rejection Ratio . . . 267
- 15.3 JFET Tail Supply . . . 267
- 15.4 Preliminary Derivations . . . 269
- 15.5 Preliminary Calculations . . . 270
- 15.6 Preliminary SPICE Simulations . . . 270
- 15.7 Experimental Procedures . . . 271
  - 15.7.1 Preparation . . . 271
  - 15.7.2 Transistor Matching . . . 271
  - 15.7.3 Assembly . . . 271
  - 15.7.4 Function Generator . . . 271
  - 15.7.5 Single Ended Input Small-Signal Gain . . . 272
  - 15.7.6 Differential Input Small-Signal Gain . . . 272
  - 15.7.7 Transfer Characteristic . . . 272
  - 15.7.8 Common Mode Gain . . . 272
  - 15.7.9 Current Source Design . . . 272
  - 15.7.10 Differential Amplifier with JFET Current Source Tail Supply . . . 273
  - 15.7.11 Square Wave Response . . . 273
- 15.8 Laboratory Report . . . 273
- 15.9 References . . . 273

16 MOSFET Amplifier 275
- 16.1 Object . . . 275
- 16.2 Device Equations . . . 275
- 16.3 Bias Equation . . . 277
- 16.4 Small-Signal Models . . . 278
  - 16.4.1 Hybrid-π Model . . . 278
  - 16.4.2 T Model . . . 279
- 16.5 Small-Signal Equivalent Circuits . . . 280
  - 16.5.1 Simplified T Model . . . 280
  - 16.5.2 Norton Drain Circuit . . . 280
  - 16.5.3 Thévenin Source Circuit . . . 282
  - 16.5.4 Summary of Models . . . 282
  - 16.5.5 Common Source Amplifier . . . 282
  - 16.5.6 Common-Drain Amplifier . . . 284
  - 16.5.7 Common-Gate Amplifier . . . 285
  - 16.5.8 MOSFET Differential Amplifier with Resistive Load . . . 286
  - 16.5.9 MOSFET Differential Amplifier with Active Load . . . 289
- 16.6 Preliminary Derivations . . . 289
- 16.7 Preliminary Calculations . . . 289
- 16.8 Preliminary SPICE Simulations . . . 290
- 16.9 Experimental Procedures . . . 290
  - 16.9.1 Preparation . . . 290
  - 16.9.2 Parameter Measurement . . . 290
  - 16.9.3 Common-Source Amplifier . . . 291
  - 16.9.4 MOSFET Differential Amplifier with Resistive Load . . . 291
  - 16.9.5 MOSFET Differential Amplifier with Current Mirror Active Load . . . 292

CONTENTS

    16.10 Laboratory Report 293
    16.11 References 293

17 Feedback Amplifiers   295
    17.1 Object 295
    17.2 Theory 295
        17.2.1 Basic Description of Feedback 295
        17.2.2 Signal-Flow Graphs 296
        17.2.3 Review of Background Theory 297
        17.2.4 Series-Shunt Feedback 299
        17.2.5 Shunt-Shunt Feedback 302
        17.2.6 Series-Series Feedback 304
        17.2.7 Shunt-Series Feedback 306
    17.3 Design Examples 310
        17.3.1 Series-Shunt 310
        17.3.2 Shunt-Shunt 311
        17.3.3 Series-Series 312
        17.3.4 Shunt-Series 312
    17.4 Preliminary Derivations 313
        17.4.1 Series-Shunt 313
        17.4.2 Shunt-Shunt 313
        17.4.3 Series-Series 313
        17.4.4 Shunt-Series 313
    17.5 Preliminary Calculations 313
        17.5.1 Series-Shunt 313
        17.5.2 Shunt-Shunt 314
        17.5.3 Series-Series 314
        17.5.4 Shunt-Series 314
    17.6 Preliminary SPICE Simulations 314
        17.6.1 Series-Shunt 314
        17.6.2 Shunt-Shunt 314
        17.6.3 Series-Series 314
        17.6.4 Shunt-Series 315
    17.7 Procedure 315
        17.7.1 Preparation 315
        17.7.2 Series Shunt 315
        17.7.3 Shunt Shunt 316
        17.7.4 Series-Series 316
        17.7.5 Shunt-Series Feedback Amplifier 317
    17.8 Laboratory Report 318
    17.9 References 318

18 Linear Op-Amp Oscillators   319
    18.1 Object 319
    18.2 Theory 319
        18.2.1 The Wien-Bridge Oscillator 320
        18.2.2 The Phase-Shift Oscillator 321
        18.2.3 The Quadrature Oscillator 322
    18.3 Preliminary Derivations 323
    18.4 Preliminary Calculations 323
    18.5 Preliminary SPICE Simulations 323
    18.6 Experimental Procedures 324

          18.6.1 Preparation 324
          18.6.2 Component Measurement 324
          18.6.3 Wien-Bridge Oscillator 324
          18.6.4 Phase Shift Oscillator 324
          18.6.5 Quadrature Oscillator 325
    18.7 Laboratory Report 325
    18.8 References 325

19 Switched-Capacitor Filters   327
    19.1 Object 327
    19.2 Theory 327
          19.2.1 Switched-Capacitor Integrator 328
          19.2.2 Second Order Filter Categories 329
    19.3 Devices 333
    19.4 Preliminary Derivations 336
    19.5 Preliminary Calculations 336
    19.6 Preliminary SPICE Simulations 336
    19.7 Experimental Procedures 337
          19.7.1 Preparation 337
          19.7.2 Clock 337
          19.7.3 Second-Order Filter 338
          19.7.4 Fourth-Order Filter 338
    19.8 Laboratory Report 339
    19.9 References 339

20 A Voltage Regulator   341
    20.1 Object 341
    20.2 Theory 341
          20.2.1 Block Diagram 341
          20.2.2 Differential Amplifier 342
          20.2.3 BJT Equivalent Circuits 342
          20.2.4 Small-Signal Output Impedance 344
          20.2.5 Current Limit 348
    20.3 Preliminary Derivations 349
    20.4 Preliminary Calculations 349
    20.5 Preliminary SPICE Simulations 349
    20.6 Experimental Procedures 349
          20.6.1 No Feedback 349
          20.6.2 Feedback 351
    20.7 Laboratory Report 351
    20.8 References 352

21 The Operational Amplifier   353
    21.1 Object 353
    21.2 Notation 353
    21.3 Op-Amp Model 353
    21.4 Voltage-Gain Transfer Function 353
    21.5 Gain-Bandwidth Product 354
    21.6 Slew Rate 355
    21.7 Relations between Slew Rate and Gain-Bandwidth Product 356
    21.8 Closed-Loop Transfer Function 356
    21.9 Transient Response 357

CONTENTS

- 21.10 Input Stage Overload 358
- 21.11 The BiFet Op Amp 359
- 21.12 Sine-Wave Response 361
- 21.13 Full Slewing Response 361
- 21.14 Intermediate Circuits 362
- 21.15 Completed Op-Amp Circuit 366
- 21.16 Preliminary Derivations 368
- 21.17 Preliminary Calculations 368
- 21.18 Preliminary SPICE Simulations 368
- 21.19 Procedure Part 1, A Discrete Op Amp 368
  - 21.19.1 Determination of Emitter Resistor and Compensation Capacitor 368
  - 21.19.2 Design of JFET Current Source 368
  - 21.19.3 BJT Transistor Parameter Measurement 369
  - 21.19.4 Differential Amplifier 371
  - 21.19.5 Laboratory Report 373
- 21.20 Procedure Part 2, A Discrete Op Amp 373
  - 21.20.1 Amplifier Loading 373
  - 21.20.2 Clipping 373
  - 21.20.3 Push-Pull Output Stage 374
  - 21.20.4 V_{BE} Multiplier 374
- 21.21 Procedure Part 3, A Discrete Op Amp 374
  - 21.21.1 Current Mirror 374
  - 21.21.2 Current Mirror Load on Diff-Amp 375
  - 21.21.3 Laboratory Report 375
- 21.22 References 376

Chapter 1

A Single Stage Common-Emitter BJT Amplifier

1.1 Object

The object of this experiment is to use a bipolar junction transistor, or BJT, to design a single-stage, common-emitter amplifier. The biasing, frequency response, and large-signal transient response are investigated.

1.2 Theory

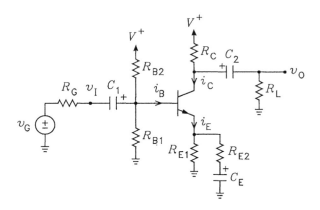

Figure 1.1: Common-emitter amplifier.

Figure 1.1 shows the circuit diagram of a single-stage, common-emitter, or CE, amplifier. A voltage source with an open circuit output voltage v_G and an output resistance R_G drives the input node labeled v_I. The input voltage is capacitively coupled through C_1 to the base of the transistor. The output is taken from the collector. It is capacitively coupled through C_2 to the load resistor R_L. Because the emitter is neither an input nor an output, the circuit is called a common-emitter amplifier. The parameter of greatest interest is the small-signal voltage gain given by

$$A_v = \frac{v_o}{v_i} \tag{1.1}$$

where v_i and v_o, respectively, are the small-signal ac components of v_I and v_O. The voltage gain must be measured with a sinusoidal input signal that sufficiently small so that the output voltage waveform is not distorted.

Capacitors C_1 and C_2 act as dc blocking capacitors to block the dc bias voltages from the source and the load. Their values are chosen so that they are effectively short circuits at the signal frequencies of interest. C_E is called an emitter-bypass capacitor. For dc, it is an open circuit and all of the emitter current must flow through R_{E1}. As frequency is increased, C_E becomes a short circuit and the emitter current flows through the parallel combination of R_{E1} and R_{E2}. This causes the gain of the circuit to increase. The value of the gain can be controlled by varying R_{E2}. For good stability of the bias current, R_{E1} is usually chosen so that the quiescent voltage across it is greater than the base-emitter voltage V_{BE}. If polar electrolytic capacitors are used for C_1, C_2, and/or C_E, the proper capacitor polarity must be observed. The polarities are marked with a plus sign in the figure.

1.2.1 Voltage Gain

For the BJT to operate as an amplifier, it must be biased in its forward active mode. This means that the base-emitter junction is forward biased and the base-collector junction is reverse biased. The term small signal means that the amplitude of the input voltage is small enough so that the output voltage waveform does not appear to be distorted. If the input amplitude is too large, the BJT can be driven into saturation and/or cutoff, causing the output signal to be clipped, i.e. the peaks are clipped off. If the positive peaks of the input voltage are too large, the base-collector junction of the transistor can be forward biased, causing the transistor to be driven into saturation. This causes the negative peaks at the output to be clipped. If the negative peaks of the input voltage are too large, the base-emitter junction can be reverse biased, causing the transistor to be driven into cutoff. This causes the positive peaks at the output to be clipped.

To solve for the small-signal voltage gain, we assume that capacitors C_1, C_2, and C_E are signal short circuits, i.e. that each appears as a dc battery in the circuit having a voltage equal to the quiescent voltage on the capacitor. The small-signal gain can be written as a product of terms as follows:

$$A_v = \frac{v_o}{v_i} = \frac{i_e}{v_i} \times \frac{i_c}{i_e} \times \frac{v_o}{i_c} \tag{1.2}$$

To solve for i_e/v_i, we use the emitter equivalent circuit given in Fig. 1.2(a), where r_{ie} is the small-signal resistance seen looking into the emitter with $v_i = 0$. It is given by

$$r_{ie} = \frac{r_x}{1+\beta} + r_e = \frac{r_x}{1+\beta} + \frac{V_T}{I_E} \tag{1.3}$$

where r_x is the base spreading resistance, β is the base-collector current gain, $r_e = V_T/I_E$ is the intrinsic emitter resistance, V_T is the thermal voltage, and I_E is the dc emitter bias current. The thermal voltage is given by $V_T = kT/q$, where $k = 1.38 \times 10^{-23}$ J/K is Boltzmann's constant, T is the Kelvin temperature, and $q = 1.6 \times 10^{-19}$ C is the electronic charge. The default value for T in SPICE is 300 K for which $V_T = 0.0259$ V.

For the second product term in Eq. 1.2, we have $i_c/i_e = \alpha$, where $\alpha = \beta/(1+\beta)$ is the emitter-collector current gain. The third term can be written from the collector equivalent circuit in Fig. 1.2(b) to obtain

$$\frac{v_o}{i_c} = -(r_{ic} \| R_C \| R_L) \tag{1.4}$$

where r_{ic} is the small-signal collector output resistance given by

$$r_{ic} = \frac{r_0 + r_{ie} \| R_{te}}{1 - \alpha R_{te}/(r_{ie} + R_{te})} \tag{1.5}$$

In this equation, r_0 is the small-signal BJT collector-emitter resistance and R_{te} is the resistance seen looking out of the emitter to ground given by $R_{te} = R_{E1} \| R_{E2}$. The collector-emitter resistance is given by

$$r_0 = \frac{V_A + V_{CB}}{I_C} \tag{1.6}$$

1.2. THEORY

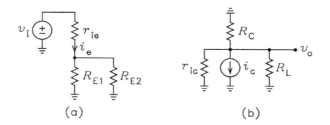

Figure 1.2: (a) Emitter equivalent circuit. (b) Collector equivalent circuit.

where V_A is the Early voltage, V_{CB} is the collector-base bias voltage, and I_C is the collector bias current.

The above equations can be combined to write the voltage gain as follows:

$$A_v = \frac{1}{r_{ie} + R_{E1} \| R_{E2}} \times \alpha \times [-(r_{ic} \| R_C \| R_L)] \tag{1.7}$$

This equation is often approximated by assuming that $\beta \to \infty$ and $V_A \to \infty$. In this case, $\alpha = 1$, $r_{ie} = r_e$ and $r_{ic} \to \infty$. Thus the approximate expression for the gain is

$$A_v \simeq -\frac{R_C \| R_L}{r_e + R_{E1} \| R_{E2}} \tag{1.8}$$

If $r_e \ll R_{E1} \| R_{E2}$, the gain can be further approximated by

$$A_v = -\frac{R_C \| R_L}{R_{E1} \| R_{E2}} \tag{1.9}$$

This equation says that the gain is approximately equal to the negative ratio of the resistance seen looking out of the collector to the resistance seen looking out of the emitter.

The negative sign in the gain equation means that the CE amplifier has an inverting gain. If the input voltage is increasing, the output voltage is decreasing, and vice versa. For a sinusoidal input, the output voltage waveform would appear to be 180° out of phase from the input waveform.

1.2.2 Biasing

The bias equivalent circuit of the CE amplifier is shown in Fig. 1.3(a). This circuit is obtained by replacing C_1, C_2, and C_E with open circuits. Fig. 1.3(b) shows the bias circuit with the circuit seen looking out of the base replaced with a Thévenin equivalent circuit, where

$$V_{BB} = V^+ \frac{R_{B1}}{R_{B1} + R_{B2}} \qquad R_{BB} = R_{B1} \| R_{B2} \tag{1.10}$$

From this circuit, we can write

$$V_{BB} = I_B R_{BB} + V_{BE} + I_E R_{E1} = \frac{I_C}{\beta} R_{BB} + V_{BE} + \frac{I_C}{\alpha} R_{E1} \tag{1.11}$$

This equation can be solved for I_C to obtain

$$I_C = \frac{V_{BB} - V_{BE}}{R_{BB}/\beta + R_{E1}/\alpha} \tag{1.12}$$

The collector bias current and the resistor values in the circuit set the output clipping voltages. Suppose v_I is driven negative to the point where the transistor just cuts off, i.e. such that $i_C = 0$. Fig. 1.4(a) shows

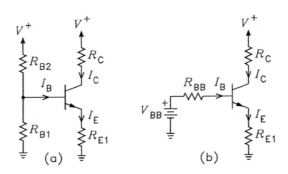

Figure 1.3: Bias equivalent circuits.

the equivalent circuit. Capacitor C_2 is represented by a dc battery with a voltage equal to the quiescent voltage across C_2. This voltage is $V_{C_2} = V^+ - I_C R_C$. The positive clipping voltage v_O^+ can be calculated from the circuit as follows:

$$v_O^+ = \frac{V^+ - V_{C_2}}{R_C + R_L} \times R_L = \frac{I_C R_C}{R_C + R_L} \times R_L = I_C \times (R_C \| R_L) \tag{1.13}$$

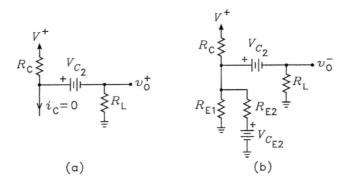

Figure 1.4: (a) Circuit for calculating v_O^+. (b) Circuit for calculating v_O^-.

Next, suppose that v_I is driven negatively so that the BJT just saturates, i.e. $v_{CE} \simeq 0$, where v_{CE} is the collector-emitter voltage. Fig. 1.4(b) shows the equivalent circuit, where the base current is assumed to be small enough so that it can be neglected which makes $\alpha \approx 1$. Capacitors C_2 and C_E are represented by dc batteries with voltages equal to the quiescent voltage across the capacitors, where $V_{C_2} = V^+ - I_C R_C$ and $V_{C_{E2}} = I_E R_{E1} = I_C R_{E1}$. By superposition of V^+, V_{C_2}, and $V_{C_{E2}}$, it follows from the circuit that the negative clipping voltage v_O^- is given by

$$\begin{aligned}v_O^- &= V^+ \frac{R_L \| R_{E1} \| R_{E2}}{R_C + R_L \| R_{E1} \| R_{E2}} - V_{C_2} \frac{R_L}{R_L + R_C \| R_{E1} \| R_{E2}} + V_{C_{E2}} \frac{R_L \| R_C \| R_{E1}}{R_{E2} + R_L \| R_C \| R_{E1}} \\ &= -\left[V^+ - I_C (R_C + R_{E1})\right] \frac{R_C \| R_L}{R_C \| R_L + R_{E1} \| R_{E2}}\end{aligned} \tag{1.14}$$

Symmetric clipping requires that $v_O^+ = -v_O^-$. This condition can be used to solve for the collector bias current I_C to obtain

$$I_C = \frac{V^+}{R_C\|R_L + R_C + R_{E1} + R_{E1}\|R_{E2}} \tag{1.15}$$

A CE amplifier biased at this collector current exhibits symmetric clipping on the positive and negative peaks of the output waveform.

The design of the CE amplifier involves the selection of the resistors and capacitors to meet specifications. For example, the load resistor R_L and midband voltage gain might be specified. The resistors R_C, R_{E1}, and R_{E2} would be selected so that the midband voltage gain meets the design criteria. Next the quiescent collector current would be selected to meet some criteria, such as symmetric clipping. Once a value for I_C is determined, the quiescent base voltage can be calculated from

$$V_B = V_{BE} + I_E R_{E1} = V_{BE} + I_C R_{E1} \tag{1.16}$$

where the base-emitter voltage is typically in the range of 0.6 V to 0.7 V. An appropriate value to assume for a design is the average of the two values, i.e. $V_{BE} = 0.65$ V. The base bias resistors R_{B1} and R_{B2} must be chosen to set the base voltage to the required value. For example, the current through R_{B2} might be specified. Denote it by $I_{R_{B2}}$. The current through R_{B1} is then $I_{R_{B2}} - I_B = I_{R_{B2}} - I_C/\beta$. The base bias resistors are then given by

$$R_{B2} = \frac{V^+ - V_B}{I_{R_{B2}}} \qquad R_{B1} = \frac{V_B}{I_{R_{B2}} - I_C/\beta} \tag{1.17}$$

When using this procedure, $I_{R_{B2}}$ is usually chosen to be $10 I_C/\beta_{\min}$ or larger, where $\beta_{\min}$ is the minimum specified value of β for the BJT. In this case, the I_C/β term in the equation for R_{B1} can usually be neglected.

1.3 Example Design

A CE amplifier is to be designed for the gain $A_v = -50$ with the load resistance $R_L = 20$ kΩ. The power supply voltage is to be $V^+ = 15$ V. The collector bias current it to be chosen for symmetric clipping. Let us choose $R_C = 10$ kΩ and $R_{E1} = 2$ kΩ. We expect $R_{E2} \ll R_{E1}$ so that Eq. (1.15) for I_C can be approximated by

$$I_C = \frac{V^+}{R_C\|R_L + R_C + R_{E1}} = 0.8 \text{ mA} \tag{1.18}$$

where we have assumed that $\alpha \simeq 1$. For this current, $r_e = 25.9/0.8 = 32.4$ Ω. For the specified gain, Eq. (1.8) requires

$$50 \simeq \frac{10\text{k}\|20\text{k}}{32.4 + 2\text{k}\|R_{E2}} \tag{1.19}$$

This equation can be solved for R_{E2} to obtain $R_{E2} = 106$ Ω. Although the closest 5% value is 110 Ω, we will choose $R_{E2} = 100$ Ω to round off in the direction of increasing gain. This tends to correct for the approximations in Eq. (1.8).

To calculate the base bias voltage, we assume $V_{BE} = 0.65$ V. Thus $V_B = 0.8 \times 2 + 0.65 = 2.25$ V. If we assume that $\beta_{\min} = 100$, it follows that the minimum current through R_{B2} is $10 \times 0.8/100 = 0.08$ mA. For this current, we have $R_{B2} = (15 - 2.25)/0.08 = 159$ kΩ and $R_{B1} = 2.25/0.08 = 28.1$ kΩ, where base current has been neglected. We will choose $R_{B2} = 150$ kΩ and $R_{B1} = 27$ kΩ. If base current is neglected, the base voltage should be $15 \times 27/(150 + 27) = 2.29$ V which is very close to the desired 2.25 V value. For these values, the current the current through R_{B2} is larger than the minimum value of 0.08 mA.

The lower cutoff frequency is a function of C_1, C_2, and C_E. It is beyond the intent of this experiment to cover the equations that relate the cutoff frequency to the capacitor values. For good response down to 100 Hz, the values $C_1 = 0.22$ μF, $C_2 = 10$ μF, and $C_E = 100$ μF should be adequate.

1.4 SPICE Simulation

The CE amplifier can be simulated with the following SPICE input deck:

```
CE AMPLIFIER
VI 1 0 AC 1 SIN(0 0.04 1E3)
RB1 2 0 27E3
RB2 6 2 150E3
RE1 4 0 2E3
RE2 4 7 100
RC 6 3 10E3
RL 5 0 20E3
C1 1 2 0.22E-6 IC=-2.2022
C2 3 5 10E-6 IC=7.2652
CE 7 0 100E-6 IC=1.5545
Q1 3 2 4 Q2N3904
VPLUS 6 0 DC 15
.MODEL Q2N3904 NPN(IS=1E-14 BF=200 VA=170 CJC=3.6E-12 TF=0.3E-9 RB=10)
.OP
.AC DEC 30 10 100E6
.TRAN 1E-6 1E-3 UIC
.FOUR 1E3 V(5)
.PROBE
.END
```

The node labels should be obvious from the circuit. The BJT is assumed to be a 2N3904 for which representatives parameter values are given in the .MODEL statement. An ac analysis can used to obtain the frequency response, i.e. the small-signal gain versus frequency. This is shown in Fig. 1.5. The figure shows that the magnitude of the gain at midband frequencies is 51, or 2% higher than the design value. The lower -3 dB frequency is set by the external capacitors C_1, C_2, and C_E. The upper -3 dB frequency is set by the internal capacitors inside the transistor, i.e. c_π and c_μ in the small-signal model. These capacitors are calculated by SPICE from the forward transit time τ_F (TF) and the zero-bias base collector capacitance c_{jco} (CJC).

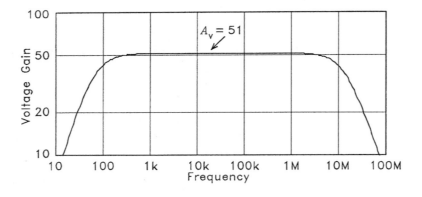

Figure 1.5: Gain magnitude versus frequency.

A transient analysis by SPICE can be used to examine the clipping response of the amplifier. However, the voltages on the capacitors in the circuit must first be set to their initial values. These voltages can be obtained from the .OUT file created by SPICE. They are listed at the end of the line for each capacitor in

1.5. PROCEDURE

the deck. The UIC in the .TRAN line means "use initial conditions." Fig. 1.6 shows the output voltage versus time for sinusoidal input signals having peak voltages of 0.04 V, 0.1 V, and 0.25 V. The output is undistorted for the 0.04 V signal, somewhat distorted for the 0.1 V signal, and severely distorted for the 0.25 V signal. The figure shows the positive and negative clipping levels to be approximately the same.

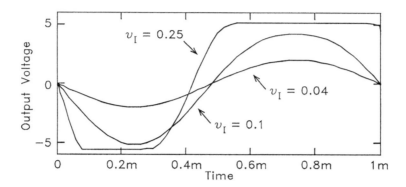

Figure 1.6: Clipping response.

SPICE can be used to calculate the amplitude of the Fourier series harmonic components in the output voltage waveform as part of the transient analysis. The .FOUR line in the deck controls this. Fig. 1.7 shows the components for the fundamental through the ninth harmonic for the three input signal amplitudes. Each is normalized so that the fundamental component is unity. The percent distortion is 2.5% for the 0.04 V signal, 8.3% for the 0.1 V signal, and 41% for the 0.25 V signal.

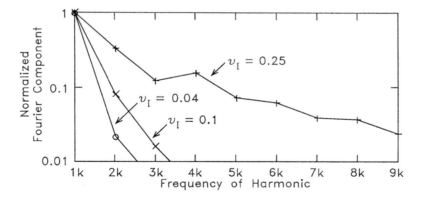

Figure 1.7: Normalized Fourier harmonic components.

1.5 Procedure

1.5.1 Breadboard Preparation

Before beginning the experiment, the solderless breadboard must be configured properly. This consists of establishing buses or rails for the common or ground, the positive power supply, and the negative power supply. Although a negative power supply voltage is not used in this experiment, it is good practice to configure the breadboard for subsequent experiments.

The laboratory bench power supply and the breadboard have female binding posts for interconnecting leads with male banana plugs. A suggested color code for the power supply wires is black for common or ground, red for positive, and yellow for negative. The binding posts on the breadboard must be mechanically secured to the breadboard with the proper mounting hardware.

There is a hole in each binding post which can be revealed by unscrewing the plastic knob. A jumper wire can be inserted into this hole and the knob screwed down on it to secure it. The wire can then be used to connect the terminal post to its respective bus on the breadboard. The wire will not make contact to the binding post if it is inserted so that the insulation around the wire is in the hole. Prepare the breadboard as shown in Fig. 1.8. The 100 μF capacitors are power supply decoupling capacitors. They serve the function of establishing an ac ground connection between each power supply rail and the ground bus. Without the capacitors, the amplifier could oscillate. Because the capacitors are electrolytic types, they must be inserted with the polarity shown or they will not function properly. Indeed, **they may explode**, causing irreversible damage to the capacitors and angst to the experimenter.

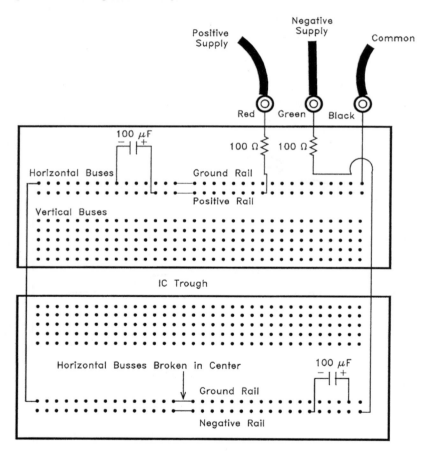

Figure 1.8: Breadboard layout.

The 100 Ω, 1/4 W resistors shown in series with the positive and negative binding posts serve an important purpose. They act to limit the current if a mistake is made on the breadboard. If the current drawn by the

1.5. PROCEDURE

circuit becomes excessive, one of these resistors may emit smoke. If it does, **do not touch the resistor**, for you can be burned. Instead, turn the bench power supply off and find your error.

Use the Digital Multimeter (DMM) as a ohmmeter to verify that the breadboard has been properly wired. Make sure that the resistance from the center of each binding post to its respective rail on the breadboard measures zero.

1.5.2 Transistor Leads

The transistor used in this experiment is the 2N3904. This is a NPN BJT. Its lead configuration is shown in Fig. 1.9. The figure also shows the configuration for the 2N3906 PNP BJT, which is the complement to the 2N3904.

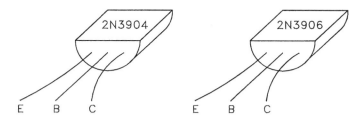

Figure 1.9: Transistor lead labels.

1.5.3 Circuit Assembly

Figure 1.10 shows the CE amplifier circuit with the element values calculated above. Resistor values with the k suffix are in kΩ. Capacitor values with the u suffix are in μF. The circuit is modified by the addition of three resistors. The 1 kΩ and 20 Ω resistors at the input form a voltage divider to attenuate the output of the function generator to prevent overload of the amplifier. The voltage division ratio is approximately 1/50. Thus 1 V output from the function generator should produce 0.02 V at the v_I node and 1 V at the output. A 100 Ω resistor is shown in series with the output. This resistor isolates the high-frequency capacitance of any test lead connected to the output from the circuit. Without this resistor, the circuit could exhibit high-frequency oscillations.

Electrolytic capacitors must be inserted with the proper polarity. The positive terminal is labeled on the circuit diagram. Use banana-plug leads to connect the positive output (use a red lead) and the common output (use a black lead), respectively, on the bench power supply to the positive rail and common rail binding posts on the breadboard. The common output on the power supply is usually called the ground because it connects to circuit ground on the breadboard. Do not use the terminal on the power supply that is labeled with a ground symbol. This terminal connects to the power supply chassis and to the third prong of the ac line cord. It does not connect to the internal power supply circuits.

1.5.4 Quiescent dc Bias Point

Turn on the dc power supply. Use the digital multimeter (DMM) to measure the voltage drop from the positive rail on the breadboard to the breadboard ground. Adjust the positive power supply voltage so that the DMM reads +15 V. After the rail voltage is set, use the DMM to measure the dc voltages at the collector, the base, and the emitter of the BJT. Each of these are measured with respect to breadboard ground. Record the measured values in the data section. Calculate and record the quiescent collector current given by

$$I_C = \frac{V^+ - V_C}{R_C} \quad (1.20)$$

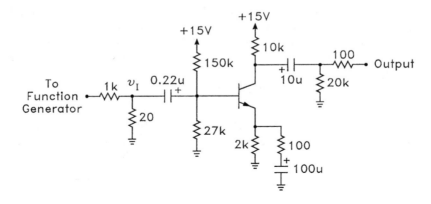

Figure 1.10: Circuit to be assembled in the laboratory.

A collector current of approximately 0.8 mA should have been obtained. If the circuit is not properly biased, the rest of the experiment is meaningless.

1.5.5 Small-Signal Amplifier

Turn on the function generator. Connect its output to the voltage divider on the breadboard and set it to produce a sine wave with a frequency of 1 kHz and a peak voltage of 1 V. Use the oscilloscope to measure the ac voltages at the output of the voltage divider, i.e. at the v_I node, and at the output of the circuit. Calculate the voltage gain from the measurements. The voltage at the output of the voltage divider should be approximately 0.02 V peak. The voltage at the output of the amplifier should be approximately 1 V peak. The gain should be approximately -50, i.e. an inverting gain of 50.

1.5.6 Frequency Response

Vary the frequency of the function generator and record the input and output sine wave amplitudes so that a plot of gain versus frequency can be made. Take data over a large enough range of frequencies so that the upper and lower -3 dB frequencies may be determined. The frequency axis on the plot should have a log scale. The measurement frequencies should be chosen to give approximately equal spacings between points on a log scale. For example, for 4 equal spacings between 1 kHz and 10 kHz, the frequencies are 1 kHz, 1.78 kHz, 3.16 kHz, 5.62 kHz, and 10 kHz. Each frequency is obtained from the preceding frequency by multiplying by $10^{0.25}$. In the lab, these might be rounded off to 1 kHz, 2 kHz, 3 kHz, 6 kHz, and 10 kHz.

1.5.7 Large-Signal Clipping Behavior

Set the function generator frequency to 1 kHz. Increase its output level until the waveform at the output of the amplifier just starts to show clipping on either its positive or its negative peak. Sketch or print the oscilloscope display.

Further increase the function generator level until both the positive and negative peaks exhibit hard clipping. Measure and record the positive and negative clipping levels. Sketch or print the display.

Set the oscilloscope to the X-Y mode. Connect the X input to the v_I node on the amplifier and the Y input to the output of the amplifier. Set the X axis sensitivity so that the slope of the straight line in the linear mode can be measured. Adjust the oscilloscope cursors so that the slope can be measured. Sketch or print the display.

1.6. LABORATORY REPORT

Replace resistor R_{E2} with a 10 kΩ potentiometer connected as a variable resistor. (Connect the wiper or center pin of the potentiometer to one of the outer pins. Use the two outer pins for the terminals of the variable resistor.) Vary the potentiometer and note the effect on the oscilloscope display.

Set the oscilloscope to the time display mode. Reduce the function generator output signal to 1 V. Observe the effect of varying the potentiometer on the output voltage waveform.

Set the potentiometer to a value that results in a gain of approximately −50.

1.5.8 Spectral Analysis

The following assumes that a **Tektronix 3012B** oscilloscope or equivalent is available which permits a FFT analysis of a signal.

Turn the oscilloscope on and wait for it to boot. Connect the input of the circuit to $CH1$ and the output of the circuit to $CH2$. Turn both $CH1$ and $CH2$ on. Press *AUTOSET*.

The oscilloscope will now be configured to display the frequency spectrum of the waveform connected to the $CH2$ input. Turn $CH1$ off and use the Vertical Scale and Position controls to position the waveform on $CH2$ to the upper half of the screen. Manually set the Time/Div to 4 ms which should display about 40 cycles of the output waveform. Press the *MATH* button and select FFT. Set the source for the FFT operation to $CH2$ the output of the circuit. A spike one horizontal division to the right of the left most graticule line should be visible. If the spike isn't visible look at the waveform record at the top of the display and change the horizontal scale so that the brackets are about one tenth of the record. Use the horizontal position control to position the brackets on the waveform record to the far left (the moving of the bracket can be speeded up by pressing the COARSE button). Vary the vertical positions of the two displays until they do not overlap. Do not position the waveform on $CH2$ off the screen; this will clip it and produce a severely distorted spectrum.

There should be a spike at $f = 1$ kHz which is the amplified input 1 kHz sine wave. This is the fundamental. The other spectral components are distortion components. Use the cursors to measure the amplitude of the spectral components. Vary the setting of the pot and observe the spectrum. Adjust the pot so that all distortion components are below the digital noise baseline (the grainy display at the bottom).

Turn the MATH function off and turn $CH1$ on. Press *AUTOSET* and measure the gain of the amplifier.

Turn the dc power supply off, remove the potentiometer and measure and record the value of the resistance.

When finished, turn all instruments off and return all leads to the wall rack.

1.6 Laboratory Report

1.6.1 Bias Values

Collector Voltage $V_C =$ ~~~~ 7.8178 V
Base Voltage $V_B =$ ~~~~ 2.1129 V
Emitter Voltage $V_E =$ ~~~~ 1.4521 V
Collector Current $I_C =$ ~~~~ 0.718 mA

1.6.2 Small-Signal Gain

$A_v = v_o/v_i =$ ~~~~ 2.75V / 0.065 V = 42.31

1.6.3 Frequency Response

A_v(midband) = −0.75 dB
f_{-3dB} (lower) = 6.3.11 Hz
f_{-3dB} (upper) = 143.30 kHz

1.6.4 Large-Signal Clipping Behavior

Measured Positive Clipping Level $v_O^+ =$ ~~4.80V 4.75V~~ 5.10V
Measured Negative Clipping Level $v_O^- =$ −6.30V
Computed Positive Clipping Level $v_O^+ =$ 4.7867 ✓
Computed Negative Clipping Level $v_O^- =$ −6.294 ✓

Use the measured value of I_C, $\alpha = 0.99$, ,and $R_{E2} = 100\ \Omega$ when using Eqs. 1.13 and 1.14 to compute the clipping levels.

Include the sketches or printouts of the large-signal clipping level that were made.

1.6.5 Spectral Analysis

Include the sketches or printouts of the large-signal clipping level that were made.

The value of the small-signal gain and corresponding resistance for R_{E2} that just resulted in a distortion that was too small to measure.

$R_{\text{pot}} =$ 15885 Ω
$A_v = v_o/v_i =$ 1.52V/0.046V = 33.0434

Answer any supplementary questions which may have been posed by the laboratory instructor.

1.7 References

1. E. J. Angelo, *Electronics: BJTs, FETs, and Microcircuits*, McGraw-Hill, 1969.
2. W. Banzhaf, *Computer-Aided Circuit Analysis Using SPICE*, 2nd ed., Prentice-Hall, 1992.
3. T. E. Brewer, *An Introduction to Electrical Measurements*, Kendall-Hunt, 1997.
4. T. C. Hayes and P. Horowitz, *Student Manual for The Art of Electronics*, Cambridge, 1989.
5. M. N. Horenstein, *Microelectronics Circuits and Devices*, 2nd ed., Prentice-Hall, 1996.
6. P. Horwitz and W. Hill, *The Art of Electronics,* 2nd edition, Cambridge University Press, 1989.
7. P. Horowitz and I. Robinson, *Laboratory Manual for The Art of Electronics*, Cambridge University Press, 1981.
8. J. H. Krenz, *An Introduction to Electrical and Electronic Devices*, Prentice-Hall, 1987.
9. R. Mauro, *Engineering Electronics,* Prentice-Hall, 1989.
10. F. H. Mitchell and F. H. Mitchell, *Introduction to Electronic Design*, Prentice-Hall, 1988.
11. Motorola, *Small-Signal Semiconductors*, DL 126, Motorola, 1987.
12. D. A. Neamen, *Electronic Circuit Analysis and Design*, Irwin, 1996.
13. C. J. Savant, M. S. Roden, and G. L. Carpenter, *Electronic Circuit Design,* 3rd ed., Discovery, 1997.
14. A. S. Sedra and K. C. Smith, *Microelectronic Circuits*, 4th edition, Oxford, 1998.
15. D. L. Schilling and C. Belove, *Electronic Circuits: Discrete and Integrated*, McGraw-Hill, 1968.
16. R. T. Howe and C. G. Sodini, *Microelectronics*, Prentice Hall, 1997.
17. P. W. Tuinenga, *SPICE*, 3rd ed., Prentice-Hall, 1995.

Chapter 2

Basic Op-Amp Circuits 1

2.1 Object

The object of this experiment is to assemble and evaluate several introductory op-amp circuits. These are the non-inverting amplifier, the inverting amplifier, the inverting amplifier with a T feedback network, the integrator, and the differentiator.

2.2 Theory

2.2.1 Voltage and Current Notation

The notation for voltages and currents used in this experiment and in following experiments is defined as follows. Total quantities (dc plus small-signal ac) are denoted by a lower-case letter with an upper-case subscript, e.g. v_I, i_O, etc. dc quantities are denoted by an upper-case letter with an upper-case subscript, e.g. V_I, I_O, etc. Small-signal ac quantities are denoted by a lower-case letter with a lower-case subscript, e.g. v_i, i_o, etc. Phasor and complex quantities (functions of s or $j\omega$) are denoted by an upper-case letter and a lower-case subscript, e.g. V_i, I_o, etc. Root-mean-square or rms quantities are also denoted by an upper-case letter and a lower-case subscript.

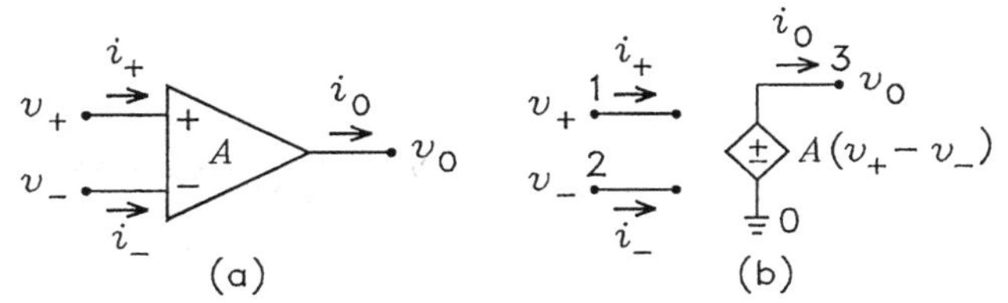

Figure 2.1: (a) Circuit symbol of the op amp (b) Controlled source model of the ideal op amp with SPICE node numbers labeled.

2.2.2 The Ideal Op-Amp

The circuit symbol for the op amp is given in Fig. 2.1(a). The output voltage v_O is given by

$$v_O = A(v_+ - v_-) \tag{2.1}$$

where v_+ is the voltage at the non-inverting input, v_- is the voltage at the inverting input, and A is the open-loop voltage gain. Most circuits containing op amps are designed under the assumption that the op amps are ideal. An ideal op amp satisfies the following three conditions:

- The input currents i_+ and i_- are zero. This means that the impedance seen looking into either input is infinite.
- The output voltage v_O is independent of the output current i_O. This means that the impedance seen looking into the output is zero.
- The open-loop gain A approaches infinity in the limit. Because $v_+ - v_- = v_O/A$, this condition means that $v_+ - v_- = 0$ or $v_+ = v_-$ when v_O is finite.

The latter condition is usually assumed to hold when negative feedback is applied from the op-amp output to its inverting input. It is said that a *virtual short circuit* exists between the two inputs. A virtual short circuit is a short circuit branch in which no current flows. It is often indicated on a circuit diagram with a dashed line. The controlled source model of the ideal op amp is given in Fig. 2.1(b).

2.2.3 Voltage-Gain Transfer Function of Physical op amps

For a sine-wave input signal, the voltage gain of a physical op amp depends on the frequency. For general purpose op amps, the voltage-gain transfer function can usually be approximated by a single-pole transfer function of the form

$$A(s) = \frac{V_o}{V_+ - V_-} = \frac{A_0}{1 + s/\omega_0} \tag{2.2}$$

where s is the complex frequency, A_0 is the dc gain, and ω_0 is the radian pole frequency. The pole frequency in Hz is given by $f_0 = \omega_0/2\pi$. Typical general purpose op amps have a dc gain $A_0 \simeq 2 \times 10^5$ and a pole frequency $f_0 \simeq 5$ Hz. For $s = j\omega$ and $\omega >> \omega_0$, the transfer function in Eq. (2.2) can be approximated by

$$A(s) \simeq \frac{A_0}{(s/\omega_0)} = \frac{A_0 \omega_0}{s} = \frac{\omega_x}{s} \tag{2.3}$$

where ω_x is the radian *gain-bandwidth product* given by

$$\omega_x = A_0 \omega_0 \tag{2.4}$$

The gain-bandwidth product in Hz is given by $f_x = \omega_x/2\pi = A_0 f_0$. Note that since the gain is dimensionless the units on the gain-bandwidth product is Hertz. Typical general purpose op amps have a gain-bandwidth product $f_x \simeq 1$ MHz. Because $|A(j2\pi f_x)| \simeq 1$, f_x is also called the op-amp *unity gain frequency*.

2.2.4 Non-Inverting Amplifier

Fig. 2.2 shows the circuit diagram of a non-inverting amplifier. The feedback network consists of resistors R_F and R_1. Resistor R_I sets the input resistance. Resistor R_O sets the output resistance. To solve for the voltage gain,

$$v_+ - v_- = \lim_{A \to \infty} \frac{v_O}{A} = 0 \tag{2.5}$$

$$v_+ = v_I \tag{2.6}$$

$$v_- = v_O \frac{R_1}{R_1 + R_F} \tag{2.7}$$

These equations can be solved for the voltage gain to obtain

$$A_v = A_{0f} = \frac{v_O}{v_I} = 1 + \frac{R_F}{R_1} \tag{2.8}$$

can be written. The input resistance is R_I and the output resistance is R_O.

2.2. THEORY

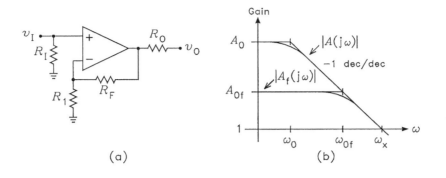

Figure 2.2: (a) Non-inverting amplifier. (b) Bode magnitude plot.

2.2.5 Voltage-Gain Transfer Function of Non-Inverting Amplifier

When the op amp is modeled as having the voltage-gain transfer function in Eq. (2.3), then

$$V_+ - V_- = \frac{V_o}{A(s)} = V_o \frac{s}{\omega_x} \tag{2.9}$$

$$V_+ = V_i \tag{2.10}$$

$$V_- = V_o \frac{R_1}{R_1 + R_F} \tag{2.11}$$

These equations can be solved for the voltage-gain transfer function to obtain

$$A_f(s) = \frac{V_o}{V_i} = \left[1 + \frac{R_F}{R_1}\right] \frac{1}{1 + (1 + R_F/R_1)s/\omega_x} = \frac{A_{0f}}{1 + s/\omega_{0f}} \tag{2.12}$$

This is the product of the gain given by Eq. (2.8) and a single-pole low-pass transfer function. The radian pole frequency ω_{0f} in the transfer function is given by

$$\omega_{0f} = \frac{\omega_x}{1 + R_F/R_1} \tag{2.13}$$

The pole frequency in Hz is given by $f_{0f} = \omega_{0f}/2\pi = f_x/(1 + R_F/R_1)$. The gain-bandwidth product for the amplifier is given by

$$A_{0f} f_{0f} = \left(1 + \frac{R_F}{R_1}\right) \frac{f_x}{1 + R_F/R_1} = f_x \tag{2.14}$$

This is the same as for the op amp without feedback. Fig. 2.2(b) shows the Bode magnitude plots for both $A(j\omega)$ and $A_f(j\omega)$. The upper half-power or -3 dB radian cutoff frequency for $A(j\omega)$ is ω_0 while that for $A_f(j\omega)$ is ω_{0f}.

2.2.6 Inverting Amplifier

Fig. 2.3(a) shows the circuit diagram of the inverting amplifier. R_1 sets the input resistance, R_F is the feedback resistor, and R_O sets the output resistance. To solve for the voltage gain, write

$$v_- = \lim_{A \to \infty} \frac{v_O}{A} = 0 \tag{2.15}$$

$$i_1 = \frac{v_I}{R_1} \tag{2.16}$$

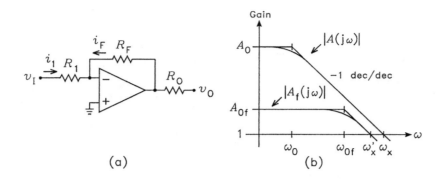

Figure 2.3: (a) Circuit diagram of the inverting amplifier. (b) Bode magnitude plots for $A(j\omega)$ and $A_f(j\omega)$.

$$i_F = \frac{v_O}{R_F} \qquad (2.17)$$

$$i_1 + i_F = 0 \qquad (2.18)$$

These equations can be solved for the voltage gain to obtain

$$A_v = A_{0f} = \frac{v_O}{v_I} = -\frac{R_F}{R_1} \qquad (2.19)$$

Because $v_- = 0$, the input resistance is R_1. The output resistance is R_O.

2.2.7 Voltage-Gain Transfer Function of Inverting Amplifier

If the op amp is modeled as having the voltage-gain transfer function in Eq. (2.3), the inverting input is no longer a virtual ground. To solve for the voltage-gain transfer function, write

$$V_- = -\frac{V_o}{A(s)} = -V_o \frac{s}{\omega_x} \qquad (2.20)$$

$$I_1 = \frac{V_1 - V_-}{R_1} \qquad (2.21)$$

$$I_f = \frac{V_o - V_-}{R_F} \qquad (2.22)$$

$$I_1 + I_f = 0 \qquad (2.23)$$

These equations can be solved for the voltage-gain transfer function to obtain

$$A_f(s) = \frac{V_o}{V_i} = -\frac{R_F}{R_1} \times \frac{1}{1 + (1 + R_F/R_1)s/\omega_x} = \frac{A_{0f}}{1 + s/\omega_{0f}} \qquad (2.24)$$

This is the product of the gain given by Eq. (2.19) and a single-pole low-pass transfer function. The radian pole frequency ω_{0f} is the same as for the non-inverting amplifier and is given by Eq. (2.13). The gain-bandwidth product of the amplifier is given by

$$|A_{0f}| f_{0f} = \frac{R_F}{R_1} \frac{f_x}{1 + R_F/R_1} = f_x \frac{R_F}{R_F + R_1} \qquad (2.25)$$

This is less than the gain-bandwidth product of the non-inverting amplifier by the factor $R_F/(R_1 + R_F)$. It can be concluded that the inverting amplifier has a lower bandwidth than the non-inverting amplifier when

2.2. THEORY

the two amplifiers have the same gain magnitude. Conversely, if the bandwidths of an inverting amplifier and a non-inverting amplifier are equal, the non-inverting amplifier has the greater gain magnitude. In the case that $R_F \gg R_1$, the bandwidths of the two amplifiers are approximately the same.

The Bode magnitude plots for $A(j\omega)$ and $A_f(j\omega)$ are given in Fig. 2.3. The radian gain-bandwidth product with feedback is labeled ω'_x and is given by $\omega'_x = \omega_x R_F / (R_1 + R_F)$.

2.2.8 Inverting Amplifier with T Feedback Network

To reduce the sensitivity of op-amp circuits to the non-ideal characteristics of op amps, very small or very large resistor values should be avoided. A good rule of thumb is to use resistor values in the range from 1 kΩ to 100 kΩ if possible. However, most circuits give no problems if resistors are in the range of 100 Ω to 1 MΩ. To achieve a very high gain with the inverting amplifier, either R_1 must be very small, R_F must be very large, or both. This problem can be overcome with a T feedback network as shown in Fig. 2.4. To solve for the voltage gain, write

$$v_- = \lim_{A \to \infty} \frac{v_O}{(-A)} = 0 \tag{2.26}$$

$$i_1 = \frac{v_I}{R_1} \tag{2.27}$$

$$i_F = \frac{v_O \left(\frac{R_3}{R_3 + R_4}\right)}{R_2 + R_3 \| R_4} \tag{2.28}$$

$$i_1 + i_F = 0 \tag{2.29}$$

These equations can be solved for the voltage gain to obtain

$$A_v = A_{0f} = \frac{v_O}{v_I} = -\frac{R_4}{R_1}\left(1 + \frac{R_2}{R_3 \| R_4}\right) \tag{2.30}$$

The input resistance is R_1 and the output resistance is R_O.

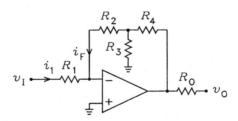

Figure 2.4: Inverting amplifier with T feedback network.

2.2.9 Voltage-Gain Transfer Function of Inverting Amplifier with T Feedback Network

If the op amp is modeled as having the voltage-gain transfer function in Eq. (2.3), the inverting input is no longer a virtual ground. In this case, write

$$V_- = \frac{V_o}{-A(s)} = -V_o \frac{s}{\omega_x} \tag{2.31}$$

$$I_1 = \frac{V_i - V_-}{R_1} \tag{2.32}$$

$$I_f = \frac{V_o\left(\frac{R_3}{R_3+R_4}\right) - V_-}{R_2 + R_3\|R_4} \tag{2.33}$$

$$I_1 + I_f = 0 \tag{2.34}$$

These equations can be solved for the voltage gain transfer function to obtain

$$A_f(s) = \frac{V_o}{V_i} = -\frac{R_4}{R_1}\left(1 + \frac{R_2}{R_3\|R_4}\right) \times \frac{1}{1 + \frac{R_4}{R_1}\left(1 + \frac{R_1+R_2}{R_3\|R_4}\right)\frac{s}{\omega_x}} \tag{2.35}$$

This is the product of the gain given by Eq. (2.30) and a single-pole low-pass transfer function. The radian pole frequency ω_{0f} in the transfer function is given by

$$\omega_{0f} = \frac{\omega_x}{\frac{R_4}{R_1}\left(1 + \frac{R_1+R_2}{R_3\|R_4}\right)} \tag{2.36}$$

This is also the radian upper half-power or -3 dB cutoff frequency of the amplifier. The Bode magnitude plot for $A_f(j\omega)$ is identical to the one for the inverting amplifier in Fig. 2.3(b), where the gain-bandwidth product is given by

$$\omega_x' = |A_f(j0)|\omega_{0f} = \frac{R_4}{R_1}\left(1 + \frac{R_2}{R_3\|R_4}\right)\omega_{0f} = \omega_x \frac{R_2 + R_3\|R_4}{R_1 + R_2 + R_3\|R_4} \tag{2.37}$$

2.2.10 Integrator

Fig. 2.5(a) shows the circuit diagram of the integrator. If the op amp is ideal, the voltage-gain transfer function of the circuit can be obtained from Eq. (2.19) by replacing R_F with the complex impedance of the capacitor C_F given by

$$Z_F = \frac{1}{C_F s} \tag{2.38}$$

The voltage-gain transfer function is

$$\frac{V_o}{V_i} = -\frac{1}{R_1 C_F s} \tag{2.39}$$

Because a division by s in the frequency domain is equivalent to an integration in the time domain, it follows that this is the transfer function of an inverting integrator with a gain constant of $1/R_1 C_F$. Because it has the units of time, the quantity $R_1 C_F$ is called the *time constant* of the integrator. The input resistance to the circuit is R_1 and the output resistance is R_O. The Bode magnitude plot for V_o/V_i is given in Fig. 2.5(b). It can be seen from the plot that the dc gain of the circuit approaches infinity. This is not possible with a physical op amp.

Because the integrator has a gain that is inversely proportional to the frequency, the gain at high frequencies is lower than the gain at low frequencies. This has the effect of decreasing the amplitude of rapid or fast variations in the output signal. For example, an integrator converts a square wave at its input into a triangular wave at its output.

2.2. THEORY

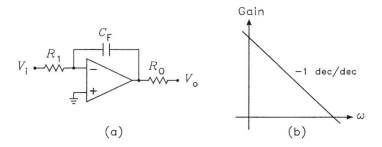

Figure 2.5: (a) Inverting integrator. (b) Bode magnitude plot.

2.2.11 Practical Integrator

Unless negative dc feedback is provided to the input from a following stage, dc offset problems at the op-amp output make it difficult to use physical op amps in integrator circuits. This problem can be solved by connecting a resistor R_F in parallel with the feedback capacitor C_F. This is illustrated in the circuit of Fig. 2.6(a). If the op amp is considered to be ideal, the voltage-gain transfer function for this circuit can be obtained from Eq. (2.19) by replacing R_F with the complex impedance

$$Z_F = R_F \parallel \frac{1}{C_F s} = \frac{R_F}{1 + R_F C_F s} \tag{2.40}$$

The voltage-gain transfer function is

$$\frac{V_o}{V_i} = -\frac{R_F}{R_1} \times \frac{1}{1 + R_F C_F s} = -\frac{1}{R_1 C_F s} \times \frac{R_F C_F s}{1 + R_F C_F s} \tag{2.41}$$

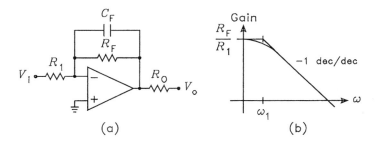

Figure 2.6: (a) Practical integrator. (b) Bode magnitude plot.

This is of the form of the transfer function of an inverting integrator with a time constant $R_1 C_F$ multiplied by a high-pass transfer function which has the upper radian cutoff frequency

$$\omega_1 = \frac{1}{R_F C_F} \tag{2.42}$$

For $s = j\omega$ and $\omega \gg \omega_1$, the high-pass transfer function has a gain magnitude of unity so that the transfer function of Eq. (2.41) is approximately equal to that of Eq. (2.39). Thus for $\omega \gg \omega_1$, the circuit of Fig. 2.6(a) performs as an integrator. The Bode magnitude plot for V_o/V_i is shown in Fig. 2.6(b). For $\omega \ll \omega_1$, it can be seen that the gain magnitude shelves at the value R_F/R_1.

2.2.12 Differentiator

Fig. 2.7(a) shows the circuit diagram of the differentiator. If the op amp is ideal, the voltage-gain transfer function of the circuit can be obtained from Eq. (2.19) by replacing R_1 with the complex impedance of the capacitor C_1 given by

$$Z_1 = \frac{1}{C_1 s} \tag{2.43}$$

The voltage-gain transfer function is

$$\frac{V_o}{V_i} = -R_F C_1 s \tag{2.44}$$

Because a multiplication by the complex frequency s in the frequency domain is equivalent to a differentiation in the time domain, it follows that this is the transfer function of an inverting differentiator with a gain constant $R_F C_1$. Because it has the units of time, the gain constant is also called the *time constant* of the differentiator. The output resistance of the circuit is R_O. The input impedance is the impedance of the capacitor C_1 to virtual ground, i.e. $Z_{in} = 1/C_1 s$. For $s = j\omega$, it follows that the input impedance approaches zero as frequency is increased. The Bode magnitude plot for V_o/V_i is shown in Fig. 2.7(b). It can be seen from the plot that the gain approaches infinity as $\omega \to \infty$. This is impossible with a physical op amp.

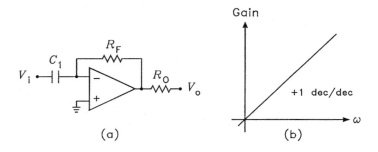

Figure 2.7: (a) Inverting differentiator. (b) Bode magnitude plot.

2.2.13 Practical Differentiator

Because the input impedance to the differentiator consists of a capacitor to a virtual ground, the circuit which drives its V_i input node can be unstable, i.e. it can oscillate. Therefore, a resistor R_1 is often used in series with C_1 as shown in Fig. 2.8(a) to prevent the input impedance from being too low at high frequencies. Because the gain of the differentiator is high at high frequencies, the amplification of out-of-band high-frequency noise can be a problem. To minimize this, a capacitor C_F can be connected in parallel with the feedback resistor R_F as shown in Fig. 2.8(a) to decrease the high-frequency gain. If the op amp is ideal, the voltage-gain transfer function of the circuit can be obtained from Eq. (2.19) by replacing R_F with the complex impedance Z_F given by Eq. (2.40) and R_1 with the complex impedance Z_1 given by

$$Z_1 = R_1 + \frac{1}{C_1 s} = \frac{1 + R_1 C_1 s}{C_1 s} \tag{2.45}$$

The voltage-gain transfer function is

$$\frac{V_o}{V_i} = -R_F C_1 s \times \frac{1}{1 + R_1 C_1 s} \times \frac{1}{1 + R_F C_F s} \tag{2.46}$$

2.3. SPICE OP-AMP MACROMODEL CIRCUITS

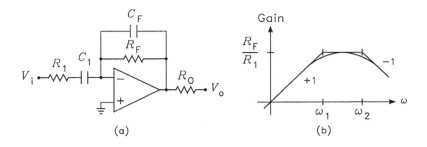

Figure 2.8: (a) Practical differentiator. (b) Bode magnitude plot.

This is of the form of the transfer function of an inverting differentiator with a time constant $R_F C_1$ multiplied by two low-pass filter transfer functions having the upper radian cutoff frequencies

$$\omega_1 = \frac{1}{R_1 C_1} \qquad (2.47)$$

$$\omega_2 = \frac{1}{R_F C_F} \qquad (2.48)$$

For $s = j\omega$ and $\omega \ll \min(\omega_1, \omega_2)$, the circuit acts as a differentiator with a gain that is approximately equal to that given Eq. (2.44). The Bode magnitude plot for V_o/V_i is shown in Fig. 2.8(b). The plot assumes that $\omega_1 < \omega_2$. For $\omega_1 < \omega < \omega_2$, it can be seen from the plot that the asymptotic gain magnitude shelves at the value R_F/R_1.

2.3 SPICE Op-Amp Macromodel Circuits

2.3.1 Quasi-Ideal Macromodel

Macromodel circuits are used to represent the op amp in SPICE simulations. The simplest macromodel is a voltage-controlled voltage source having a voltage gain equal to the dc gain of the op amp. Such a circuit is illustrated in Fig. 2.1(b). In SPICE, a voltage-controlled voltage source is modeled by an "E amplifier". The SPICE subcircuit code for such a macromodel with a gain $A = 2 \times 10^5$ is given below. This is referred to as the *quasi-ideal macromodel subcircuit* in the following. The non-inverting input is node 1, the inverting input is node 2, and the output is node 3. This subcircuit model is useful in SPICE simulations of most op-amp circuits where the op amp is considered to be ideal in the design of the circuit.

```
*QUASI-IDEAL OP-AMP SUBCIRCUIT
.SUBCKT OPAMP 1 2 3
EOUT 3 0 1 2 2E5
.ENDS OPAMP
```

The node numbers in the op-amp subcircuit code are "dummy" node numbers. The actual node numbers in a circuit may are used in the subcircuit call. For example, the code line

```
X1 0 4 9 OPAMP
```

refers to op amp X1 with non-inverting input node 0, inverting input node 4, and output node 9.

2.3.2 Small-Signal AC Macromodel

A more elaborate macromodel is required if the ac frequency response behavior of the op amp is to be modeled. Such a model is given in Fig. 2.9. This circuit models the small-signal input resistance, output impedance, and voltage-gain transfer function of the op amp. Only a single pole is modeled in the voltage-gain transfer function. The SPICE subcircuit codes for the 741 and the TL071 op amps are given below. These are referred to as the *small-signal ac macromodel subcircuits* in the following. The non-inverting input is node 1, the inverting input is node 2, and the output is node 3. The ac macromodel is useful in SPICE simulations of most op-amp circuits where the small-signal frequency response of the op amp is to be modeled.

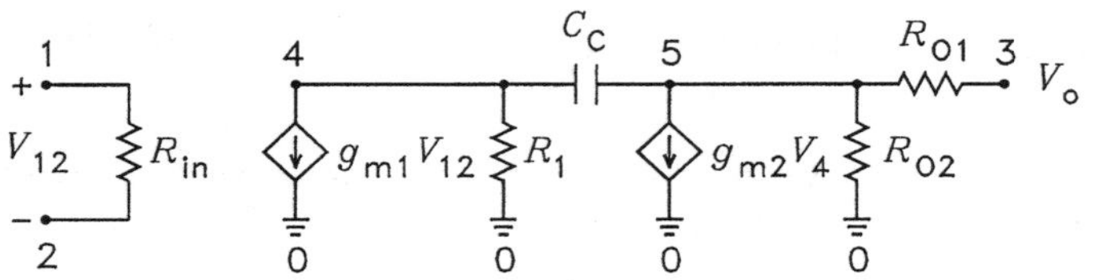

Figure 2.9: Op-amp macromodel.

```
*741 OP-AMP SUBCIRCUIT      *TL071 OP-AMP SUBCIRCUIT
.SUBCKT OA741 1 2 3         .SUBCKT TL071 1 2 3
RIN 1 2 2E6                 RIN 1 2 2E12
GM1 4 0 1 2 1.38E-4         GM1 4 0 1 2 2.83E-4
R1 4 0 1E5                  R1 4 0 1E5
CC 4 5 20E-12               CC 4 5 15E-12
GM2 5 0 4 0 106             GM2 5 0 4 0 283
RO1 3 5 150                 RO1 3 5 50
RO2 5 0 150                 RO2 5 0 25
.ENDS OA741                 .ENDS TL0711
```

The resistor R_O shown connected in series with the output of the op amps in this experiment is included for practical reasons, viz. to reduce the possibility that the circuit will oscillate when a capacitive load such as an oscilloscope probe is connected to it. It should not be included in the theoretical derivations, calculations, or simulations. For these analyses the output of the circuit should be taken directly at the output node of the of the appropriate op amp.

2.4 Preliminary Derivations

(a) Derive Eq. (2.8) for the voltage-gain transfer function for the non-inverting amplifier. **(b)** Derive Eq. (2.19) for the voltage-gain transfer function of the inverting amplifier. **(c)** Derive Eq. (2.30) for the voltage gain of the inverting amplifier with a T feedback network. **(d)** An op amp with a gain bandwidth product $f_x = 1$ MHz is used to realize an inverting amplifier and a non-inverting amplifier each having a midband gain magnitude of 10. What is the bandwidth of each amplifier? Repeat the calculations for gain magnitudes of 5, 2, and 1. **(e)** Derive Eq. (2.41) for the voltage-gain transfer function of a practical integrator. **(f)** Derive Eq. (2.46) for the voltage-gain transfer function of a practical differentiator.

2.5 Preliminary Calculations

2.5.1 Non-Inverting Amplifier

For the circuit of Fig. 2.2a, it is desired to have a voltage gain of $+A_v$, an input resistance of R_i, and an output resistance of R_O. The current drawn by the feedback network is specified to be 0.1 mA when $v_O = 10$ V. Calculate the values of the resistors in the circuit. The 741 op amp has a gain-bandwidth product of 1 MHz. The TL071 op amp has a gain-bandwidth product of 3 MHz. Calculate the pole frequency in Hz for the amplifier for each op-amp type.

2.5.2 Inverting Amplifier

For the circuit of Fig. 2.3a, it is desired to have a voltage gain of $-A_v$, an input resistance of R_I, and an output resistance of R_O. Calculate the required values for R_I, R_F, and R_O. Calculate the pole frequency in Hz for the amplifier if a 741 op amp is used. Repeat the calculation for the TL071.

2.5.3 Inverting Amplifier with T Feedback Network

For the circuit of Fig. 2.4, it is desired to have a voltage gain of $-A_v$, an input resistance of R_I, and an output resistance of R_O. Calculate suitable values for the resistors in the circuit. Calculate the pole frequency in Hz for the amplifier if a 741 op amp is used. Repeat the calculation for the TL071.

2.5.4 Integrator

For the practical integrator in Fig. 2.6a having the voltage-gain transfer function given by Eq. (2.41), determine the frequency range for which the circuit approximates an ideal integrator for the circuit element values $R_1 = 100$ kΩ, $R_F = 3$ MΩ, and $C_F = 0.1$ μF. Use the criterion that the radian frequency should be at least 10 times the radian pole frequency of the high-pass transfer function in Eq. (2.42).

2.5.5 Differentiator

For the practical differentiator in Fig. 2.8a having the voltage-gain transfer function given by Eq. (2.46), determine the frequency range for which the circuit approximates an ideal differentiator for the circuit element values $R_1 = 1$ kΩ, $C_1 = 0.01$ μF, $R_F = 47$ kΩ, and $C_F = 100$ pF. Use the criterion that the radian frequency should one-tenth the lowest radian pole frequency for the two low-pass transfer functions in Eq. (2.46).

2.6 Preliminary SPICE Simulations

2.6.1 Non-Inverting Amplifier

Use *PSpice* to calculate and graphically display the small-signal ac gain versus frequency for the non-inverting amplifier designed in part 1 of the Preliminary Calculations. Use the small-signal ac macromodel of the TL071 for the simulation. The line of code in the main program which "calls" the subcircuit is of the form XNAME N1 N2 N3 OPAMP, where NAME is any name for the op amp (e.g. OA1, OA2, etc.), N1 is the non-inverting input node, N2 is the inverting input node, and N3 is the output node. Because R_O is a floating resistor in the circuit, it should be omitted from the simulation. The SPICE "dot command" for the calculations is the .AC command which has the form .AC DEC N F1 F2, where N is the number of points per decade, F1 is the start frequency, and F2 is the stop frequency. Calculate the gain over the frequency interval from 100 Hz to 1 MHz with 20 points per decade. Use the *PSpice* Probe Graphics Processor to display the gain with a log scale on the vertical axis. (The log scale option can be invoked through the y-axis menu in Probe.) Use the Probe cursor to determine the upper cutoff frequency at which the gain is reduced by a factor of $1/\sqrt{2}$.

Compare this frequency to the pole frequency calculated for the TL071 op amp in part 1 of the Preliminary Calculations.

2.6.2 Inverting Amplifier

Use *PSpice* to calculate the dc small-signal gain, input resistance, and output resistance of the inverting amplifier designed in part 2 of the Preliminary Calculations. Because R_O is a floating resistor in the circuit, it should be omitted from the simulation. The SPICE "dot command" for the calculations is the .TF command. For the simulation, model the op amp with the quasi-ideal macromodel subcircuit. After running *PSpice* with the .TF command, the output data are in the .OUT file.

2.6.3 Inverting Amplifier with T Feedback Network

Use *PSpice* to calculate and graphically display the small-signal ac gain versus frequency for the inverting amplifier with T feedback network designed in part 3 of the Preliminary Calculations. Because R_O is a floating resistor in the circuit, it should be omitted from the simulation. For the simulations, use 20 points per decade over the frequency interval from 100 Hz to 1 MHz. The gain should be displayed with a log scale on the vertical axis. Use the ac small-signal model for the 741 op amp for the simulation. Use the Probe cursor to determine the upper cutoff frequency at which the gain is reduced by a factor of 0.707. Compare this frequency to the pole frequency calculated for the 741 op amp in step 3 of the Preliminary Calculations.

2.6.4 Integrator

Use *PSpice* to calculate and graphically display the small-signal ac gain versus frequency for the practical integrator of step 4 of the Preliminary Calculations. Use the small-signal ac macromodel of the TL071 for the simulation. The Bode magnitude plot should be displayed with a log scale on the vertical axis with 20 points per decade. From the Bode plot, determine the frequency range for which the slope of the magnitude plot is −1 decade/decade on log-log scales. This is the frequency range for which the circuit acts as an integrator. Compare this to the frequency range determined in part 4 of the Preliminary Calculations.

2.6.5 Differentiator

Use *PSpice* to calculate and graphically display the small-signal ac gain versus frequency for the practical differentiator of part 5 of the Preliminary Calculations. Use the small-signal ac macromodel of the 741 for the simulation. The gain should be displayed with a log scale on the vertical axis using 20 points per decade. From the plot, determine the frequency range for which the slope of the Bode magnitude plot is +1 decade/decade on log-log scales. This is the frequency range for which the circuit acts as a differentiator. Compare this to the frequency range determined in step 4 of section 5.

2.7 Experimental Procedures

2.7.1 Preparation

The op-amp types specified for this experiment are the 741 and TL071. In the case of the 741, different manufacturers use different prefixes and suffixes with the part number, but it is universally known as the 741. Both the 741 and the TL071 are fabricated in 8-pin dual in-line packages called DIP's. The 8-pin DIP is illustrated in Fig. 2.10, where the view is with the pins pointing away from the viewer. Pin 2 is the inverting input, pin 3 is the non-inverting input, pin 4 is the negative dc power supply input, pin 6 is the output, pin 7 is the positive dc power supply input, and pin 8 is not used. Pins 1 and 5 are used only if it is desired to null a dc offset voltage that can appear at the output due to non-ideal op-amp characteristics. In this case, the outside leads of a 10 kΩ potentiometer are connected to pins 1 and 5 and the potentiometer

2.7. EXPERIMENTAL PROCEDURES

wiper is connected to the negative dc power supply rail. The potentiometer wiper can be adjusted to null, i.e. reduce to zero, the dc offset voltage at the op-amp output.

Dual in line package (DIP) with pin numbers for the 741 and TL071 op amps.

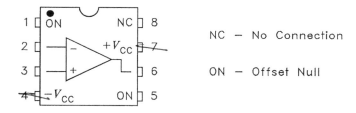

Figure 2.10: Dual in line package (DIP) with pin numbers for the 741 and LF351 op amps.

For the circuits specified in the procedures, each power supply rail is decoupled with a 100 Ω 1/4 W resistor and a 100 µF 25 V (or greater) capacitor. Because the capacitors are electrolytic types, they must be connected with the proper polarity. The purpose of the resistors in the decoupling networks is to limit the current to the circuit in the event of a mistake. The capacitors provide a signal short circuit to ground on the rails to prevent oscillations which can be induced by positive feedback from the power supply rails into the op amp. A 100 Ω resistor is specified in series with the output of each amplifier circuit. This makes each circuit have an output resistance of 100 Ω. Without this resistor, an op amp can oscillate with a capacitive load such as the capacitance of the oscilloscope and oscilloscope leads.

2.7.2 Non-Inverting Amplifier

(a) Assembly

Assemble the non-inverting amplifier circuit shown in Fig. 2.11 on the breadboard. Use a 741 op amp. The circuit element values are $R_I = 10$ kΩ, $R_1 = 1$ kΩ, $R_F = 3$ kΩ, and $R_O = 100$ Ω.

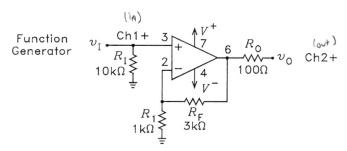

Figure 2.11: Non-inverting amplifier. gain: $1 + \frac{R_F}{R_1} = 4$

(b) Power Supply Adjustment

Set the dual power supply to produce dc rail voltages of $+15$ V and -15 V. The common or ground on the power supply is connected to the circuit ground on the breadboard.

(c) Function Generator Adjustment

Set the function generator for a minimum amplitude sine wave output signal with a frequency of 1 kHz and a dc offset (the default value) of zero. Connect the output of the function generator to the op-amp circuit. Connect channel 1 of the dual channel oscilloscope to the circuit input and channel 2 to its output. Adjust the function generator to produce a sine wave input signal with an ac peak amplitude of 1 V. Record the peak amplitudes of the input and output voltage waveforms and calculate the voltage gain.

(d) Bandwidth Measurement

Set the function generator to produce a 1 V ac peak-to-peak signal at the op-amp output, increase the generator frequency until the op-amp output voltage drops by a factor of $1/\sqrt{2}$. Record the frequency as the upper half-power or -3 dB cutoff frequency of the amplifier.

(e) Clipping

Set the generator frequency to 100 Hz. Increase the amplitude of the input voltage until the op-amp output is clipped. Record (sketch or print) the output waveform when the amplifier is just at the verge of clipping (just beginning to distort) and when it is in hard clipping (the upper and lower peaks are straight lines).

2.7.3 Inverting Amplifier

(a) Assembly

Assemble the inverting amplifier circuit shown in Fig. 2.12 on the breadboard. Use a 741 op amp. The circuit element values are $R_1 = 1$ kΩ, $R_F = 6.8$ kΩ, and $R_O = 100$ Ω. The power supply connections and power supply decoupling networks are not shown in the circuit diagram but must be included. Be certain that pin 7 is connected to the positive rail and pin 4 to the negative rail.

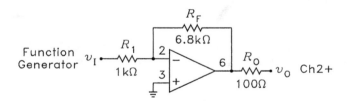

Figure 2.12: Inverting amplifier.

(b) Bandwidth and Clipping Measurements

Repeat steps (b) through (e) of part 2.7.2.

2.7.4 Inverting Amplifier with T Feedback Network

(a) Assembly

Assemble the inverting amplifier with a T feedback network shown in Fig. 2.13. The power supply connections and decoupling networks are not shown but must be included. Make certain that pin 4 is connected to the negative rail and pin 7 to the positive rail. Resistors R_A and R_B form a L-pad input attenuator which limits the maximum signal voltage applied to the v_I input node. (This attenuator is not necessary if the function generator has a stepped attenuator output and should be omitted and the function generator

2.7. EXPERIMENTAL PROCEDURES

directly connected to the v_i node.) The gain of the amplifier is V_o/V_i; the L pad attenuator is merely being used to reduce the output of the function generator to a level that will not result in clipping for this high gain amplifier. The output resistance of the L-pad is $R_A \| R_B$. Use a 741 op amp. The circuit element values are $R_A = R_1 = R_3 = 1$ kΩ, $R_B = R_0 = 100$ Ω, $R_2 = R_4 = 10$ kΩ.

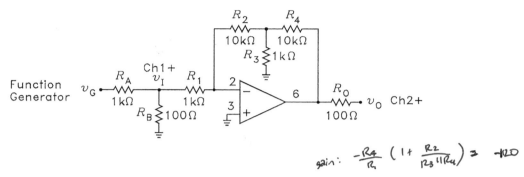

gain: $-\frac{R_4}{R_1}\left(1 + \frac{R_2}{R_3 \| R_4}\right) = -120$

Figure 2.13: Inverting amplifier with T feedback network.

(b) Function Generator Adjustment

Apply a sine wave with a peak voltage of 1 V, a dc level of zero (the default value), and a frequency of 1 kHz to the circuit input.

(c) Bandwidth and Clipping Measurements

Repeat steps (b) through (e) of part 2.7.2 with the dual channel oscilloscope set to observe the waveforms at the v_I and the v_0 nodes.

2.7.5 Integrator

(a) Assembly

Assemble the practical inverting integrator circuit shown in Fig. 2.14. The power supply connections and power supply decoupling networks are not shown but must be included. Use a 741 op amp. The circuit element values are $R_1 = 100$ kΩ, $R_F = 3$ MΩ, $R_0 = 100$ Ω, and $C_F = 0.1$ μF. Make certain that pin 7 is connected to the positive rail and pin 4 to the negative rail. Use the dual channel oscilloscope to observe the input and output voltage waveforms.

(b) Function Generator Adjustment

Apply a sine wave with a dc level of zero (the default value), a peak amplitude of 5 V, and a frequency of 100 Hz to the input of the circuit. Record the peak amplitudes of the input and output voltage waveforms and calculate the gain of the circuit.

(c) Frequency Response

Vary the frequency of the generator over the range from 10 Hz to 1 kHz and measure the gain as a function of frequency. If the output signal is clipped, reduce the level of the input signal until an undistorted output is obtained. Use the measured data to plot the gain versus frequency of the circuit. For the plot, use log scales for both the vertical axis and the horizontal axis.

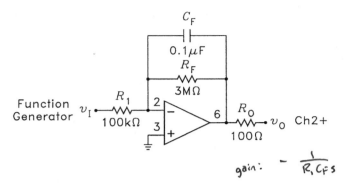

Figure 2.14: Inverting integrator.

(d) Square Wave Response

Change the input to a 100 Hz square wave with a dc level of zero and a peak amplitude of 5 V. Sketch or print the input and output time domain waveforms. Manually vary the frequency of the function generator from 10 Hz to 1 kHz and note the change in the output signal shape in the time domain as the frequency frequency of the source is changed. If the output signal is clipped, reduce the level of the input signal until an undistorted output is obtained.

2.7.6 Differentiator

(a) Assembly

Assemble the practical differentiator circuit shown in Fig. 2.15. The power supply connections and power supply decoupling networks are not shown but must be included. Use a TL071 op amp. Be certain that pin 7 is connected to the positive rail and pin 4 to the negative rail. The circuit element values are $R_1 = 1$ kΩ, $C_1 = 0.01$ μF, $R_F = 47$ kΩ, $C_F = 100$ pF, and $R_0 = 100$ Ω.

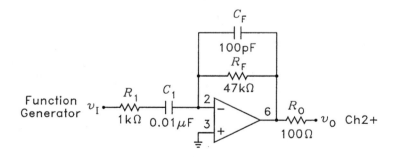

Figure 2.15: Differentiator circuit.

(b) Sine-Wave Response

Apply a sine wave to the circuit input with a dc level of zero (the default value), a peak amplitude of 1 V, and a frequency of 100 Hz. Sketch or print the output voltage waveform (time domain display) showing any distortion which appears on the waveform. (The distortion would not occur if the input signal were a pure sine wave. Some function generators do not generate pure sine waves. Its output signal may be corrupted by high-frequency harmonics of the fundamental frequency.)

(c) Frequency Response

Vary the frequency of the generator over the range from 10 Hz to 1 kHz and measure the gain of the circuit as a function of frequency. If the output signal is clipped, reduce the level of the input signal until the peaks are no longer clipped. Use the measured data to plot the gain versus frequency of the circuit. For the plots, use log scales for both the vertical axis and the horizontal axis.

(d) Triangular-Wave Response

Change the input to a triangular wave and repeat step (b). Namely, observe the output of the circuit (time domain display) when the input is a triangular wave. Manually vary the frequency of the function generator from 10 Hz to 1 kHz and note any change in the shape of the output signal in the time domain as the frequency of the source is changed. If the output signal is clipped, reduce the level of the input signal until the peaks are no longer clipped.

(e) Square-Wave Response

Change the input to a square wave and repeat step (b). Manually vary the frequency of the function generator from 10 Hz to 1 kHz and note the change in the shape output signal in the time domain as the frequency of the source is changed.

2.8 Laboratory Report

The laboratory report should be include all appropriate sketches and tables. The verification sheet must be included. Answer any supplementary questions posed by the laboratory instructor.

2.9 References

1. R. F. Coughlin and F. F. Driscoll, *Operational Amplifiers and Linear Integrated Circuits*, Prentice-Hall, 1987.
2. P. Horowitz and W. Hill, *The Art of Electronics,* 2nd edition, Cambridge University Press, 1989.
3. P. Horowitz and I. Robinson, *Laboratory Manual for The Art of Electronics,* Cambridge University Press, 1981.
4. E. J. Kennedy, *Operational Amplifier Circuits*, Holt Rhinehart and Winston, 1988.
5. J. H. Krenz, *An Introduction to Electrical and Electronic Devices,* Prentice-Hall, 1987.
6. J. M. McMenamin, *Linear Integrated Circuits*, Prentice-Hall, 1985.
7. A. S. Sedra and K. C. Smith, *Microelectronic Circuits*, 4th edition, Holt, Oxford, 1998.
8. W. D. Stanley, *Operational Amplifiers with Linear Integrated Circuits*, 2nd edition, Merrill, 1989.

Chapter 3

Elementary Op-Amps 2

3.1 Object

This experiment is a continuation of Experiment 2. Several op-amp circuits are assembled and evaluated. In addition, performance limitations imposed by the non-ideal characteristics of physical op amps are covered. The non-ideal characteristics that are examined are peak clipping, output current limiting, dc offset, input bias current, finite open-loop gain and bandwidth, and slew rate limiting. The op amps used in the circuits are the 741 and the TL071. The op-amp macromodel specified for the computer simulations simulates a finite gain and bandwidth, non-infinite input resistance, a non-zero output impedance, peak clipping at the output, output current limiting, and slew rate limiting.

3.2 Theory

3.2.1 Peak Clipping

The peak clipping voltage of an op amp is the maximum amplitude output voltage that it can put out. An op amp is driven into peak clipping when amplitude of the input signal is too large. If the input signal is a sine wave, the output signal appears as a sine wave with its peaks "flattened out", i.e. clipped off. The 741 op amp has a peak clipping voltage that is approximately 1.9 V lower than the power supply rails. The TL071 clips at a voltage that is approximately 1.5 V lower than the rails. For the usual ±15 V rails, it follows that the 741 clips at approximately ±13.1 V and the TL071 clips at approximately ±13.5 V.

Fig. 3.1 shows the circuit diagram of an inverting amplifier which may be used to measure the op-amp clipping voltage. With a signal generator connected to the input and an oscilloscope connected to the output, the amplitude of the generator is increased until peak clipping is observed. A sine-wave test signal is normally used with a typical frequency of 100 Hz.

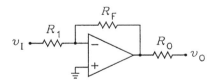

Figure 3.1: Inverting amplifier for measuring the peak clipping level, output current limit, and slew rate of an op-amp.

3.2.2 Output Current Limiting

The internal transistors in an op amp which supply output current to an external load can fail if the output current becomes too large. To protect these transistors, protection circuits are included inside the op amp which limit the maximum output current. For the 741, the output current is limited to approximately 26 mA. For the TL071, it is limited to approximately 31 mA. When the current limit circuits are activated, the op-amp output voltage will appear to be peak clipped. For example, with a 100 Ω load resistor on the op amp, the maximum output voltage from a 741 is approximately $0.026 \times 100 = 2.6$ V while it is approximately $0.031 \times 100 = 3.1$ V for the LF 351.

The circuit of Fig. 3.1 can be used to measure the current limit of an op amp. With a typical value for R_O of 100 Ω and the v_O terminal grounded, a signal generator is connected to the input and an oscilloscope is connected to the output. The generator amplitude is increased until the output voltage appears to be peak clipped. The peak voltage divided by R_O is the current limit. A sine-wave test signal is normally used with a typical frequency of 100 Hz.

3.2.3 Slew Rate Limiting

General purpose op amps have internal capacitors which prevent the circuits from oscillating when feedback is added. When a circuit oscillates, it puts out a signal with no input signal. The capacitors are known as frequency compensation capacitors because they affect the frequency response of the op amp, i.e. the gain versus frequency. With some op amps, the frequency compensation capacitors are connected to external terminals. Both the 741 and the TL071 have internal compensation capacitors which are fabricated on the integrated circuit. For the 741, the capacitor has a value of 30 pF. For the TL071, it is 10 pF.

The slew rate of an op amp is the maximum time rate of change (time derivative) of its output voltage. The slew rate is set by the compensation capacitor and the maximum current available to charge the capacitor. It is given by

$$SR = \left|\frac{dv_O}{dt}\right|_{\max} = \frac{I}{C_c} \quad (3.1)$$

where C_c is the compensation capacitor and I is the maximum current available to charge it. The current I is normally set by a constant current source in the integrated circuit. The slew rate of an op amp is usually specified with the units *volts per microsecond*.

An op amp exhibits slew rate limiting or slewing when the rate of change of the input voltage multiplied by the voltage gain of the amplifier is greater than the slew rate of the op amp. It is normally measured with a square wave applied at the op-amp input. The nominal value of the slew rate for the 741 is 0.5 V/μs. For the TL071, it is 13 V/μs.

The inverting amplifier circuit shown in Fig. 3.1 can be used to measure the slew rate of an op amp. A square wave input signal with a typical frequency of 10 kHz is applied to the circuit and the output voltage is observed on an oscilloscope. The amplitude of the input signal is increased until the maximum slope of the oscilloscope waveform does not change with a further increase in input voltage. The maximum slope of the waveform is the op-amp slew rate. It is common to observe slight differences when the slew rate is measured on the rising and falling portions of the output voltage waveform.

3.2.4 DC Offset and Bias Currents

If both the inverting and non-inverting terminals of an ideal op amp are connected to ground, the op-amp output voltage is zero. For physical op amps, however, the dc output voltage that is present when the inputs are grounded is not zero due to unavoidable and unpredictable mismatches between the active devices in the integrated circuit. An undesirable dc output voltage is called a dc offset voltage. When negative feedback is used with the op amp, the dc offset is reduced by the feedback and is usually in the range of several millivolts. In many applications, this does not pose a significant performance limitation.

An ideal op amp has no current flowing in either the inverting or the non-inverting input leads. In contrast, physical op amps have bias currents that flow in the input leads. The order of magnitude of the

3.2. THEORY

bias currents is determined by the type of active devices used in the op-amp input stage. The 741 has a bipolar junction transistor (BJT) input stage while the TL071 has a junction field effect transistor (JFET) input stage. Thus the input bias currents will be much larger for the 741 than for the TL071. These currents flow through the external resistive networks connected to the op-amp inputs and cause a dc voltage which is amplified and appears as a dc offset voltage at the op-amp output. The component of the dc voltage present at the output due to the bias currents can be minimized by making the external resistances seen looking out of the inverting and the non-inverting terminals equal, where the resistances are calculated with both the generator and the op-amp output voltages zeroed. This assumes that the bias currents flowing in the two input terminals are approximately equal. For the 741, a nominal value for the input bias currents is 81 nA. For the LF 351, a nominal value is 13 pA.

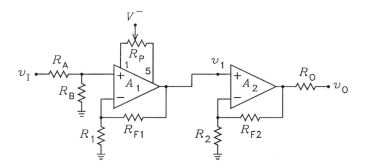

Figure 3.2: Two-stage amplifier for illustrating the effects of a non-zero input currents in the op-amp.

Many op amps have a provision for nulling the dc offset voltage that appears at the output. For the 741 and TL071 op amps, the two end terminals of a potentiometer (or "pot" for short) are connected to pins 1 and 5 of the op amp and the wiper of the pot is connected to the negative power supply rail. (The pin layout for the op amps is given in Experiment 1). By adjusting the wiper, the dc offset voltage can be reduced to zero. This assumes that the input voltage to the op amp does not have a dc component.

Whether the dc offset voltage appearing at the output of an op amp poses a significant performance limitation depends on the application. Consider the two-stage dc coupled amplifier shown in Fig. 3.2. The dc offset voltage at the output of A1 is amplified by A2. If the gain of A2 is high, it follows that the predominant component of the dc offset at V_O is caused by the dc offset at the output of op amp A1. Therefore, offset compensation is most effective if it is applied to A1. If A1 is a 741 or a TL071, this can be accomplished with the pot R_P as shown in the figure.

3.2.5 Band-Pass Inverting Amplifier

The circuit of Fig. 3.3 is an inverting amplifier that has a voltage-gain transfer function that is of the form of a band-pass filter. That is, the amplifier passes signals inside a frequency band and rejects signals outside that band. (Although the circuit is similar to the practical differentiator circuit in Experiment 2, the pole frequencies in the voltage-gain transfer function are chosen differently.) To calculate the voltage-gain transfer function, define the complex impedances Z_1 and Z_F as follows:

$$Z_1 = R_1 + \frac{1}{C_1 s} = \frac{1 + R_1 C_1 s}{C_1 s} \tag{3.2}$$

$$Z_F = R_F \parallel \left(\frac{1}{C_F s}\right) = \frac{R_F}{1 + R_F C_F s} \tag{3.3}$$

If the op amp is assumed to be ideal, the voltage-gain transfer function is given by

$$\frac{V_o}{V_i} = -\frac{Z_F}{Z_1} = -\frac{R_F}{R_1} \times \frac{R_1 C_1 s}{1 + R_1 C_1 s} \times \frac{1}{1 + R_F C_F s} \quad (3.4)$$

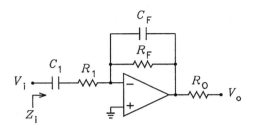

Figure 3.3: Inverting amplifier having a band-pass voltage-gain transfer function.

This is of the form of the product of three terms – a gain constant $(-R_F/R_1)$ which sets the midband gain, high-pass transfer function which sets the lower half-power frequency, and a low-pass transfer function which sets the upper half-power frequency. The radian pole frequency of the high-pass transfer function, which is the radian lower half-power frequency, is $\omega_l = \omega_1 = 1/(R_1 C_1)$. The radian pole frequency of the low-pass transfer function, which is the radian upper half-power frequency, is $\omega_u = \omega_2 = 1/(R_F C_F)$. Assume that $R_1 C_1 \gg R_F C_F$ so that the pole in the high-pass function occurs at a much lower frequency than the pole in the low-pass function. This makes the gain in the band between these two frequencies approximately a constant given by $V_o/V_i = -R_F/R_1$ which is the midband gain. The minus sign makes this circuit an inverting amplifier since in the midband frequencies the input and output signals are 180° out of phase. For this circuit the midband frequencies are those for which the capacitor C_1 is a short circuit and the capacitor C_F as well as the internal capacitors inside the op amp are open circuits.

Because $V_- = 0$, the input impedance transfer function for the circuit of Fig. 3.3 is given by Eq. (3.2). This is of the form of a constant (R_1) divided by a high-pass transfer function having the time constant $R_1 C_1$. At low frequencies, the impedance is capacitive while at high frequencies it is resistive.

3.2.6 Non-inverting Amplifier with 100% DC Feedback

The circuit shown in Fig. 3.4 is a non-inverting amplifier which has a capacitor in the feedback network that gives 100% feedback at dc. Audio circuits are often designed with 100% dc feedback to minimize dc offset problems at the output. If the op amp is ideal, the voltage gain transfer function is calculated as follows:

$$\frac{V_o}{V_i} = 1 + \frac{R_F}{R_1 + 1/C_1 s} = \frac{1 + (R_1 + R_F) C_1 s}{1 + R_1 C_1 s} \quad (3.5)$$

This transfer function is called a shelving transfer function because the gain shelves at different non-zero values at low frequencies and at high frequencies. At low frequencies, it predicts a gain of unity. At high frequencies, it predicts a gain of $1 + R_F/R_1$. (The mid-band frequency range for this circuit is taken to be those frequencies for which capacitor C_1 is a short circuit but the capacitors inside the op amp are open circuits. This makes the mid-band gain $1 + R_F/R_1$.) The transfer function has the radian pole frequency $\omega_p = 1/(R_1 C_1)$ and the radian zero frequency $\omega_z = 1/[(R_1 + R_F) C_1]$. Because the high-frequency gain is larger than the low-frequency gain, the transfer function is called a high-pass shelving transfer function.

3.2.7 Single-Power Supply Amplifiers

In applications such as automobile audio circuits, op amps are operated from a single power supply voltage. Fig. 3.5 shows the circuit diagrams of a non-inverting amplifier and an inverting amplifier which are operated

3.2. THEORY

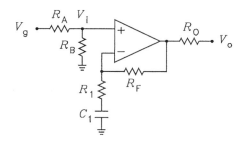

Figure 3.4: Non-inverting amplifier with 100% dc feedback.

from a single power supply. At dc, the capacitors in both circuits are open circuits. By voltage division, the dc voltage at each non-inverting input is $V^+/2$. Each op amp forces the dc voltage at its inverting input to be $V^+/2$. With 100% dc feedback, the output voltage must equal the voltage at the inverting input. It follows that the quiescent output voltage is $V^+/2$. Thus each op amp is biased at a dc voltage that allows maximum symmetrical output voltage swing, i.e. $+V^+/2$ and $-V^+/2$.

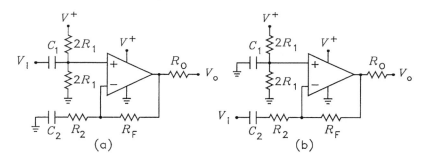

Figure 3.5: Single power supply amplifiers. (a) Non inverting. (b) Inverting.

For the non-inverting amplifier of Fig. 3.5(a), write

$$\frac{V_+}{V_i} = \frac{R_1}{R_1 + 1/C_1 s} = \frac{R_1 C_1 s}{1 + R_1 C_1 s} \tag{3.6}$$

$$\frac{V_o}{V_+} = 1 + \frac{R_F}{R_2 + 1/C_2 s} = \frac{1 + (R_2 + R_F) C_2 s}{1 + R_2 C_2 s} \tag{3.7}$$

It follows that the voltage-gain transfer function of the circuit is given by

$$\frac{V_o}{V_i} = \frac{V_+}{V_i} \times \frac{V_o}{V_+} = \frac{R_1 C_1 s}{1 + R_1 C_1 s} \times \frac{1 + (R_2 + R_F) C_2 s}{1 + R_2 C_2 s} \tag{3.8}$$

The transfer function is of the form of a high-pass function multiplied by a high-pass shelving function. It has the radian pole frequencies $\omega_{p1} = 1/(R_1 C_1)$ and $\omega_{p2} = 1/(R_2 C_2)$ and the radian zero frequencies $\omega_{z1} = 0$ and $\omega_{z2} = 1/[(R_2 + R_F) C_2]$. If $R_1 C_1 = (R_2 + R_f) C_2$, the pole in the high-pass function cancels the zero in the high-pass shelving function which makes the overall transfer function a high-pass function. At midband, the capacitors in the circuit are signal short circuits while those inside the op amp are open circuits and the midband voltage gain is $1 + R_F/R_2$. The midband input resistance is R_1. The output resistance is R_O.

For the inverting amplifier of Fig. 3.5(b), the voltage-gain transfer function can be written

$$\frac{V_o}{V_i} = -\frac{R_F}{R_2 + 1/C_2 s} = -\frac{R_F}{R_2} \times \frac{R_2 C_2 s}{1 + R_2 C_2 s} \tag{3.9}$$

This is of the form of a constant $(-R_F/R_2)$ multiplied by a high pass transfer function having the radian pole frequency $\omega_p = 1/(R_2 C_2)$. At midband, the capacitors in the circuit are signal short circuits while those inside the op amp are open circuits and the voltage gain is $V_o/V_{i1} = -R_F/R_2$. The midband input resistance is R_2. The output resistance is R_O.

3.3 SPICE Macromodels

Macromodel circuits are used to represent the op amp in SPICE simulations. In Experiment 2, the *quasi-ideal macromodel* and the *small-signal ac macromodel* are introduced. For simulations which must predict peak clipping, slew-rate limiting, and current limiting, a more advanced macromodel is required. Such models are given in Fig. 3.6 for op amps with a BJT input stage such as the 741 and in Fig. 3.7 for op amps with a JFET input stage such as the TL071. The SPICE subcircuit codes for the 741 and the TL071 op-amp types are given below.

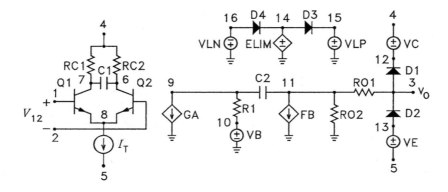

Figure 3.6: Macromodel for op amp with BJT diff amp.

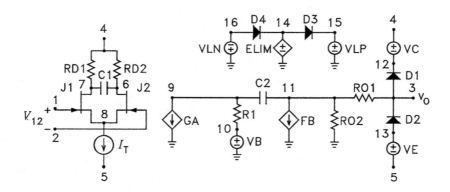

Figure 3.7: Macromodel for op amp with JFET diff amp.

```
*741 OP-AMP SUBCIRCUIT      *TL071 OP-AMP SUBCIRCUIT
.SUBCKT OA741 1 2 3         .SUBCKT TL071 1 2 3
VCC 4 0 DC 15               VCC 4 0 DC 15
VEE 0 5 DC 15               VEE 0 5 DC 15
IT 8 5 1.016E-5             IT 8 5 1.95E-4
VB 10 0 DC 0                VB 10 0 DC 0
VC 4 12 DC 2.6              VC 4 12 DC 2.2
VE 13 5 DC 2.6              VE 13 5 DC 2.2
VLP 15 0 DC 25              VLP 15 0 DC 30
VLN 0 16 DC 25              VLN 0 16 DC 30
Q1 7 1 8 QX                 J1 7 1 8 JX
Q2 6 2 8 QX                 J2 6 2 8 JX
D1 3 12 DX                  D1 3 12 DX
D2 13 3 DX                  D2 13 3 DX
D3 14 15 DX                 D3 14 15 DX
D4 16 14 DX                 D4 16 14 DX
RC1 4 7 7597                RD1 4 7 3536
RC2 4 6 7597                RD2 4 6 3536
R1 9 10 1E5                 R1 9 10 1E5
RO1 3 11 150                RO1 3 11 50
RO2 11 0 150                RO2 11 0 25
C1 6 7 4.664E-12            C1 6 7 3.498E-12
C2 9 11 20E-12              C2 9 11 15E-12
GA 9 0 6 7 1.377E-4         GA 9 0 6 7 2.828E-4
FB 11 0 POLY(5) VB VC VE    FB 11 0 POLY(5) VB VC VE
+VLP VLN 0 10.61E6 -10E6    +VLP VLN 0 28.29E6 -30E6
+10E6 10E6 -10E6            +30E6 30E6 -30E6
ELIM 14 0 11 3 6.667        ELIM 14 0 11 3 20
.MODEL QX NPN(IS=8E-16      .MODEL JX NJF(IS=12.5E-12
+BF=62.5)                   +BETA=250.1E-6 VTO=-1)
.MODEL DX D(IS=8E-16)       .MODEL DX D(IS=8E-16)
.ENDS OA741                 .ENDS TL071
```

In the circuits of Figs. 3.6 and 3.7, the dc gain is set by the transconductance of the input diff-amp, the load resistances for the diff amp (R_{C1} and R_{C2}), and the gain constant of the current-controlled current source FB for the current component that is controlled by the current through VB. The first or dominant pole in the voltage-gain transfer function is set by C_2. The second pole is set by C_1. The slew rate is set by the ratio of the tail current I_T to C_2. When diodes D_1 and D_2 conduct, the output voltage is clipped. The voltage sources VC and VD determine the output clipping voltage. When diodes D_3 and D_4 conduct, the output current is limited. The voltage sources VLP and VLN determine the output current limit.

The resistor R_O shown connected to the output of the op amps in the circuits in this experiment is included for practical reasons, viz. to reduce the possibility that the circuit will oscillate when a capacitive load such as an oscilloscope probe is connected to it. It should not be included in the theoretical derivations, calculations, or simulations. For these analyses the output of the circuit should be taken directly at the output node of the of the appropriate op amp.

3.4 Preliminary Derivations

1. An op amp is required to put out a sine-wave signal with a peak level of 10 V. **(a)** What is the highest frequency that the op amp will put out the signal without slew rate limiting if it is a 741 with a slew rate of 0.5 V/μsec? **(b)** What is the highest frequency that the op amp will put out the signal without slew rate limiting if it is a TL071 with a slew rate of 13 V/μsec?

2. (a) Derive Eq. (3.4) for the voltage gain transfer function and Eq. (3.2) for the input impedance transfer function of an inverting band-pass amplifier. **(b)** Derive Eq. (3.5) for the voltage gain transfer function of the non-inverting amplifier with 100% dc feedback. **(c)** Derive Eq. (3.8) for the voltage-gain transfer function of the single power supply non-inverting amplifier. **(d)** Derive Eq. (3.9) for the voltage gain transfer function of the single power supply inverting amplifier.

3.5 Preliminary Calculations

1. For the circuit of Fig. 3.3, denote the lower half-power cutoff frequency by f_ℓ, the upper half-power cutoff frequency by f_u, and the midband gain by $V_o/V_i = -A_v$. Numerical values for these parameters should have been specified by the laboratory instructor in the preceding lab period. If $C_1 = 0.1$ μF, what must be the values of R_1, R_F, and C_F? Assume the op amp to be ideal.

2. For the non-inverting amplifier of Fig. 3.4, denote the pole frequency in the voltage-gain transfer function by f_p and the midband voltage gain by $V_o/V_i = A_v$. Numerical values for these parameters should have been specified by the laboratory instructor in the preceding lab period. If $C_1 = 0.1$ μF, calculate the required values of R_1 and R_F. If the overall midband gain is to be unity and $R_B = 100$ Ω, calculate the required value for R_A.

3. (a) For the non-inverting amplifier of Fig. 3.5(a), denote the midband voltage gain by $V_o/V_i = A_v$ and the lower half-power cutoff frequency by f_ℓ. Numerical values for these parameters should have been specified by the laboratory instructor in the preceding lab period. If $C_1 = C_2 = 0.1$ μF, calculate the required values for R_1, R_2, and R_F if there is to be a pole-zero cancellation in the voltage gain transfer function of Eq. (3.8), i.e. $R_1 C_1 = (R_2 + R_F) C_2$. **(b)** For the non-inverting amplifier of Fig. 3.5(b), denote the midband voltage gain by $V_o/V_i = -A_v$ and the lower half-power cutoff frequency by f_ℓ. Numerical values for these parameters should have been specified by the laboratory instructor in the preceding lab period. If $C_2 = 0.1$ μF, determine values of R_2 and R_F that meet this design specification. Why are the values of C_1 and R_1 immaterial for this circuit?

3.6 Preliminary SPICE Simulations

The op-amp macromodels and SPICE subcircuit codes for the 741 and the TL071 op amps given in Section 3.3 are to be used in the simulations. In using SPICE subcircuits, the "call" from the main SPICE program for the subcircuit named OAXYZ is of the form X N1 N2 N3 OAXYZ, where N1 is the node number of the non-inverting input, N2 is the node number of the inverting input, and N3 is the node number of the output. In simulating circuits that contain more than one op amp of the same type, it is necessary to use only one subcircuit for that op-amp type.

3.6.1 Nonideal Properties of Op-Amp

The following simulations are to be performed for both the 741 and the TL071 op amps. **(a)** The circuit of Fig. 3.1 has the element values $R_1 = 1$ kΩ and $R_F = 10$ kΩ. The resistor R_O is to be omitted, i.e. consider it to be a short circuit. Write the SPICE code to perform a .dc sweep on the circuit for V_G in the range -2 V $\leq V_G \leq +2$ V with an increment of $\Delta V_G = 0.1$ V. Use the PROBE graphics post processor to display the output voltage V_O versus the input voltage V_G. From the output, determine the positive and negative peak clipping voltages of the op amp. **(b)** Modify the SPICE code to add a resistor $R_L = 100$ Ω from the op-amp output terminal to ground. Repeat the preceding part and use the graph of V_O versus V_G to determine the positive and negative current limits of the op amp. **(c)** Remove the resistor R_L from the SPICE code. Modify the code to display slewing on the output voltage waveform. For v_I, use an independent PULSE source to define a 10 kHz square wave having an initial value of -0.1 V, a pulsed value of $+0.1$ V, a time delay of 0, rise and fall times of 1 ns, and a pulse width of one-half the period. The appropriate "dot command" in

SPICE is the `.TRAN` command. Display the output voltage waveform as a function of time for two cycles of the input square wave. Use the cursor feature of the `PROBE` graphics processor to solve for the positive and negative slew rates of the op amp.

3.6.2 Inverting and Noninverting Amplifiers

Simulate the amplifiers designed in the Preliminary Calculations. Use the simplest model of the op amp—the quasi ideal op amp. Perform an `AC` analysis. Use an `AC` source for v_i with an amplitude of 1 and plot the magnitude of the gain versus frequency. Use a start frequency that is less that one tenth of the smallest critical frequency (a critical frequency is either a pole or zero) and a stop frequency that is at least ten times the largest critical frequency.

3.7 Experimental Procedures

3.7.1 Preparation

(a) Prepare the electronic breadboard to provide buses for the positive and negative power supply rails and the circuit ground. Each power supply rail should be decoupled with a 100 Ω 1/4 W resistor, and a 100 μF 25 V (or greater) capacitor. The resistors are connected in series with the external power supply leads and the capacitors are connected from power supply rail to ground on the circuit side of the resistors. The capacitors must be installed with proper polarity to prevent reverse polarity breakdown. (b) The op amps specified for this experiment are the 741 and the TL071. The pin configurations for these are given in Experiment 1. (When viewed from the top with the pins away from the viewer, pin 2 is the inverting input, pin 3 is the non-inverting input, pin 4 is $-V_{CC}$, pin 7 is $+V_{CC}$, pin 6 is the output, and pins 1 and 5 are used for the optional offset null.) The power supply voltages are $+15$ V and -15 V for all steps in this procedure except for the single power supply circuits. In this case, the power supply voltage is $+15$ V.

3.7.2 Peak Clipping

This part is to be performed using the 741 op amp. (a) Assemble the circuit of Fig. 3.1. The element values are $R_1 = 1$ kΩ, $R_f = 10$ kΩ, and $R_O = 100$ Ω. (b) Set the function generator to produce a minimum amplitude 100 Hz sine wave with zero dc offset. Connect the generator output to the circuit input and to channel 1 of the dual channel oscilloscope. Connect the output of the circuit to channel 2 of the oscilloscope. (c) Slowly increase the generator output until peak clipping is observed. Record the positive and negative peak clipping amplitudes for the verge of clipping. (d) Increase the function generator output level until the op amp is driven into hard clipping. Record the positive and negative peak amplitudes for hard clipping. Note the change in the signal for hard clipping versus the verge of clipping.

soft clipping @ 1.2

3.7.3 Current Limiting

This part is to be performed using the 741 op amp. The circuit and element values are those of the preceding part. (a) Ground the v_O output terminal of the circuit to the power supply ground bus on the breadboard. (Do not remove the 100 Ω resistor connected to pin 6 of the op amp. Ground the other side of this 100 Ω resistor so that the resistance to ground from pin 6 of the op amp is 100 Ω.) Connect channel 2 of the dual channel oscilloscope to the op-amp output terminal, i.e. to the left of resistor R_O in Fig. 3.1. Channel 1 should be connected to the function generator output which is the circuit input. (b) Set the function generator to produce a minimum amplitude 100 Hz sine wave with zero dc offset. (c) Slowly increase the generator output until current limiting is observed, i.e. the waveform appears clipped. Calculate and record the positive and negative peak currents for the verge of current limit. (d) Increase the function generator output level until the op amp is driven into hard current limiting. Calculate and record the positive and negative peak currents for hard current limit.

3.7.4 Slewing

This part is to be performed using the 741 op amp. The circuit and element values are those of the preceding part. **(a)** Remove the ground wire from resistor R_O and connect channel 2 to the circuit output. Set the generator for a 1 V peak amplitude 10 kHz square wave with a dc offset of zero. **(c)** While observing the output voltage waveform on the oscilloscope, increase the frequency of the function generator until the time required for the lower to upper transition of the waveform is approximately 1/4 of the period of the waveform. The transitions between the upper and lower voltage levels should be approximately straight line segments. The slope of the straight line transition from the lower to the upper level is the positive slew rate. Similarly, the slope of the straight line transition from the upper to the lower level is the negative slew rate. Measure and record both slew rates. Although these are not usually identical, they should have the same order of magnitude. It may be necessary to use the sweep magnifier on the oscilloscope to accurately measure the slope of the lines.

3.7.5 DC Offset and Bias Currents

This part is to be performed using 741 op amps. **(a)** Assemble the circuit in Fig. 3.2. The the pot R_P is omitted for this step. The circuit element values are $R_A = 10$ kΩ, $R_B = 100$ Ω, $R_{f1} = 300$ kΩ, $R_1 = 300$ kΩ, $R_{F2} = 47$ kΩ, $R_2 = 1$ kΩ, and $R_o = 100$ Ω. **(b)** Connect the circuit input to the generator and to channel A of the dual channel oscilloscope. Set the function generator to produce a sine wave with a dc level of zero, a frequency of 1 kHz, and a peak-to-peak amplitude of 2 V. Connect the circuit output to channel 2 of the oscilloscope. Both channels of the oscilloscope should be dc coupled. Use the digital multimeter to measure the dc voltages at the output of both op amps. **(c)** Connect the pot R_P as shown in Fig. 3.2. The value of R_P is 10 kΩ. One end connects to pin 5 of op amp A1 and the other end connects to pin 1. The wiper of the pot is connected to the -15 V power supply rail. **(e)** Vary the setting of the wiper on the pot and observe the effect on v_O. Adjust the wiper so that v_O is a sine wave with a dc level of 0 V. Measure the dc output voltage of both op amps with the digital multimeter.

3.7.6 Inverting Amplifier

This part is to be performed with a 741 op amp. **(a)** Assemble the circuit shown in Fig. 3.3 using the element values calculated in step 1 of the Preliminary Calculations. For the resistors, use the nearest 5% values to those calculated. The resistor R_O is specified to be 100 Ω. **(b)** Set the function generator for a minimum amplitude sine wave output with a dc level of zero. Connect the circuit input to the generator and to channel 1 of the dual channel oscilloscope. Connect the circuit output to channel 2 of the oscilloscope. **(c)** Slowly increase the output of the generator until a 1 V peak-to-peak signal is obtained at the circuit output. Use the oscilloscope to measure the frequency response of the circuit over the frequency range from 10 Hz to 100 kHz, i.e. measure the ratio of the amplitude of the output sine wave to amplitude of the input sine and the phase shift from the input to the output. If it appears that the output sine wave is distorted, the amplifier may be slewing or clipped and the amplitude of V_i should be decreased and the measurements repeated. The gain of the circuit is to be plotted as a function of frequency of the source. The gain versus frequency is plotted on log-log graph paper.

3.7.7 Non-Inverting Amplifier

This part is to be performed with a TL071 op amp. **(a)** Assemble the non-inverting amplifier shown in Fig. 3.4 using the element values calculated in step 2 of the Preliminary Calculations. (The resistors R_A and R_B may be omitted if the function generator voltage can be reduced to a level so that the output does not clip.) The resistor R_O is specified to be 100 Ω. Connect the function generator as V_g, Ch1+ to V_i, and Ch2+ to V_o. **(b)** Repeat the measurements described in part 6 for this amplifier.

3.7.8 Single Power Supply Amplifiers

This part is to be performed with a TL071 op amp. **(a)** Assemble the circuit of Fig. 3.5(a) using the element values calculated in step 3 of the Preliminary Calculations. The resistor R_O is specified to be 100 Ω. **(b)** Measure the dc offset at the circuit output. This should be within 10% of $V^+/2$. **(c)** Repeat the measurements described in part 6 for this amplifier. **(d)** Modify the circuit to obtain the circuit of Fig. 3.5(b). Repeat the measurements of the preceding step.

3.8 Laboratory Report

The laboratory report should contain appropriate sketches, plots, and tables of the experimental data taken including all SPICE assignments and derivations and calculations. Also, answer any supplementary questions posed by the laboratory instructor. The verification sheet should be included.

3.9 References

1. R. F. Coughlin and F. F. Driscoll, *Operational Amplifiers and Linear Integrated Circuits*, Prentice-Hall, 1987.
2. P. Horowitz and W. Hill, *The Art of Electronics,* 2nd edition, Cambridge University Press, 1989.
3. P. Horowitz and I. Robinson, *Laboratory Manual for The Art of Electronics,* Cambridge University Press, 1981.
4. E. J. Kennedy, *Operational Amplifier Circuits*, Holt Rhinehart and Winston, 1988.
5. J. H. Krenz, *An Introduction to Electrical and Electronic Devices,* Prentice-Hall, 1987.
6. J. M. McMenamin, *Linear Integrated Circuits*, Prentice-Hall, 1985.
7. Motorola, *Linear and Interface Circuits*, Series F, Motorola, 1988.
8. National Semiconductor, *Linear Databook 1,* National Semiconductor, 1988.
9. A. S. Sedra and K. C. Smith, *Microelectronic Circuits,* 4th edition, Oxford, 1998.
10. W. D. Stanley, *Operational Amplifiers with Linear Integrated Circuits*, 2nd edition, Merrill, 1989.
11. Texas Instruments, *Linear Circuits Operational Amplifier Macromodels*, Texas Instruments, 1990.
12. Texas Instruments, *Linear Circuits Amplifiers, Comparators, and Special Functions,* Texas Instruments, 1989.

Chapter 4

Differential and Instrumentation Amplifier

4.1 Object

This experiment is a continuation of the previous two. Additional applications of op amps are examined such as a differential amplifier, an op-amp instrumentation amplifier, and a curve tracer.

4.2 Theory

The differential or balanced-input amplifier is used in applications where a low level signal from a transducer such as a strain gauge or microphone is to be amplified. In such applications, the differential amplifier is superior to single-ended or unbalanced input amplifiers in rejecting unwanted noise signals at the input that are called common-mode or longitudinal noise. In this experiment, a single op-amp circuit commonly called a differential amplifier is examined. In addition, a three op-amp differential amplifier called an instrumentation amplifier is examined.

4.2.1 Differential Amplifier

Fig. 4.1 shows the circuit diagram of the differential amplifier with a signal source modeled as a Thevénin generator connected between its inputs. Resistor R_0 sets the output resistance of the circuit. Superposition can be used to express the open-circuit output voltage as a function of the two node voltages v_1 and v_2 as follows:

$$v_O = \frac{R_2}{R_1 + R_2}\left[1 + \frac{R_2}{R_1}\right]v_1 - \frac{R_2}{R_1}v_2 = \frac{R_2}{R_1}(v_1 - v_2) \tag{4.1}$$

It follows from this expression that the output is zero if $v_1 = v_2$. Therefore, if a noise signal is present equally on both inputs, i.e. a common-mode noise, it will be canceled at the output of the amplifier.

For a floating differential source such as the one shown in Fig. 4.1, the differential input resistance seen by the source is $2R_1$. This follows because the op amp forces a virtual short circuit to exist between its two inputs. By voltage division, it follows that $v_1 - v_2 = v_G \times 2R_1/(R_G + 2R_1)$ so that the output signal from the amplifier can be written

$$v_O = \frac{R_2}{R_1} \times \frac{2R_1}{R_G + 2R_1}v_G \tag{4.2}$$

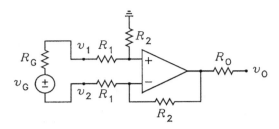

Figure 4.1: Differential amplifier.

This equation shows that $2R_1$ should be large compared to R_G to minimize loss of the input signal due to current flowing through R_G. Because this condition may be difficult to meet in practice, the differential amplifier is not suitable in all applications.

4.2.2 Instrumentation Amplifier

The disadvantages of the differential amplifier can be overcome with the instrumentation amplifier circuit of Fig. 4.2. This figure shows a differential amplifier consisting of op amp A_3 with two op amps added to the circuit between the generator and the differential amplifier. The added op amps are operated as non-inverting amplifiers so that a very high input resistance can be achieved. This resistance is set by resistors labeled R_1 These are necessary in practical circuits to provide a path to ground for the leakage currents which flow in op-amp input leads. Superposition of the input voltages v_1 and v_2 can be used to solve for the output voltages of op amps A_1 and A_2 as follows:

$$v_3 = \left(1 + \frac{R_3}{R_2}\right) v_1 - \frac{R_3}{R_2} v_2 \tag{4.3}$$

$$v_4 = \left(1 + \frac{R_3}{R_2}\right) v_2 - \frac{R_3}{R_2} v_1 \tag{4.4}$$

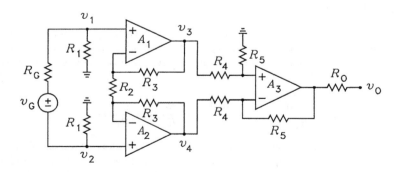

Figure 4.2: Instrumentation amplifier.

op amp A_3 is a differential amplifier which has an output voltage proportional to the difference between v_3 and v_4. The open-circuit output voltage is given by

4.2. THEORY

$$v_O = \frac{R_5}{R_4}(v_3 - v_4) = \frac{R_5}{R_4}\left(1 + \frac{2R_3}{R_2}\right)(v_1 - v_2) \tag{4.5}$$

Because $v_1 - v_2 = 2R_1/(R_G + 2R_1)\,v_G$, it follows that the voltage output from the circuit can be written

$$v_O = \frac{R_5}{R_4}\left(1 + \frac{2R_3}{R_2}\right)\frac{2R_1}{R_G + 2R_1}v_G \tag{4.6}$$

For the instrumentation amplifier, it is simple to impose the design constraint that $2R_1$ be much larger that R_G so that the loss of input voltage due to current flowing through R_G is not significant. This is accomplished in practice if $2R_1 \gg R_G$, e.g. $2R_1 \geq 100 R_G$. In this case v_O can be approximated by

$$v_O \cong \frac{R_5}{R_4}\left(1 + \frac{2R_3}{R_2}\right)v_G \tag{4.7}$$

4.2.3 Op-Amp Curve Tracer

As an application of differential amplifiers an op-amp curve tracer will be considered. A curve tracer is an electronic circuit that provides voltages that are proportional to the terminal voltages and/or currents for a semiconductor device. Such a circuit is shown in Fig. 4.3.

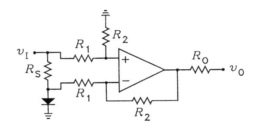

Figure 4.3: Differential amplifier curve tracer.

The differential amplifier is used to measure the voltage across the resistor R_S which serves as a current sampling resistor. The output of the op amp is connected to the horizontal axis of an oscilloscope that is being used in the X-Y mode. The anode of the diode is then connected to the vertical axis of the oscilloscope and then a plot of diode current versus diode voltage is obtained. The diode current is given by the voltage on the vertical axis divided by R_S. Elaborate laboratory instruments known as curve tracers use these same principles to display the terminal characteristics of solid state devices.

A better curve tracer can be obtained by using an instrumentation amplifier as shown in Fig. 4.4. The circuit that senses the diode current has a much high input impedance than the differential amplifier employed in Fig. 4.3.

Note: The resistor R_O shown connected to the output of the op amps in the circuits in this experiment is included for practical reasons, viz. to reduce the possibility that the circuit will oscillate when a capacitive load such as an oscilloscope probe is connected to it. It should not be included in the theoretical derivations, calculations, or simulations. For these analyses the output of the circuit should be taken directly at the output node of the of the appropriate op amp.

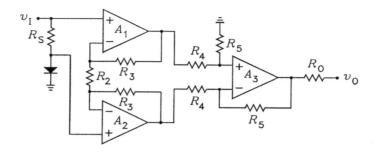

Figure 4.4: Instrumentation amplifier curve tracer.

4.3 Preliminary Derivations

4.3.1 Differential Amplifier

(a) Use superposition of the node voltages v_1 and v_2 in Fig. 4.1 to derive Eq. 4.1. (b) For the circuit of Fig. 4.1, use the concept of a virtual short between the + and − input terminals of the op amp to show that $v_1 - v_2 = v_G \times 2R_1/(R_G + 2R_1)$. (c) For the circuit of Fig. 4.1, show that v_1 and v_2 are given by

$$v_1 = \frac{R_1 - R_2}{2R_1 + R_G} v_G \qquad (4.8)$$

$$v_2 = -\frac{R_1 + R_2}{2R_1 + R_G} v_G \qquad (4.9)$$

It can be concluded from this result that the differential amplifier does not have balanced input voltages with respect to ground, i.e. $v_1 \neq -v_2$.

4.3.2 Instrumentation Amplifier

(a) In the circuit of Fig. 4.2, let $v_1 = v_2 = v_{CM}$, where v_{CM} is a common-mode input voltage. Show that $v_3 = v_4 = v_{CM}$. To show this, use symmetry to demonstrate that the current through R_2 must be zero. Therefore, op amps A_1 and A_2 must have unity gain for a common-mode input signal. (b) In the circuit of Fig. 4.2, let $v_1 = -v_2 = v_D/2$, where v_D is a differential voltage. Show that $v_3 = -v_4 = (1 + 2R_3/R_2)(v_D/2)$. To show this, consider resistor R_2 to be broken into two series resistors of value $R_2/2$ each. Use symmetry to reason that the common node between these two resistors is a virtual ground for the differential signal. Therefore, op amps A_1 and A_2 must have a gain of $1 + R_3/(R_2/2) = 1 + 2R_3/R_2$. (c) Use the results of the preceding two steps to derive Eq. 4.5. To do this, break the input voltages v_1 and v_2 into the common-mode and differential components given by $v_{CM} = (v_1 + v_2)/2$ and $v_D = (v_1 - v_2)$.

4.4 Preliminary Calculations

4.4.1 Differential Amplifier

(a) In the circuit of Fig. 4.1, solve for R_1 and R_2 which will give a differential voltage gain $A_D = v_O/(v_1 - v_2)$ and differential input resistance $R_{in} = (v_1 - v_2)/i_1$ specified by the laboratory instructor. (b) If the generator is removed from the circuit of Fig. 4.1 and the v_2 node is grounded, calculate the input resistance to the v_1 node for the element values of the preceding step. Calculate the input resistance to the v_2 node if the v_1 node is grounded. If v_1 is shorted to v_2, calculate the input resistance to the circuit. (c) If

$v_G = 1$ V and $R_G = 1$ kΩ, calculate the output voltage v_O for the circuit of Fig. 4.1 for the element values of step (a). Calculate v_O if v_1 is grounded. Calculate v_O if v_2 is grounded. Why is v_O different for these three conditions?

4.4.2 Instrumentation Amplifier

(a) For the circuit of Fig. 4.2, determine R_1, R_2, R_3, R_4, and R_5 to yield the gain and input impedance that were specified for the differential amplifier in the previous problem. (b) If the generator is removed from the circuit and the v_2 node is grounded, calculate the input resistance to the v_1 node for the element values of the preceding step. Calculate the input resistance to the v_2 node if the v_1 node is grounded. If v_1 is shorted to v_2, calculate the input resistance to the circuit. (c) If $v_G = 1$ V and $R_G = 1$ kΩ, calculate the output voltage v_O for the circuit for the element values of step (a). Calculate v_O if v_1 is grounded. Calculate v_O if v_2 is grounded. Why is v_O the same for these three conditions?

4.5 Preliminary SPICE Simulations

4.5.1 Differential Amplifier

The SPICE macromodel for the 741 op amp given in Experiment 3 is to be used for the simulations. (a) Write the SPICE code to analyze the circuit of Fig. 4.1. For the element values, use those calculated in part 4.4.1 of the preliminary calculations. The generator output resistance is $R_G = 1$ kΩ. Using a 1 V ac generator voltage, perform a ".ac" analysis of the circuit over the range from 1 kHz to 1 MHz with 20 points per decade. From the SPICE output, determine the low-frequency gain of the circuit, the upper half-power or -3 dB cut-off frequency, and the impedance seen by the generator. (b) Repeat the preceding step if v_1 is grounded. (c) Repeat step (a) if v_2 is grounded. (d) Remove the SPICE code for v_G and R_G from the circuit. Connect v_1 and v_2 together and drive the input with a 1 V ac source. Perform an ".AC" analysis to determine the circuit gain and the impedance seen by the generator over the same frequency range used for step (a).

4.5.2 Instrumentation Amplifier

(a) Write the SPICE code to analyze the circuit of Fig. 4.2. For the element values, use those calculated in part 4.4.2 of the preliminary calculations. The generator output resistance is $R_G = 1$ kΩ. Using a 1 V ac generator voltage, perform a ".AC" analysis of the circuit over the range from 1 kHz to 1 MHz with 20 points per decade. From the SPICE output, determine the low-frequency gain of the circuit, the upper half-power or -3 dB cut-off frequency, and the impedance seen by the generator. (b) Repeat the preceding step if v_1 is grounded. (c) Repeat step (a) if v_2 is grounded. (d) Remove the SPICE code for v_G and R_G from the circuit. Connect v_1 and v_2 together and drive the input with a 1 V ac source. Perform an AC analysis to determine the circuit gain and the impedance seen by the generator over the same frequency range used for step (a).

4.6 Experimental Procedures

4.6.1 Preparation

Prepare the electronic breadboard to provide busses for the positive and negative power supply rails and the circuit ground. Each power supply rail should be decoupled with a 100 Ω, 1/4 W resistor and a 100 μF, 25 V (or greater) capacitor. The resistors are connected in series with the external power supply leads and the capacitors are connected from power supply rail to ground on the circuit side of the resistors. The capacitors must be installed with proper polarity to prevent reverse polarity breakdown.

4.6.2 Differential Amplifier

Single-Ended Gain—Inverting Input Grounded

(a) Assemble the differential amplifier of Fig. 4.1 with $R_O = 100\ \Omega$ and the values for R_1 and R_2 to produce the gain specified by the laboratory instructor. (b) Ground the v_2 input node and connect the output of the function generator to the v_1 input. The function generator should be set up for a 1 kHz sine wave output of minimum amplitude with a dc offset of zero. (c) Connect one channel of the dual channel oscilloscope to the v_1 input and the other channel to the v_O output. Both channels of the oscilloscope should be dc coupled. (d) Use the digital multimeter to adjust the decoupled power supply voltages to $+15$ V and -15 V (on the circuit side of the 100 Ω decoupling resistors). (e) Increase the voltage output of the function generator until v_O measures 1 V peak. Note that v_O is in phase with v_1. Record v_1 and calculate the gain of the circuit. (f) Increase the frequency of the function generator until v_O drops by 3 dB (a factor of approximately 0.71). Record the upper half-power or -3 dB cut-off frequency for the v_1 input. (g) Reset the function generator to 1 kHz and adjust its output until v_O measures 1 V peak. Record the peak value of v_1. Connect a 10 kΩ, 1/4 W resistor in series between the generator and the v_1 input. Set the function generator level until v_O measures 1 V peak. Use the dual channel oscilloscope to measure the voltage on each side of the 10 kΩ resistor. Use the ratio of the two voltages to calculate the input resistance to the v_1 input.

Single-Ended Gain—Non-Inverting Input Grounded

(a) For the circuit of Fig. 4.1, repeat the gain measurement made in the previous step with the v_1 input grounded and the signal generator connected to the v_2 input. Note that the output voltage is inverted with respect to the input for this connection. (b) From the data obtained in this step and in the preceding part, calculate the two single-ended gains v_O/v_1 and v_O/v_2.

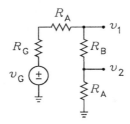

Figure 4.5: Circuit that simulates a floating-signal source.

Differential Gain—Floating Source

(a) The circuit shown in Fig. 4.5 simulates a floating-input source. Assemble the circuit using the element values $R_A = 10$ kΩ and $R_B = 1$ kΩ. Connect the v_1 and v_2 outputs of the circuit to the v_1 and v_2 inputs of the differential amplifier. (b) Connect the dual channel inputs of the oscilloscope to the v_1 and v_2 inputs and set the oscilloscope to display the difference between the two voltages. Set the function generator to produce a differential input voltage of $v_1 - v_2 = 0.2$ V. (c) Connect the output of the differential amplifier to an unused channel of the oscilloscope. Use the oscilloscope to measure the output voltage. Calculate the differential gain of the amplifier as

$$A_D = \frac{v_O}{v_1 - v_2}$$

4.6. EXPERIMENTAL PROCEDURES

Common-Mode Gain

(a) For the circuit of Fig. 4.1, connect the v_1 and v_2 inputs together and connect the output of the function generator to the common input. Measure the common-mode gain of the circuit at a frequency of 1 kHz (the output voltage divided by the common input). If the equal value resistors in the circuit are well matched the gain will be very small and difficult to measure. (Ideally the gain is zero.) It may be necessary to increase the function generator level above the value used in the preceding steps. If a distorted sine wave appears at v_O, the circuit is being overloaded and the level of the generator should be decreased. (b) Calculate the common-mode rejection ratio of the circuit as the ratio of the differential gain obtained in previous steps to the common-mode gain. Express it in decibels. (c) Increase the function generator frequency and observe the variation of the common-mode output voltage as a function of frequency. The generator level should be decreased if a distorted output waveform is obtained.

Common-Mode Gain Minimization

Replace the resistor R_2 which connects from v_+ to ground in Fig. 4.1 with a fixed resistor in series with a potentiometer connected as a variable resistor; the value of the fixed resistor and potentiometer should be selected so that the sum of the resistance of the potentiometer and the fixed resistor can be set equal to the design value of R_2. Set the function generator frequency to 1 kHz. (a) Repeat the measurements described in the preceding part while adjusting the potentiometer to obtain a minimum common-mode gain. Use the ohmmeter to measure the value of R_2 which minimizes v_O. (b) Set the function generator to produce a square-wave output. Determine the setting of the potentiometer which minimizes the v_O. Record the output voltage waveform. Measure the value of R_2. Is it the same value found in the preceding step?

4.6.3 Instrumentation Amplifier

Single-Ended Gain—Inverting Input Grounded

(a) Assemble the instrumentation amplifier of Fig. 4.2 with $R_O = 100\ \Omega$ and the values for R_1 through R_5 required to produce the gain specified by the laboratory instructor. Since this circuit requires three op amps, any combination of 741s and 351s may be used. Alternatively, a 347 may be used; this IC has four op amps (pin connections are given at the end of the laboratory manual). (b) Ground the v_2 input node and connect the output of the function generator to the v_1 input. The function generator should be set up for a 1 kHz sine output of minimum amplitude with a dc offset or level of zero. (c) Connect one channel of the dual channel oscilloscope to the v_1 input and the other channel to the v_O output. Both channels of the oscilloscope should be dc coupled. (d) Use the multimeter to adjust the decoupled power supply voltages to $+15$ V and -15 V (circuit side of the 100 Ω decoupling resistors). (e) Increase the voltage output of the function generator until v_O measures 1 V peak. Note that v_O is in phase with v_1. Record v_1 and calculate the gain of the circuit. (f) Increase the frequency of the function generator until v_O drops by 3 dB. Record the upper half-power or $-3\ dB$ cut-off frequency for the v_1 input. (g) Reset the function generator to 1 kHz and adjust its output until v_O again measures 1 V peak. Record the peak value of v_1. Connect a 100 kΩ, 1/4 W resistor in series between the generator and the v_1 input. Increase the generator level until v_O again measures 1 V peak. Record the peak value of the generator output voltage. Use the data obtained to calculate the input resistance to the v_1 input.

Single-Ended Gain—Non-Inverting Input Grounded

(a) For the circuit of Fig. 4.2, repeat the measurements described in the preceding part with the v_1 input grounded and the signal generator connected to the v_2 input. Note that the output voltage is inverted with respect to the input for this connection. (b) From the data obtained in this part and in the preceding part, calculate the single-ended gains, v_O/v_1 and v_O/v_2.

Differential Gain—Floating Source

(a) Assemble the circuit of Fig. 4.5 using the element values $R_A = 10$ kΩ and $R_B = 1$ kΩ. Connect the v_1 and v_2 outputs of the circuit to the v_1 and v_2 inputs of the instrumentation amplifier. **(b)** Connect the dual channel inputs of the oscilloscope to the v_1 and v_2 inputs and set the oscilloscope to display the difference between the two voltages. Set the function generator to produce a peak-to-peak differential input voltage of $v_1 - v_2 = 0.4$ V. **(c)** Connect the output of the differential amplifier to an unused channel of the oscilloscope. Use the oscilloscope to measure the output voltage. Calculate the differential gain of the amplifier

$$A_D = \frac{v_O}{v_1 - v_2}$$

Common-Mode Gain

(a) For the circuit of Fig. 4.2, connect the v_1 and v_2 inputs together and connect the output of the function generator to the common input. Measure the common-mode gain of the circuit at a frequency of 1 kHz. If the equal value resistors in the circuit are well matched, the gain will be very small and difficult to measure. (Ideally, the gain is zero.) It may be necessary to increase the function generator level above the value used in the preceding steps. If a distorted sine wave appears at v_O, the circuit is being overloaded and the level of the generator should be decreased. **(b)** Calculate the common-mode rejection ratio of the circuit as the ratio of the differential gain to the common-mode gain. **(c)** Increase the generator frequency and observe the variation of the common-mode output voltage as a function of frequency. The generator level should be decreased if a distorted output waveform is obtained.

Common-Mode Gain Minimization

Replace the resistor R_5 which connects from the v_+ terminal of A_3 to ground with a resistor in series with a potentiometer; the resistor and potentiometer should be selected so that total resistance can be varied above and below the design value. Set the function generator frequency to 1 kHz. **(a)** Repeat the measurements described in the preceding part while adjusting the potentiometer to obtain a minimum common-mode gain. Use the ohmmeter to measure the value of R_5 which minimizes v_O. **(b)** Set the function generator to produce a square-wave output. Determine the setting of the potentiometer which minimizes the v_O. Record the output voltage waveform. Measure the value of R_5. Is it the same value found in the preceding step?

4.6.4 Curve Tracer—Differential Amplifier

Assemble the circuit shown in Fig. 4.3. The band on the diode is the cathode corresponding to the straight line on the circuit symbol. The voltage source v_G is a 100 Hz triangular wave with a dc level of zero, and a peak-to-peak value of 10 V. The element values are $R_O = 100$ Ω, $R_S = 1$ kΩ, $R_1 = R_2 = 470$ kΩ. Connect the v_O to the vertical axis of the oscilloscope and the anode of the diode to the horizontal axis. Note that neither oscilloscope lead is connected to the function generator output. Set the oscilloscope to the X-Y mode. Set the VOLTS/DIV for the vertical and horizontal axis to suitable values that display the diode current and voltage. Sketch or print the display. (If a **Tektronix 3012B** oscilloscope is being used, the vertical axis is $CH2$ and the horizontal axis is $CH1$. The $X - Y$ mode display is obtained by pressing the DISPLAY button and then the XY, Triggered XY softkey. The sensitivity for the X axis ($CH1$) should be set to 0.1 Volts/div and the vertical axis ($CH2$ should be set to 2 Volts/div).

4.7 Laboratory Report

The laboratory report should include all appropriate sketches and tables of experimental data including all SPICE assignments, derivations, and calculations. The verification sheet should be included. Also, answer any supplementary questions posed by the laboratory instructor.

4.8 References

1. R. F. Coughlin and F. F. Driscoll, *Operational Amplifiers and Linear Integrated Circuits*, Prentice-Hall, 1987.
2. P. Horowitz and W. Hill, *The Art of Electronics,* 2nd edition, Cambridge University Press, 1989.
3. P. Horowitz and I. Robinson, *Laboratory Manual for The Art of Electronics,* Cambridge University Press, 1981.
4. E. J. Kennedy, *Operational Amplifier Circuits*, Holt Rhinehart and Winston, 1988.
5. J. H. Krenz, *An Introduction to Electrical and Electronic Devices,* Prentice-Hall, 1987.
6. J. M. McMenamin, *Linear Integrated Circuits*, Prentice-Hall, 1985.
7. Motorola, *Linear and Interface Circuits*, Series F, Motorola, 1988.
8. National Semiconductor, *Linear Databook 1,* National Semiconductor, 1988.
9. A. S. Sedra and K. C. Smith, *Microelectronic Circuits,* 4th edition, Oxford, 1998.
10. W. D. Stanley, *Operational Amplifiers with Linear Integrated Circuits*, 2nd edition, Merrill, 1989.
11. Texas Instruments, *Linear Circuits Operational Amplifier Macromodels*, Texas Instruments, 1990.
12. Texas Instruments, *Linear Circuits Amplifiers, Comparators, and Special Functions,* Texas Instruments, 1989.

Chapter 5

Non-Linear Op-Amp Circuits

5.1 Object

This experiment examines some elementary nonlinear op-amp circuits such as active rectifiers, peak detectors, and comparators.

5.2 Theory

5.2.1 Half-Wave Precision Rectifier

Figure 5.1(a) shows the circuit diagram of a passive half-wave rectifier consisting of a diode and a load resistor. To explain the operation of the circuit, let us consider the diode to be ideal. That is, it is a short circuit if a voltage is applied to cause a current to flow in the direction of the arrow and an open circuit if a voltage is applied to cause a current to flow against the direction of the arrow. In this case, the output voltage is given by

$$\begin{aligned} v_O &= v_I \quad \text{for} \quad v_I \geq 0 \\ &= 0 \quad \text{for} \quad v_I < 0 \end{aligned} \tag{5.1}$$

The circuit is called a half-wave rectifier because it passes the signal only on its positive cycle. If the diode is reversed, it pass the signal on its negative cycle. The voltage across a physical diode is not zero when it is forward biased but varies logarithmically with the current. This degrades the performance of the rectifier unless the applied voltage is large compared to the voltage drop across the diode.

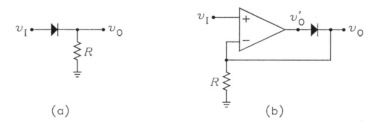

Figure 5.1: (a) Half-wave rectifier circuit. (b) Precision half-wave rectifier circuit.

Many of the limitations of physical diodes in rectifier circuits can be overcome with op amps. Such a circuit is shown in Fig. 5.1(b). For $v_I > 0$, the op amp output voltage v'_O is positive. This forward biases the diode and causes v_O to go positive. Because the op amp has negative feedback, the difference voltage between its two inputs is forced to zero, making $v_O = v_I$. For $v_I < 0$, the voltage v'_O is negative. This reverse biases the diode. The current flow through it is essentially zero, making $v_O = 0$. It follows that the output voltage v_O is given by Eq. (5.1).

The circuit of Fig. 5.1(b) has the disadvantage that the op amp loses feedback when $v_I < 0$. When this happens, the op amp gain increases to its open-loop value and v'_O falls to the negative saturation voltage $-V_{SAT}$. When v_I again goes positive, v'_O must increase from $-V_{SAT}$ to a positive voltage before the diode becomes forward biased. Because the op amp slew rate is not infinite, v'_O cannot change instantaneously. This causes a time delay before the diode becomes forward biased, thus degrading the operation of the circuit. This degradation becomes worse as the frequency of the input signal is increased.

Figure 5.2 shows an improved circuit for which the op amp does not saturate. For $v_I > 0$, v'_O goes negative. This causes D_1 to be reverse biased and D_2 to be forward biased. The negative feedback through R_F forces the voltage at the inverting input to be zero. To solve for v_O, we can write $i_1 + i_F = 0$, $i_1 = v_I/R_1$, and $i_F = v_O/R_F$. Solution of these equations yields $v_O = (-R_F/R_1)\,v_I$. For $v_I < 0$, v'_O goes positive. This causes D_1 to be forward biased and D_2 to be reverse biased, making $i_F = 0$. The op amp has negative feedback through D_1 which forces the voltage at the inverting input to be zero. It follows that $v_O = 0$. Thus the circuit operates as a precision half-wave rectifier. The voltage gain is $-R_F/R_1$. The negative sign means that the gain is inverting. Because the op amp does not saturate when $v_O = 0$, the circuit does not have the limitations of the one in Fig. 5.1(b). If the direction of each diode is reversed, $v_O = 0$ for $v_I > 0$ and $v_O = (-R_F/R_1)\,v_I$ for $v_I < 0$. Again, the circuit has an inverting voltage gain.

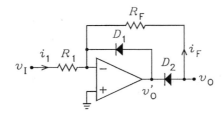

Figure 5.2: Improved precision half-wave rectifier.

5.2.2 Full-Wave Precision Rectifiers

In many applications, a full-wave rectifier is preferred over a half-wave rectifier. The output voltage from a full-wave rectifier is given by

$$v_O = +k\,|v_I| \quad \text{or} \quad v_O = -k\,|v_I| \tag{5.2}$$

where k is the gain. The plus sign is used for a non-inverting gain and the minus sign is used for an inverting gain. This section describes four precision full-wave rectifier circuits.

Circuit 1

Figure 5.3 shows the circuit diagram of a commonly used precision full-wave rectifier circuit. It consists of the half-wave circuit of Fig. 5.2 followed by an inverting summer circuit. The output voltage is given by $v_O = (-R_F/R_2)\,v_I + (-2R_F/R_2)\,v_{O1}$. For $v_I > 0$, $v_{O1} = -v_I$ so that $v_O = (-R_F/R_2)\,v_I + (2R_F/R_2)\,v_I = (R_F/R_2)\,v_I$. For $v_I < 0$, $v_{O1} = 0$ so that $v_O = (-R_F/R_2)\,v_I$. These results can be combined to obtain

$$v_O = \frac{R_F}{R_2} \times |v_I| \tag{5.3}$$

If the direction of each diode is reversed, the output is multiplied by -1.

5.2. THEORY

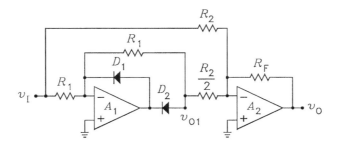

Figure 5.3: First precision full-wave rectifier.

Circuit 2

A second precision full-wave rectifier is shown in Fig. 5.4. For $v_I > 0$, D_1 is forward biased and D_2 is reverse biased. Because the voltage at the inverting input to A_1 is zero, it follows that $v_B = 0$. To solve for v_A, we can write $i_1 + i_3 = 0$, $i_1 = v_I/R_1$, and $i_3 = v_A/R$. Solution for v_A yields $v_A = (-R/R_1)\, v_I$. Because $v_B = 0$, A_2 operates as an inverting amplifier with an output voltage given by $v_O = (-R/R)\, v_A = (R/R_1)\, v_I$. For $v_I < 0$, D_1 is reverse biased and D_2 is forward biased. For this case, we can write $i_1 + i_2 + i_3 = 0$, $i_1 = v_I/R_1$, $i_2 = v_B/R$, and $i_3 = v_C/2R$. But the negative feedback around A_2 makes $v_B = v_C$. These equations can be solved for v_B to obtain $v_B = (-2R/3R_1)\, v_I$. Because $v_B \neq 0$, A_2 operates as a non-inverting amplifier with v_B as its input, a resistor R from its output to its inverting input, and a resistance $2R$ from its inverting input to virtual ground at the inverting input to A_1. Thus A_2 has the voltage gain $(1 + R/2R) = 3/2$. It follows that $v_O = (3/2)\, v_B = (-R/R_1)\, v_I$. These results can be combined to obtain

$$v_O = \frac{R}{R_1} \times |v_I| \qquad (5.4)$$

If the diodes are reversed, the output is multiplied by -1.

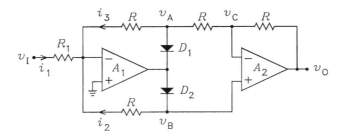

Figure 5.4: A second precision full-wave rectifier.

Circuit 3

Figure 5.5 shows the circuit diagram of a third precision full-wave rectifier. Because the input connects to the non-inverting inputs of A_1 and A_2, this circuit exhibits a much higher input resistance than the circuits of Figs. 5.3 and 5.4. For $v_I > 0$, D_1 is forward biased and D_2 is reverse biased. The negative feedback around A_1 forces $v_A = v_I$. To solve for v_O, we can use superposition of v_I and v_A. For $v_A = 0$, A_2 operates as a non-inverting amplifier with the gain $[1 + R/(R + R/2)] = 5/3$. For $v_I = 0$, A_2 operates as an inverting amplifier with the gain $-[R/(R+R/2)] = -2/3$. It follows that $v_O = (5/3)v_I + (-2/3)v_A = (5/3)v_I + (-2/3)v_I = v_I$. For $v_I < 0$, D_1 is reverse biased and D_2 is forward biased. By voltage division, we can write $v_A = v_B \times R/(R+R) = v_B/2$, or $v_B = 2v_A$. But $v_A = v_I$, thus $v_B = 2v_I$. To solve for v_O, we use superposition of v_I and v_B. For $v_B = 0$, A_2 operates as a non-inverting amplifier with a gain $[1 + R/(R/2)] = 3$. For $v_I = 0$, A_2 operates as an inverting amplifier with a gain $-R/(R/2) = 2$. It follows that v_O is given by $v_O = 3v_I + (-2)v_B = 3v_I + (-2)(2v_I) = -v_I$. These results can be combined to obtain

$$v_O = |v_I| \tag{5.5}$$

If the diodes are reversed, the output is multiplied by -1.

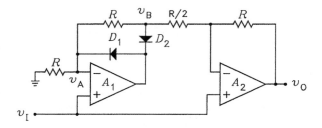

Figure 5.5: Third precision full-wave rectifier circuit.

Circuit 4

The circuit diagram of a fourth precision full-wave rectifier is shown in Fig. 5.6. For $v_I > 0$, D_1 is forward biased and D_2 is reverse biased. The negative feedback around A_1 forces $v_A = 0$. Because there is no current through R_2, the voltage at the non-inverting input to A_2 is v_I. A_2 operates as a non-inverting amplifier with the gain $(1 + R_1/R)$. Thus the output voltage is $v_O = (1 + R_1/R)v_I$. For $v_I < 0$, D_1 is reverse biased and D_2 is forward biased. The current i_1 is given by v_I/R_1. Because there is no current through D_1, i_1 must flow into the output of A_2. Because $v_A = 0$, it follows that $v_O = -i_1(R + R_1) = -(1 + R_1/R)v_I$. Thus the general expression for v_O is

$$v_O = \left(1 + \frac{R_1}{R}\right) \times |v_I| \tag{5.6}$$

If the diodes are reversed, the output voltage is multiplied by -1. The reason R_2 is included in the circuit is to limit the current in D_2 when that diode is forward biased.

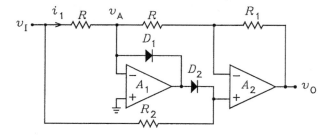

Figure 5.6: A fourth precision full-wave rectifier.

5.2. THEORY

5.2.3 Peak-Detector Circuit

Figure 5.7 shows the circuit diagram of a peak detector. To explain its operation, let us assume that R_1 is removed from the circuit. Let the initial voltage on the capacitor be zero. If v_I goes positive, v_{O1} will go positive. This forward biases D_1 and reverse biases D_2. The capacitor charges positively, thus causing v_O to increase. Because there is no current through R_2, the voltage at the inverting input of A_1 is v_O. The negative feedback around A_1 causes the voltage difference between its inputs to be zero. Thus it follows that $v_O = v_I$.

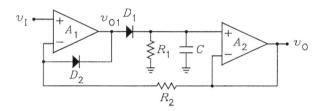

Figure 5.7: Peak-detector circuit.

Now let v_I decrease from its previous positive peak value. This causes v_{O1} to decrease, causing D_1 to cut off. When this happens, A_1 loses feedback and its output voltage decreases rapidly, causing D_2 to become forward biased. R_2 limits the current in D_2 when this occurs. The capacitor voltage cannot change because there is no discharge path with R_1 removed from the circuit. Therefore, v_O remains at the previous peak value of v_I. If v_I increases again to a value greater than v_O, D_1 will be forward biased and D_2 will be reverse biased, causing the capacitor to charge to a higher positive voltage. Thus the circuit acts as a peak detector which holds the highest positive peak voltage applied to its input. When D_1 is reverse biased, its reverse saturation current and the input bias current of A_2 can cause the capacitor voltage to change. Therefore, low leakage current diodes and op amps with a JFET input stage, i.e. a bifet op amp, should be used. If the directions of the diodes are reversed, the circuit detects the negative peaks of the input signal.

With R_1 in the circuit, the capacitor discharges with a time constant $\tau = R_1 C$ when D_1 is reverse biased, causing v_O to decrease toward zero. However, D_1 becomes forward biased again when $v_O = v_I$, thus causing v_O to follow the positive peaks in the input as opposed to holding the peaks. Very large discharge time constants require large values for R_1, for C, or for both. Large value capacitors should be avoided in this circuit because A_1 can be driven into current limiting when the capacitor is charged. This can cause the circuit not to detect the full peak value of the input. All physical capacitors exhibit a leakage resistance R_P in parallel with the capacitor. This resistor appears in parallel with R_1 in the circuit. If R_P cannot be neglected, the effective discharge time constant is $\tau = (R_1 \| R_P) C$.

5.2.4 Comparators

An op amp that is operated without feedback or with positive feedback is called a comparator. Fig. 5.8 shows the circuit diagram of an op amp operated as a non-inverting comparator. For $v_I > V_{REF}$, the output voltage is $+V_{SAT}$. For $v_I < V_{REF}$, the output voltage is $v_O = -V_{SAT}$. Thus we can write the general relation

$$v_O = V_{SAT} \text{sgn}(v_I - V_{REF}) \tag{5.7}$$

where $\text{sgn}(x)$ is the signum function defined by

$$\begin{aligned} \text{sgn}(x) &= +1 \quad \text{for} \quad x > 0 \\ &= -1 \quad \text{for} \quad x < 0 \end{aligned} \tag{5.8}$$

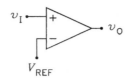

Figure 5.8: Op amp used as a non-inverting comparator.

If the inputs to the op amp are reversed, it becomes an inverting comparator and the output voltage is multiplied by -1.

Several comparators can be used to realize a bar-level indicator. When such an indicator is used to display the level of an audio signal, it is sometimes called an active VU meter, where VU stands for volume units. Fig. 5.9 shows the circuit diagram of a bar-level indicator circuit having 10 levels. The circuit is realized with the LM3914 Dot/Bar Display Driver and the LTA-1000 Bar Graph Array. The LM3914 is an 18-pin DIP (dual-in-line package) integrated circuit and the LTA-1000 is a 20-pin DIP that contains 10 LEDs (light emitting diode).

The LM3914 contains an internal voltage reference of 1.25 V that connects between pins 7 and 8. This source, in conjunction with resistors R_1 and R_2, is used to set the reference voltage V_{REF} that is applied to pin 6. Because the voltage across R_1 is 1.25 V, the current through R_1 is $1.25/R_1$. This current plus the current I_{ADJ} flows through R_2 to circuit ground. Thus the voltage V_{REF} across R_1 and R_2 is given by

$$V_{REF} = 1.25 + \left(\frac{1.25}{R_1} + I_{ADJ}\right) R_2 = 1.25\left(1 + \frac{R_2}{R_1}\right) + I_{ADJ} R_2 \qquad (5.9)$$

The current I_{ADJ} is specified to be less than 120 μA which is small enough to usually neglect in the equation. The signal input is connected to pin 5 which connects to a voltage follower buffer. The output of the buffer connects to the inverting input of each of the 10 voltage comparators.

The reference voltage for each comparator is derived from a voltage divider consisting of ten 1 kΩ resistors connected in series. For a zero signal input, the output voltage of each comparator is $+V_{SAT}$. This reverse biases all of the LEDs so that none are on. If the buffer output voltage increases to a level that is greater than the reference voltage of any comparator, the output voltage from that comparator falls to zero, thus forward biasing the LED connected to that comparator. This causes the LED to emit light. Not shown in the figure is a resistor in series with each LED that limits the current when the comparator output voltage falls to zero.

5.3 Preliminary Derivations

5.3.1 Precision Half-Wave Rectifier

Derive the transfer characteristic of the circuit shown in Fig. 5.2. Determine the effect on the transfer characteristic of reversing both of the diodes in this circuit. Determine the input impedance of this circuit.

5.3.2 First Precision Full-Wave Rectifier

Derive the transfer characteristic of the circuit shown in Fig. 5.3. Determine the effect on the transfer characteristic of reversing both of the diodes in this circuit. Determine the input impedance of this circuit.

5.3.3 Second Precision Full-Wave Rectifier

Derive the transfer characteristic of the circuit shown in Fig. 5.4. Determine the effect on the transfer characteristic of reversing both of the diodes in this circuit. Determine the input impedance of this circuit.

5.3. PRELIMINARY DERIVATIONS

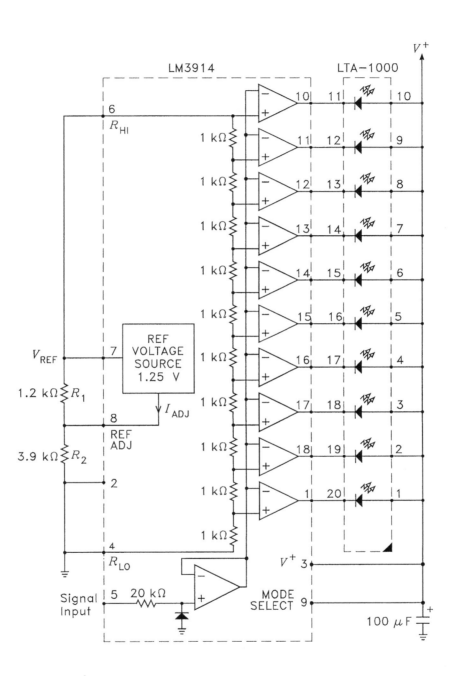

Figure 5.9: Bar-graph level indicator.

5.3.4 Third Precision Full-Wave Rectifier

Derive the transfer characteristic of the circuit shown in Fig. 5.5. Determine the effect on the transfer characteristic of reversing both of the diodes in this circuit.

5.3.5 Fourth Precision Full-Wave Rectifier

Derive the transfer characteristic of the circuit shown in Fig. 5.6. Determine the effect on the transfer characteristic of reversing both of the diodes in this circuit.

5.4 Preliminary Calculations

5.4.1 Peak-Hold Circuit

Determine the values of R_1 and C to set the release time constant of this peak hold circuit to 1 s, 10 s, and 100 s. Perform the calculation for $C = 0.1\ \mu F$, 1 μF, and 10 μF.

5.4.2 Half-Wave Rectifier

Determine the dc level of the output of the passive and active half-wave rectifiers if the input is a 1 kHz sine wave with a dc level of zero and an amplitude of 5 V.

5.4.3 Full-Wave Rectifier

Determine the dc level of the output of a precision full-wave rectifier if the input is a 1 kHz sine wave with a dc level of zero and an amplitude of 5 V.

5.5 Preliminary SPICE Simulations

5.5.1 Transient Analysis

Perform a transient analysis (.TRAN) of the circuits in:
1. Fig. 5.2
2. Fig. 5.3
3. Fig. 5.4
4. Fig. 5.5
5. Fig. 5.6

The input should be a sinusoidal with a dc level of zero, a frequency of 1 kHz, and a peak value of 5 V. Model the op amps as ideal and use the default parameters for the diodes. Pick appropriate values for the resistors and use the default values for the diode parameters. Model the op amps as simple gain blocks.

Determine the effect of reversing each diode in the circuit.

5.5.2 DC Sweep

Plot the output voltage as a function of the input voltage for the circuits in:
1. Fig. 5.2
2. Fig. 5.3
3. Fig. 5.4
4. Fig. 5.5
5. Fig. 5.6

5.6. EXPERIMENTAL PROCEDURES

Sweep the input voltage from -10 V to $+10$ V. This is the dc transfer characteristic of these circuits. Use the same models for the diodes and op amps as in the previous steps. Pick appropriate values for the resistors.

Determine the effect of reversing each diode in the circuit.

5.6 Experimental Procedures

For all circuits, a 100 Ω resistor should be used in series with the oscilloscope and voltmeter input leads when connecting the leads to any points in the circuit that are not circuit ground. Without the resistor, the capacitance of the leads can cause the op amp circuits to oscillate.

5.6.1 Preparation

(a) Prepare the electronic breadboard to provide buses for the positive and negative power supply rails and the circuit ground. Each power supply rail should be decoupled with a 100 Ω 1/4 W resistor, and a 100 μF 25 V (or greater) capacitor. The resistors are connected in series with the external power supply leads and the capacitors are connected from power supply rail to ground on the circuit side of the resistors. The capacitors must be installed with proper polarity to prevent reverse polarity breakdown. **(b)** The op amps specified for this experiment are either the 741 or the TL071. The pin assignments are identical. (When viewed from the top with the pins away from the viewer, pin 2 is the inverting input, pin 3 is the non-inverting input, pin 4 is $-V_{CC}$, pin 7 is $+V_{CC}$, pin 6 is the output, and pins 1 and 5 are used for the optional offset null.) The power supply voltages are $+15$ V and -15 V for all steps in this procedure.

5.6.2 Half-Wave Rectifier

Passive Half-Wave Rectifier

(a) Assemble the circuits shown in Fig. 5.1 and 5.2. All three circuits are to be assembled simultaneously so that they may be compared. Use a value of 3 kΩ for all of the resistors. The diodes are 1N4148 or equivalent. Use 741s for the op amps. Set the function generator to produce a sine wave with a dc level of zero (the default value), a frequency of 100 Hz, and a peak-to-peak value of $2\ V$. Connect the output of the function generator to v_i for each circuit and Ch1 of the oscilloscope, the output of the passive half-wave rectifier to Ch2. Press Auto-Scale and print or sketch the display. Set the oscilloscope for $X - Y$ mode and plot the output versus the input; set the sensitivity for the horizontal and vertical axes so that the transfer characteristic can be adequately displayed. Print or sketch the display.

(b) Change the frequency of the function generator to 10 kHz and repeat part a.

(c) Change the frequency of the function generator to 100 kHz and repeat part a.

Precision Half-Wave Rectifier

Repeat the above procedure for the precision half-wave rectifier.

Improved Precision Half-Wave Rectifier

Repeat the above procedure for the improved precision half-wave rectifier.

5.6.3 Full-Wave Rectifier

Circuit 1

Assemble the circuit shown in Fig. 5.3, with the power supply turned off. Use either 741s or TL071s for the op amps and 1N4148 or equivalents for the diodes. Use 3 kΩ for the resistors R_1, R_2 and R_F. Turn the power supply on. Set the function generator to produce a sine wave with a dc level of zero, a frequency of 100 Hz, and a peak-to-peak value of 4 V. Connect the output of the function generator to Ch 1 of the oscilloscope and v_i and Ch 2 to the output. Sketch or print the display. Set the oscilloscope to make an $X - Y$ sketch or print the display. If the output of the precision full-wave rectifiers oscillates (the output is a periodic waveform that is unrelated to the input) place a 100 pF from the output terminal to the inverting input of the first op amp (in addition to the other components, of course).

Circuit 2

Repeat the above procedure for the circuit shown in Fig. 5.4. Select all of the resistors to be 3 $k\Omega$.

Circuit 3

Repeat the above procedure for the circuit shown in Fig. 5.5. Pick $R = 3$ $k\Omega$.

Circuit 4

Repeat the above procedure for the circuit shown in Fig. 5.6. Select all of the resistors as 3 $k\Omega$.

5.6.4 Peak Hold Circuit

(a) Assemble the circuit shown in Fig. 5.7 with the power off. Use either 741s or TL071s for the op amps and 1N4148s or equivalent for the diodes. Use a value of 30 $k\Omega$ for R_2, $C = 10$ μF, and $R_1 = 1$ MΩ. Turn the power on. Set the function generator to produce a sine wave with a dc level of zero (the default value), a frequency of 100 Hz, and a peak-to-peak value of 10 V. Connect the function generator to the input of the circuit and to Ch 1 of the oscilloscope. Connect the output of the circuit to Ch 2 of the oscilloscope. Set the sensitivity to 2 V/div for each channel and use the vertical position controls so that the center horizontal grid line is zero volts for each channel. Set the time base to display several cycles of function generator voltage. The output of the circuit should be a straight line touching the positive peak of the input sine wave. Sketch or print the display. Note the effect of varying the input amplitude up and down on the function generator. Why does it go up faster than down?

5.6.5 Bar-Graph Array

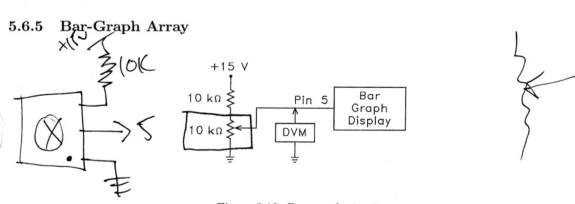

Figure 5.10: Bar-graph circuit.

Assemble the circuit shown in Fig. 5.10 with the power off. The portion of this figure which is labeled as "Bar Graph Display" is Fig. 5.9. The LM3914 is a standard 18 pin DIP IC. Note that the pin locations shown in Fig. 5.9 are grouped by function rather than physical location. The pins locations shown for the LTA-1000 DIP Bar Graph Array are grouped by their physical location. Three corners of the LTA-1000 are square while the remaining one is beveled or rounded and it is this corner that indicates pin 1. The portion of Fig. 5.9 inside the dotted lines are internal to the devices while the portion outside the dotted lines are external components or connection which must be made. Note that the anode of each LED must be directly connected positive binding post of the dc power supply; i.e. the power supply decoupling network must be modified because of the excessive current that the LEDs will draw when light. Replace the 100 Ω resistors with 7.5 Ω resistors for any steps employing the bar graph array.

Turn the dc power supply on. Vary the potentiometer and measure and record the input voltage at which each of the 10 segments of the bar graph array lights. Retain this circuit for the next step.

5.6.6 Level Indicator

Design a level indicator. The specifications for the level indicator are that an input sinusoidal voltage with an rms value of 0 V is to produce a dark or unlit bar graph and a value of V_r or larger is to light all ten segments. An input sinusoidal with an rms value of slightly less than V_r should light 9 segments. The value V_r will be specified by the laboratory instructor. The display should not flicker for small frequencies such as 10 Hz or less.

Since this is a design step, the designer must first design the circuit before it is assembled and any experimental measurements are made to determine if it meets the design criteria. Circuits sufficient to implement the design have been examined in previous steps in this experiment. The designer is free to innovate.

5.7 Laboratory Report

Turn in all sketches taken or printouts made and a tabular summary of all data taken as well as the design of the level indicator. Include the verification sheet.

5.8 References

1. R. F. Coughlin and F. F. Driscoll, *Operational Amplifiers and Linear Integrated Circuits*, Prentice-Hall, 1987.
2. P. Horowitz and W. Hill, *The Art of Electronics,* 2nd edition, Cambridge University Press, 1989.
3. P. Horowitz and I. Robinson, *Laboratory Manual for The Art of Electronics,* Cambridge University Press, 1981.
4. E. J. Kennedy, *Operational Amplifier Circuits*, Holt Rhinehart and Winston, 1988.
5. J. H. Krenz, *An Introduction to Electrical and Electronic Devices*, Prentice-Hall, 1987.
6. J. M. McMenamin, *Linear Integrated Circuits*, Prentice-Hall, 1985.
7. Motorola, *Linear and Interface Circuits*, Series *F*, Motorola, 1988.
8. National Semiconductor, *Linear Databook 1,* National Semiconductor, 1988.
9. A. S. Sedra and K. C. Smith, *Microelectronic Circuits,* 4th edition, Oxford, 1998.
10. D. H. Sheingold, *Nonlinear Circuits Handbook*, 2nd edition, Analog Devices, 1976.
11. W. D. Stanley, *Operational Amplifiers with Linear Integrated Circuits*, 2nd edition, Merrill, 1989.
12. Texas Instruments, *Linear Circuits Operational Amplifier Macromodels*, Texas Instruments, 1990.
13. Texas Instruments, *Linear Circuits Amplifiers, Comparators, and Special Functions*, Texas Instruments, 1989.

Chapter 6

Active Filters

6.1 Object

The object of this experiment is to examine the performance of several operational-amplifier or op-amp active filters. The basic theory of filter design and active filters is covered.

6.2 Introduction

Electronic filters are commonly used to either pass or reject signals with frequencies that lie in a particular band of the frequency spectrum. In some cases, a filter may be used to impart a desired phase shift as a function of frequency. Historically, early filters were designed with passive components, i.e. with resistors, inductors, and capacitors. Such filters are called passive filters. For low-frequency applications, these have the disadvantage that the inductors are physically large, lossy, and expensive. Contemporary filters that are designed for signals in the frequency band below approximately 100 kHz make use of resistors, capacitors, and op amps. Such filters are called active filters. These filters have the advantage that no inductors are required. Any filter that can be realized as a passive filter can also be realized as an active filter that has no inductors. Compared to a passive filter, the main disadvantage of an active filter is that it requires a power supply to power the op amps.

6.3 Classes of Filter Functions

The filters considered in this experiment can be divided into four classes. These are low-pass, high-pass, band-pass and band-reject. Although it is impossible to realize an ideal filter, the characteristics of the four classes of filters are simplest to describe for ideal filters. An ideal low-pass filter has a cutoff frequency below which the gain is independent of frequency and above which the gain is zero. Fig. 6.1(a) illustrates the magnitude response of an ideal low-pass filter having a gain K. The responses of two physically realizable filters are also shown. An ideal high-pass filter has a cutoff frequency above which the gain is constant and below which the gain is zero. The magnitude responses of an ideal high-pass filter and two physically realizable filters are illustrated in Fig. 6.1(b). An ideal band-pass filter has two cutoff frequencies between which the gain is constant and zero elsewhere. The magnitude responses of an ideal band-pass filter and two physically realizable filters are illustrated in Fig. 6.1(c). An ideal band-reject filter has two cutoff frequencies between which the gain is zero and constant elsewhere. The magnitude responses of an ideal band-reject filter and two physically realizable filters are illustrated in Fig. 6.1(d).

Low-pass filters are used in applications where it is desired to remove the high-frequency content of a signal. For example, aliasing distortion can occur if a signal is applied to the input of an analog-to-digital converter that has a frequency higher than one-half the sampling frequency of the converter. A low-pass

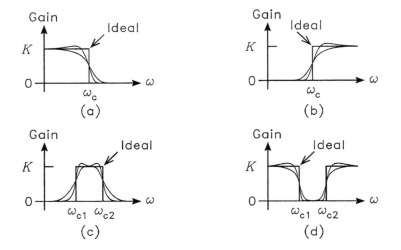

Figure 6.1: (a) Low pass. (b) High pass. (c) Band pass. (d) Band reject.

filter might be used to bandlimit the signal. Similarly, a high-pass filter is used in applications where it is desired to remove the low-frequency content of a signal. For example, the tweeter driver in a loudspeaker can be damaged by low frequencies signals. To prevent this, a high-pass filter called a crossover network must be connected to the tweeter.

A band-pass filter is used in applications where it is desired to pass only the frequencies in a band. For example, to detect a low level tone that is buried in noise, a band-pass filter might be used to pass the tone and reject the noise. A band-reject filter is used in applications where it is desired to reject a particular frequency or band of frequencies. For example, a 60 Hz hum induced in the amplifier of a public address system might be filtered out with a band-pass filter.

6.4 Transfer Functions

For sinusoidal time variations, the input voltage to a filter can be written

$$v_I(t) = \mathrm{Re}\left[V_i e^{j\omega t}\right] \tag{6.1}$$

where V_i is the phasor input voltage, i.e. it has an amplitude and a phase, and $e^{j\omega t} = \cos\omega t + j\sin\omega t$. A sinusoidal signal is the only signal in nature that is preserved by a linear system. Therefore, if the filter is linear, its output voltage can be written

$$v_O(t) = \mathrm{Re}\left[V_o e^{j\omega t}\right] \tag{6.2}$$

where V_o is the phasor output voltage. The ratio of V_o to V_i is called the voltage-gain transfer function. It is a function of frequency. Denote

$$T(j\omega) = \frac{V_o}{V_i} \tag{6.3}$$

In general, write $T(j\omega)$ as follows:

$$T(j\omega) = A(\omega) e^{j\varphi(\omega)} \tag{6.4}$$

where $A(\omega)$ and $\varphi(\omega)$ are real functions of ω. $A(\omega)$ is called the gain function and $\varphi(\omega)$ is called the phase function.

6.4. TRANSFER FUNCTIONS

As an example, consider the filter input voltage

$$v_I(t) = V_1 \cos(\omega t + \theta_1) = \text{Re}\left[V_1 e^{j\theta_1} e^{j\omega t}\right] \tag{6.5}$$

The corresponding phasor input and output voltages are

$$V_i = V_1 e^{j\theta_1} \qquad V_o = V_1 e^{j\theta_1} A(\omega) e^{j\varphi(\omega)} \tag{6.6}$$

It follows that the filter output voltage is

$$v_O(t) = \text{Re}\left[V_1 e^{j\theta_1} A(\omega) e^{j\varphi(\omega)} e^{j\omega t}\right] = A(\omega) V_1 \cos[\omega t + \theta_1 + \varphi(\omega)] \tag{6.7}$$

This equation illustrates why $A(\omega)$ is called the gain function and $\varphi(\omega)$ is called the phase function.

The complex frequency s is usually used in place of $j\omega$ in writing transfer functions. In general, most transfer functions can be written in the form

$$T(s) = K \frac{N(s)}{D(s)} \tag{6.8}$$

where K is a gain constant and $N(s)$ and $D(s)$ are the lowest order polynomials in s containing no reciprocal powers of s. The roots of $D(s)$ are called the poles of the transfer function. The roots of $N(s)$ are called the zeros.

As an example, consider the function

$$T(s) = 4 \frac{s/4 + 1}{s^2/6 + 5s/6 + 1} = 4 \frac{s/4 + 1}{(s/2 + 1)(s/3 + 1)} \tag{6.9}$$

The function has a zero at $s = -4$ and poles at $s = -2$ and $s = -3$. Note that $T(\infty) = 0$. Because of this, some texts would say that $T(s)$ has a zero at $s = \infty$. However, this is not correct because $N(\infty) \neq 0$.

Note that the constant terms in the numerator and denominator of $T(s)$ are both unity. This is one of two standard ways for writing transfer functions. Another way is to make the coefficient of the highest powers of s unity. In this case, the above transfer function would be written

$$T(s) = 6 \frac{s+4}{s^2 + 5s + 6} = 6 \frac{s+4}{(s+2)(s+3)} \tag{6.10}$$

Because it is usually easier to construct Bode plots with the first form, that form is used here.

Because the complex frequency s is the operator which represents d/dt in the differential equation for a system, the transfer function contains the differential equation. Let the transfer function above represent the voltage gain of a circuit, i.e. $T(s) = V_o/V_i$, where V_o and V_i, respectively, are the phasor output and input voltages. It follows from Eq. (6.9) that

$$\left(\frac{s^2}{6} + \frac{5s}{6} + 1\right) V_o = 4\left(\frac{s}{4} + 1\right) V_i \tag{6.11}$$

When the operator s is replaced with d/dt, this yields the differential equation

$$\frac{1}{6} \frac{d^2 v_O}{dt^2} + \frac{5}{6} \frac{dv_O}{dt} + v_O = \frac{dv_I}{dt} + 4v_I \tag{6.12}$$

where v_O and v_I, respectively, are the time domain output and input voltages. Note that the poles are related to the derivatives of the output and the zeros are related to the derivatives of the input.

6.5 Frequency Transformations

Filter transfer functions are normally derived as low-pass functions. Frequency transformations are then used to transform the low-pass functions into either high-pass, band-pass, or band-reject transfer functions. For a low-pass filter, let the normalized frequency p be defined by

$$p = \frac{s}{\omega_c} \tag{6.13}$$

where ω_c is a normalization frequency. For the case of low-pass and high-pass filters, ω_c is usually called the filter cutoff frequency. Depending on the type of filter, it is not necessarily the -3 dB cutoff frequency. To distinguish between the two in the following, ω_3 is used to denote the -3 dB cutoff frequency in cases where it differs from ω_c. In the case of a band-pass filter, ω_c is the center frequency of the band-pass response. In the case of a band-reject filter, ω_c is the center frequency of the band-reject response.

The frequency transformations are defined as follows:

$$\begin{array}{lll} \text{Low-Pass to High-Pass} & p \to & \dfrac{1}{p} \\[4pt] \text{Low-Pass to Band-Pass} & p \to & B\left(p + \dfrac{1}{p}\right) \\[4pt] \text{Low-Pass to Band-Reject} & p \to & \left[B\left(p + \dfrac{1}{p}\right)\right]^{-1} \end{array} \tag{6.14}$$

where the arrow is read "is replaced by." The parameter B determines the -3 dB bandwidth of the band-pass and band-reject functions.

6.5.1 Transformations of First-Order Functions

As an example, consider the first-order low-pass filter function

$$T_{\text{LP}}(s) = K \frac{1}{1 + s/(a\omega_c)} \tag{6.15}$$

where a is a positive constant. (Note that ω_c is equal to the -3 dB frequency of $T_{\text{LP}}(s)$ only if $a = 1$.) The function $T_{\text{LP}}(p)$ is given by

$$T_{\text{LP}}(p) = K \frac{1}{1 + p/a} \tag{6.16}$$

The high-pass, band-pass, and band-reject transfer functions, respectively, are given by

$$T_{\text{HP}}(p) = K \frac{1}{1 + 1/(ap)} = K \frac{ap}{1 + ap} \tag{6.17}$$

$$T_{\text{BP}}(p) = K \frac{1}{1 + B(p + 1/p)/a} = K \frac{ap/B}{p^2 + ap/B + 1} \tag{6.18}$$

$$T_{\text{BR}}(p) = K \frac{1}{1 + [B(p + 1/p)]^{-1}/a} = K \frac{p^2 + 1}{p^2 + p/(aB) + 1} \tag{6.19}$$

Note that the order of the transfer function is doubled for the band-pass and band-reject transformations.

6.5.2 Transformations of Second-Order Functions

Consider the second-order low-pass function

$$T_{\text{LP}}(s) = K \frac{1}{(s/a\omega_c)^2 + (1/b)(s/a\omega_c) + 1} \tag{6.20}$$

where a and b are positive constants. The function $T_{\text{LP}}(p)$ is given by

$$T_{\text{LP}}(p) = K\frac{1}{(p/a)^2 + (1/b)(p/a) + 1} \tag{6.21}$$

The high-pass, band-pass, and band-reject transfer functions, respectively, are given by

$$T_{\text{HP}}(p) = K\frac{1}{(1/ap)^2 + (1/b)(1/ap) + 1} = K\frac{(ap)^2}{(ap)^2 + (1/b)(ap) + 1} \tag{6.22}$$

$$\begin{aligned}T_{\text{BP}}(p) &= K\frac{1}{[B(p+1/p)/a]^2 + (1/b)[B(p+1/p)/a] + 1} \\ &= K\frac{(a/B)^2 p^2}{p^4 + [a/(bB)]p^3 + \left[(a/B)^2 + 2\right]p^2 + [a/(bB)]p + 1}\end{aligned} \tag{6.23}$$

$$\begin{aligned}T_{\text{BR}}(p) &= K\frac{1}{\left\{[B(p+1/p)]^{-1}/a\right\}^2 + (1/b)\left[\left\{[B(p+1/p)]^{-1}/a\right\}/a\right] + 1} \\ &= K\frac{(p^2+1)^2}{p^4 + [1/(abB)]p^3 + \left[1/(aB)^2 + 2\right]p^2 + [1/(abB)]p + 1}\end{aligned} \tag{6.24}$$

6.6 Butterworth Transfer Functions

The general form of a nth-order low-pass filter transfer function having no zeros can be written

$$T_{\text{LP}}(s) = K\frac{1}{1 + c_1(s/\omega_c) + c_2(s/\omega_c)^2 + \cdots + c_n(s/\omega_c)^n} \tag{6.25}$$

where K is the dc gain constant, ω_c is a normalization frequency, and the c_i are positive constants. The magnitude squared function is obtained by setting $s = j\omega$ and solving for $|T_{\text{LP}}(j\omega)|^2$. This function contains only even powers of ω and is of the form

$$|T_{\text{LP}}(j\omega)|^2 = K^2\frac{1}{1 + C_1(\omega/\omega_c)^2 + C_2(\omega/\omega_c)^4 + \cdots + C_n(\omega/\omega_c)^{2n}} \tag{6.26}$$

where the C_i are positive constants which are related to the c_i.

For the Butterworth filter, the constants C_i are chosen so that $|T_{\text{LP}}(j\omega)|^2$ approximates the magnitude squared function of an ideal low-pass filter in the maximally flat sense. The magnitude squared function for the ideal filter is defined by

$$\begin{aligned}|T_{\text{LP}}(j\omega)|^2 &= K^2 \text{ for } \omega \leq \omega_c \\ &= 0 \text{ for } \omega > \omega_c\end{aligned} \tag{6.27}$$

To obtain the maximally-flat approximation, the C_i are chosen to make as many derivatives as possible of $|T_{\text{LP}}(j\omega)|^2$ equal to zero at $\omega = 0$. If the derivative of a function is zero, the derivative of the reciprocal of the function is also zero. It follows that the maximally flat condition can be imposed by solving for the constants C_i which make as many derivatives as possible of $\left[|T_{\text{LP}}(j\omega)|^2\right]^{-1}$ equal to zero at $\omega = 0$. Because the denominator polynomial of $|T_{\text{LP}}(j\omega)|^2$ is an even function, all odd-order derivatives are already zero at $\omega = 0$. For the second derivative to be zero, this requires that $C_1 = 0$. For the fourth derivative to be zero, this requires that $C_2 = 0$. This procedure is repeated to obtain $C_i = 0$ for all i. However, C_n cannot set $C_n = 0$ because this would make the approximating function independent of frequency. Therefore, set $C_i = 0$ for all $1 \leq i \leq n-1$ to obtain

$$|T_{\text{LP}}(j\omega)|^2 = K^2 \frac{1}{1 + C_n (\omega/\omega_c)^{2n}} \quad (6.28)$$

The first $2n - 1$ derivatives of this function are zero at $\omega = 0$.

It is standard to choose C_n to make $|T_{\text{LP}}(j\omega_c)|^2 = K^2/2$. This forces the -3 dB frequency ω_3 to be equal to the normalization frequency ω_c. This condition requires $C_n = 1$. Thus the magnitude squared function of the nth-order Butterworth low-pass filter becomes

$$|T_{\text{LP}}(j\omega)|^2 = K^2 \frac{1}{1 + (\omega/\omega_c)^{2n}} \quad (6.29)$$

Figure 6.2 shows example plots of $|T_{\text{LP}}(j\omega)|$ for $1 \leq n \leq 5$ for the Butterworth low-pass filter. The plots assume that $K = 1$. The horizontal axis is the normalized radian frequency $v = \omega/\omega_c$. Each function has the value 1 at $v = 0$, the value 0.5 at $v = 1$, and approaches 0 as $v \to \infty$. As the order n increases, the width of the flat region in the passband is extended and the filter exhibits a sharper cutoff. The response characteristic is called maximally flat because there are no ripples in the passband response.

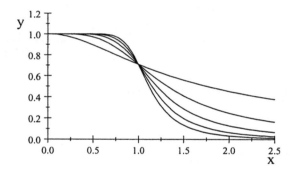

Figure 6.2: Plots of the Butterworth magnitude response for $1 \leq n \leq 5$.

To illustrate how the maximally flat condition is applied to a specific filter transfer function, consider the third-order low-pass function

$$T_{\text{LP}}(s) = K \frac{1}{1 + c_1 (s/\omega_c) + c_2 (s/\omega_c)^2 + c_3 (s/\omega_c)^3} \quad (6.30)$$

The magnitude squared function is given by

$$|T_{\text{LP}}(j\omega)|^2 = K^2 \frac{1}{1 + (c_1^2 - 2c_2)(\omega/\omega_c)^2 + (c_2^2 - 2c_1 c_3)(\omega/\omega_c)^4 + c_3^2 (\omega/\omega_c)^6} \quad (6.31)$$

For this to be maximally flat with a -3 dB cutoff frequency of ω_c, this requires that

$$c_1^2 - 2c_2 = 0 \qquad c_2^2 - 2c_1 c_3 = 0 \qquad c_3 = 1 \quad (6.32)$$

Solution for c_1 and c_2 yields $c_1 = c_2 = 2$. Thus the Butterworth third-order low-pass transfer function is

$$\begin{aligned} T_{\text{LP}}(s) &= K \frac{1}{1 + 2(s/\omega_c) + 2(s/\omega_c)^2 + (s/\omega_c)^3} \\ &= K \frac{1}{1 + (s/\omega_c)} \times \frac{1}{1 + (s/\omega_c) + (s/\omega_c)^2} \end{aligned} \quad (6.33)$$

6.6. BUTTERWORTH TRANSFER FUNCTIONS

The maximally flat filters are called Butterworth filters after S. Butterworth who described the procedure for deriving the transfer functions in his 1930 paper "On the Theory of Filter Amplifiers" which was published in *Wireless Engineer*. The resulting denominator polynomials for $T_{\text{LP}}(s)$ are called Butterworth polynomials. The first six Butterworth polynomials in factored form are

$$
\begin{aligned}
b_1(x) &= (x+1) \\
b_2(x) &= (x^2 + 1.4142x + 1) \\
b_3(x) &= (x+1)(x^2 + x + 1) \\
b_4(x) &= (x^2 + 0.7654x + 1)(x^2 + 1.8478x + 1) \\
b_5(x) &= (x+1)(x^2 + 0.6180x + 1)(x^2 + 1.6180x + 1) \\
b_6(x) &= (x^2 + 0.5176x + 1)(x^2 + 1.4142x + 1)(x^2 + 1.9319x + 1)
\end{aligned}
\tag{6.34}
$$

6.6.1 Even-Order Butterworth Filters

For an even-order Butterworth low-pass filter of order n, the transfer function can be written in the product form

$$
T_{\text{LP}}(s) = K \prod_{i=1}^{n/2} \frac{1}{(s/\omega_c)^2 + (1/b_i)(s/\omega_c) + 1}
\tag{6.35}
$$

The constants b_i are given by

$$
b_i = \frac{1}{2 \sin \theta_i}
\tag{6.36}
$$

where the θ_i are given by

$$
\theta_i = \frac{2i-1}{n} \times 90° \quad \text{for } 1 \leq i \leq \frac{n}{2}
\tag{6.37}
$$

Example 1 *Solve for the transfer functions of the second-order Butterworth low-pass and high-pass filters.*

Solution. For $n=2$, there is only one second-order transfer function. The calculations are summarized as follows:

i	θ_i	b_i
1	45°	$1/\sqrt{2}$

The low-pass transfer function is given by

$$
T(s) = K \frac{1}{(s/\omega_c)^2 + \sqrt{2}(s/\omega_c) + 1}
$$

The high-pass transfer function is obtained by replacing s/ω_c with ω_c/s to obtain

$$
\begin{aligned}
T(s) &= K \frac{1}{(\omega_c/s)^2 + \sqrt{2}(\omega_c/s) + 1} \\
&= K \frac{(s/\omega_c)^2}{(s/\omega_c)^2 + \sqrt{2}(s/\omega_c) + 1}
\end{aligned}
$$

6.6.2 Odd-Order Butterworth Filters

For an odd-order Butterworth low-pass filter of order n, the transfer function can be written in the product form

$$
T_{\text{LP}}(s) = K \frac{1}{s/\omega_c + 1} \times \prod_{i=1}^{(n-1)/2} \frac{1}{(s/\omega_c)^2 + (1/b_i)(s/\omega_c) + 1}
\tag{6.38}
$$

The constants b_i are given by
$$b_i = \frac{1}{2\sin\theta_i} \tag{6.39}$$
where the θ_i are given by
$$\theta_i = \frac{2i-1}{n} \times 90° \text{ for } 1 \leq i \leq \frac{n-1}{2} \tag{6.40}$$

Example 2 *Solve for the transfer functions of the third-order Butterworth low-pass and high-pass filters.*

Solution. For $n = 3$, each transfer function contains one first-order polynomial and one second-order polynomial. The calculations for the second-order polynomial are summarized as follows:

i	θ_i	b_i
1	30°	1

The low-pass transfer function is given by
$$T_{\text{LP}}(s) = K\frac{1}{(s/\omega_c)+1} \times \frac{1}{(s/\omega_c)^2 + (s/\omega_c) + 1}$$

The high-pass transfer function is obtained by replacing s/ω_c with ω_c/s to obtain
$$\begin{aligned}T_{\text{HP}}(s) &= K\frac{1}{(\omega_c/s)+1} \times \frac{1}{(\omega_c/s)^2 + (\omega_c/s) + 1}\\ &= K\frac{(s/\omega_c)}{(s/\omega_c)+1} \times \frac{(s/\omega_c)^2}{(s/\omega_c)^2 + (s/\omega_c) + 1}\end{aligned}$$

6.7 Chebyshev Filter Transfer Functions

6.7.1 The Chebyshev Approximation

In 1899, the Russian mathematician P. L. Chebyshev (also written Tschebyscheff, Tchebysheff, or Tchebicheff) described a set of polynomials $t_n(x)$ which have the feature that they ripple between the peak values of $+1$ and -1 for $-1 \leq x \leq +1$. His polynomials are widely used in filter approximations for frequencies that span the audio band to the microwave band. The Chebyshev polynomials are defined as

$$t_n(x) = \begin{cases} \cos[n\cos^{-1}(x)] & |x| \leq 1 \\ \cosh[n\cosh^{-1}(x)] & |x| > 1 \end{cases} \tag{6.41}$$

which leads to the recursive relationship
$$t_{n+1}(x) = 2xt_n(x) - t_{n-1}(x) \tag{6.42}$$

from which the Chebyshev polynomials can be obtained.

The first six Chebyshev polynomials are
$$\begin{aligned}t_1(x) &= x \\ t_2(x) &= 2x^2 - 1 \\ t_3(x) &= 4x^3 - 3x \\ t_4(x) &= 8x^4 - 8x^2 + 1 \\ t_5(x) &= 16x^5 - 20x^3 + 5x \\ t_6(x) &= 32x^6 - 48x^4 + 18x^2 - 1\end{aligned} \tag{6.43}$$

6.7. CHEBYSHEV FILTER TRANSFER FUNCTIONS

Figure 6.3 shows the plots of the first four of these polynomials over the range $-2 \leq x \leq +2$.

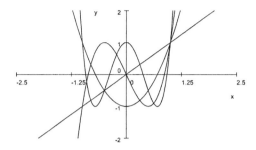

Figure 6.3: Plots of Chebyshev polynomials for $1 \leq n \leq 4$.

The Chebyshev approximation to the magnitude squared function of a low-pass filter is given by

$$|T_{\text{LP}}(j\omega)|^2 = K^2 \frac{1 + \epsilon^2 t_n^2(0)}{1 + \epsilon^2 t_n^2(\omega/\omega_c)} \tag{6.44}$$

where K is the dc gain constant and ϵ is a parameter which determines the amount of ripple in the approximation. For $\omega = 0$, it follows that $|T_{\text{LP}}(j\omega)|^2 = K^2$. For n odd, $t_n^2(0) = 0$ so that the numerator in $|T_{\text{LP}}(j\omega)|^2$ has the value 1. For $0 \leq \omega \leq \omega_c$, the denominator ripples between the values 1 and $1 + \epsilon^2$. This causes $|T_{\text{LP}}(j\omega)|^2$ to ripple between the values K^2 and $K^2/(1+\epsilon^2)$. At $\omega = \omega_c$, it has the value $K^2/(1+\epsilon^2)$. For $\omega > \omega_c$, $|T_{\text{LP}}(j\omega)|^2 \to 0$ as $\omega \to \infty$.

For n even, $t_n^2(0) = 1$ so that the numerator in $|T_{\text{LP}}(j\omega)|^2$ has the value $1 + \epsilon^2$. For $0 \leq \omega \leq \omega_c$, the denominator ripples between the values 1 and $1 + \epsilon^2$. This causes $|T_{\text{LP}}(j\omega)|^2$ to ripple between the values K^2 and $K^2(1+\epsilon^2)$. At $\omega = \omega_c$, it has the value K^2. For $\omega > \omega_c$, $|T_{\text{LP}}(j\omega)|^2 \to 0$ as $\omega \to \infty$. The major difference between the even and odd order approximations is that the odd-order functions ripple down from the zero frequency value whereas the even-order functions ripple up.

Figure 6.4 shows example plots of $|T_{\text{LP}}(j\omega)|$ for the 0.5 dB ripple 4th and 5th order filters. The plots assume that $K = 1$. The horizontal axis is the normalized radian frequency $v = \omega/\omega_c$. The figure shows the 4th order approximation rippling up by 0.5 dB from its zero frequency value. The 5th order approximation ripples down by 0.5 dB from its zero frequency value. Compared to the Butterworth filters, the Chebyshev filters exhibit a sharper cutoff at the expense of ripple in the passband. The more the ripple, the sharper the cutoff.

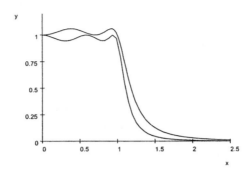

Figure 6.4: Plots of the magnitude responses of the 0.5 dB ripple 4th and 5th order Chebyshev filters.

6.7.2 The dB Ripple

The dB ripple for a Chebyshev filter is the peak-to-peak passband ripple in the Bode magnitude plot of the filter response. It is related to the parameter ϵ as follows:

$$\text{dB ripple} = 10 \log\left(1 + \epsilon^2\right) \qquad \epsilon = \sqrt{10^{\text{dB}/10} - 1} \tag{6.45}$$

6.7.3 The Cutoff Frequency

Unlike the Butterworth filter, the cutoff frequency for the Chebyshev filter is not the frequency at which the response is down by 3 dB. The cutoff frequency is the frequency at which the Bode magnitude plot leaves the "equal-ripple box" in the filter passband. For the even-order low-pass filters, the gain at the cutoff frequency is equal to the zero frequency gain. For the odd-order low-pass filters, the gain at the cutoff frequency is down from the zero frequency gain by an amount equal to the dB ripple.

For the nth-order Chebyshev low-pass filter, the -3 dB frequency can be obtained by setting $|T(j\omega)|^2 = K^2/2$ and solving for ω. This leads to the equation

$$t_n(x) - \sqrt{\frac{1}{\epsilon^2} + 2t_n^2(0)} = 0 \tag{6.46}$$

where $x = \omega/\omega_c$. The positive value of x which satisfies this equation can then be used to determine the -3 dB frequencies of the four filter types from the following relations:

$$\text{Low Pass (of order } n\text{)} \qquad \frac{\omega_3}{\omega_c} = x \tag{6.47}$$

$$\text{High Pass (of order } n\text{)} \qquad \frac{\omega_c}{\omega_3} = x \tag{6.48}$$

$$\text{Band Pass (of order } 2n\text{)} \qquad B\left|\frac{\omega_3}{\omega_c} - \frac{\omega_c}{\omega_3}\right| = x \tag{6.49}$$

$$\text{Band Reject (of order } 2n\text{)} \qquad B^{-1}\left|\frac{\omega_3}{\omega_c} - \frac{\omega_c}{\omega_3}\right|^{-1} = x \tag{6.50}$$

where ω_3 is the -3 dB radian frequency. For any x, there are two values of ω_3 which satisfy the band-pass and band-reject relations. Denote these by ω_a and ω_b. The geometric mean of these two frequencies must equal ω_c, i.e. $\omega_c = \sqrt{\omega_a \omega_b}$. In a design from specifications, the -3 dB frequencies might be specified for the band-pass and band-reject filters. In this case, the required values for the center frequency and the parameter B in the frequency transformation can be solved for.

6.7. CHEBYSHEV FILTER TRANSFER FUNCTIONS

Example 3 *An 8th-order 0.5 dB ripple band-pass filter is to have the −3 dB cutoff frequencies $f_a = 10^{1/4}$ kHz and $f_b = 10^{3/4}$ kHz. Calculate the center frequency and the required value of B in the low-pass to band-pass transformation.*

Solution. The −3 dB frequencies satisfy $f_c = \sqrt{f_a f_b}$. Solution for the center frequency yields $f_c = 10^{1/2}$ kHz. The parameter ϵ has the value $\epsilon = \sqrt{10^{0.5/10} - 1} = 0.34931$. The 8th-order band-pass filter is derived from a frequency transformation of the 4th-order low-pass transfer function. Thus it is necessary to solve for the positive real root of the equation

$$8x^4 - 8x^2 + 1 - \sqrt{\frac{1}{0.34931^2} + 2} = 0$$

The desired root is $x = 1.1063$. The value of B is thus given by

$$B = \frac{x}{|f_a/f_c - f_c/f_a|} = \frac{x}{|f_b/f_c - f_c/f_b|} = \frac{1.1063}{|10^{1/4} - 10^{-1/4}|} = 0.9098$$

6.7.4 The Parameter h

To write the Chebyshev transfer functions, the parameter h is required. Given the order n and the ripple parameter ϵ, h is defined by

$$h = \tanh\left(\frac{1}{n} \sinh^{-1} \frac{1}{\epsilon}\right) \tag{6.51}$$

6.7.5 Even-Order Chebyshev Filters

For an even-order Chebyshev low-pass filter of order n, the transfer function can be written in the product form

$$T_{\text{LP}}(s) = K \prod_{i=1}^{n/2} \frac{1}{(s/a_i\omega_c)^2 + (1/b_i)(s/a_i\omega_c) + 1} \tag{6.52}$$

The constants a_i and b_i are given by

$$a_i = \left(\frac{1}{1-h^2} - \sin^2\theta_i\right)^{1/2} \quad \text{for } 1 \leq i \leq n/2 \tag{6.53}$$

$$b_i = \frac{1}{2}\left(1 + \frac{1}{h^2 \tan^2\theta_i}\right)^{1/2} \quad \text{for } 1 \leq i \leq n/2 \tag{6.54}$$

where the θ_i are given by

$$\theta_i = \frac{2i-1}{n} \times 90° \quad \text{for } 1 \leq i \leq n/2 \tag{6.55}$$

Example 4 *Solve for the transfer function of the 0.5 dB ripple fourth-order Chebyshev low-pass and high-pass filters.*

Solution. The parameters ϵ and h are given by

$$\epsilon = \left(10^{0.5/10} - 1\right)^{1/2} = 0.3493 \qquad h = \tanh\left(\frac{1}{4}\sinh^{-1}\frac{1}{0.3493}\right) = 0.4166$$

For $n = 4$, there are two second-order polynomials in each transfer function. The calculations are summarized as follows:

i	θ_i	a_i	b_i
1	22.5°	1.0313	2.9406
2	67.5°	0.59703	0.7051

The low-pass transfer function is given by

$$T_{\text{LP}}(s) = K\frac{1}{(s/1.0313\omega_c)^2 + (1/2.9406)(s/1.0313\omega_c) + 1}$$
$$\times \frac{1}{(s/0.59703\omega_c)^2 + (1/0.7051)(s/0.59703\omega_c) + 1}$$

The high-pass transfer function is obtained by replacing s/ω_c with ω_c/s to obtain

$$T_{\text{HP}}(s) = K\frac{1}{(\omega_c/1.0313s)^2 + (1/2.9406)(\omega_c/1.0313s) + 1}$$
$$\times \frac{1}{(\omega_c/0.59703s)^2 + (1/0.7051)(\omega_c/0.59703s) + 1}$$
$$= K\frac{(1.0313s/\omega_c)^2}{(1.0313s/\omega_c)^2 + (1/2.9406)(1.0313s/\omega_c) + 1}$$
$$\times \frac{(0.59703s/\omega_c)^2}{(0.59703s/\omega_c)^2 + (1/0.7051)(0.59703s/\omega_c) + 1}$$

6.7.6 Odd-Order Chebyshev Filters

For an odd-order Chebyshev low-pass filter of order n, the transfer function can be written in the form

$$T_{\text{LP}}(s) = K\frac{1}{s/a_{(n+1)/2}\omega_c + 1} \times \prod_{i=1}^{(n-1)/2} \frac{1}{(s/a_i\omega_c)^2 + (1/b_i)(s/a_i\omega_c) + 1} \tag{6.56}$$

The constants a_i and b_i are given by

$$a_{(n+1)/2} = \frac{h}{\sqrt{1-h^2}} \text{ for } i = (n+1)/2 \tag{6.57}$$

$$a_i = \left(\frac{1}{1-h^2} - \sin^2\theta_i\right)^{1/2} \text{ for } 1 \leq i \leq (n-1)/2 \tag{6.58}$$

$$b_i = \frac{1}{2}\left(1 + \frac{1}{h^2\tan^2\theta_i}\right)^{1/2} \text{ for } 1 \leq i \leq (n-1)/2 \tag{6.59}$$

where the θ_i are given by

$$\theta_i = \frac{2i-1}{n} \times 90° \text{ for } 1 \leq i \leq (n-1)/2 \tag{6.60}$$

Example 5 *Solve for the transfer function for the 1 dB ripple third-order Chebyshev low-pass and high-pass filters.*

Solution. The parameters ϵ and h are given by

$$\epsilon = \left(10^{1/10} - 1\right)^{1/2} = 0.5088 \qquad h = \tanh\left(\frac{1}{3}\sinh^{-1}\frac{1}{0.5088}\right) = 0.4430$$

For $n = 3$, each transfer function contains one second-order polynomial and one first-order polynomial. The calculations are summarized as follows:

i	θ_i	a_i	b_i
1	30°	0.9971	2.0177
2	—	0.4942	—

6.8. ELLIPTIC FILTER TRANSFER FUNCTIONS

The low-pass transfer function is given by

$$T_{\text{LP}}(s) = K \frac{1}{(s/0.4942\omega_c) + 1} \times \frac{1}{(s/0.9971\omega_c)^2 + (1/2.0177)(s/0.9971\omega_c) + 1}$$

The high-pass function is obtained by replacing s/ω_c with ω_c/s to obtain

$$\begin{aligned} T_{\text{HP}}(s) &= K \frac{1}{(\omega_c/0.4942s) + 1} \times \frac{1}{(\omega_c/0.9971s)^2 + (1/2.0177)(\omega_c/0.9971s) + 1} \\ &= K \frac{(0.4942s/\omega_c)}{(0.4942s/\omega_c) + 1} \times \frac{(0.9971s/\omega_c)^2}{(0.9971s/\omega_c)^2 + (1/2.0177)(0.9971s/\omega_c) + 1} \end{aligned}$$

6.8 Elliptic Filter Transfer Functions

The elliptic filter is also known as the Cauer-Chebyshev filter. It is a filter which exhibits equal ripple in the passband and in the stopband. It can be designed to have a much steeper rolloff than the Chebyshev filter. The approximation to the magnitude squared function of the elliptic low-pass filter is of the form

$$|T_{\text{LP}}(j\omega)|^2 = K^2 \frac{1 + \epsilon^2 R_n^2(0)}{1 + \epsilon^2 R_n^2(\omega/\omega_c)} \qquad (6.61)$$

where K is the dc gain constant, ϵ determines the dB ripple, and $R_n(\omega/\omega_c)$ is the rational Chebychev function. This function has the form

$$\begin{aligned} R_n(x) &= \prod_{i=1}^{n/2} q_i^2 \frac{1 - (x/q_i)^2}{1 - (q_i x)^2} \qquad \text{for } n \text{ even} \\ &= x \prod_{i=1}^{(n-1)/2} q_i^2 \frac{1 - (x/q_i)^2}{1 - (q_i x)^2} \qquad \text{for } n \text{ odd} \end{aligned} \qquad (6.62)$$

where $0 < q_i < 1$. From this definition, it follows that $R_n(x)$ satisfies the following properties:

$$R_n\left(\frac{1}{x}\right) = \frac{1}{R_n(x)} \qquad R_n(q_i) = 0 \qquad R_n\left(\frac{1}{q_i}\right) = \infty \qquad (6.63)$$

$$\begin{aligned} R_n(0) &= \prod_{i=1}^{n/2} q_i^2 \qquad \text{for } n \text{ even} \\ &= 0 \qquad \text{for } n \text{ odd} \end{aligned} \qquad (6.64)$$

$$\begin{aligned} R_n(1) &= (-1)^{n/2} \qquad \text{for } n \text{ even} \\ &= (-1)^{(n-1)/2} \qquad \text{for } n \text{ odd} \end{aligned} \qquad (6.65)$$

$$\begin{aligned} R_n(\infty) &= \prod_{i=1}^{n/2} \frac{(-1)^i}{a_i^2} \qquad \text{for } n \text{ even} \\ &= \infty \qquad \text{for } n \text{ odd} \end{aligned} \qquad (6.66)$$

It is beyond the scope of this treatment to describe how the q_i are specified. For n even, the transfer function can be put into the form

$$T_{\text{LP}}(s) = \prod_{i=1}^{n/2} \frac{(s/c_i\omega_c)^2 + 1}{(s/a_i\omega_c)^2 + (1/b_i)(s/a_i\omega_c) + 1} \qquad (6.67)$$

For n odd, the form is

$$T_{\text{LP}}(s) = \frac{1}{s/a_{(n+1)/2}\omega_c + 1} \times \prod_{i=1}^{(n-1)/2} \frac{(s/c_i\omega_c)^2 + 1}{(s/a_i\omega_c)^2 + (1/b_i)(s/a_i\omega_c) + 1} \qquad (6.68)$$

These differ from the forms for the Chebyshev filter by the presence of zeros on the $j\omega$ axis. That is, $T_{\text{LP}}(j\omega) = 0$ for $\omega = c_i\omega_c$.

Figure 6.5 shows the plot of $|T_{\text{LP}}(jv)|$ for a 4th-order elliptic filter, where $v = \omega/\omega_c$ is the normalized radian frequency. The filter has a dB ripple of 0.5 dB for $v \leq 1$. Like the Chebyshev filters, the gain ripples up from its dc value for the even order filter. There are two zeros in the response, one at $v = 1.59$ and the other at $v = 3.48$. For $v > 1.5$, the gain ripples between 0 and 0.0153. As v becomes large, the gain approaches the value 0.0153 which is 36.3 dB down from the dc value. Odd order elliptic filters have the property that the gain ripples down from the dc value and approaches 0 as v becomes large.

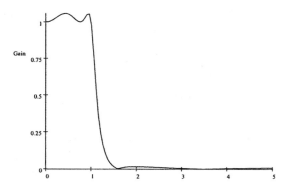

Figure 6.5: Magnitude response versus normalized frequency for the example elliptic filter.

The passband for the elliptic filter is the band defined by $\omega \leq \omega_c$. Like the Chebyschev filters, the dB ripple is defined in this band. The even-order filters ripple up from the dc gain, whereas the odd-order filters ripple down. At $\omega = \omega_c$, the gain of an even-order filter is equal to the dc gain, whereas the gain of an odd-order filter is down by an amount equal to the dB ripple. For $\omega > \omega_c$, the gain decreases rapidly and is equal to zero at the zero frequencies of the transfer function. Between adjacent zeros, the gain peaks up in an equal ripple fashion, i.e. with equal values at the peaks. Let the gain at the peaks between adjacent zeros have a dB level that is down from the dc gain by $A_{\min}$ dB. The stopband frequency ω_s is defined as the frequency between the cutoff frequency ω_c and the first zero frequency at which the gain is down by $A_{\min}$ dB. For $\omega > \omega_c$, the gain is down from the dc level by $A_{\min}$ dB or more.

The three parameters which define the alignment of an elliptic filter are the dB ripple in the passband, the ratio ω_s/ω_c of the stopband frequency to the cutoff frequency, and the minimum dB attenuation $A_{\min}$ in the stopband. The tables below give the values of the a_i, b_i, and c_i in the elliptic transfer functions for 0.5 dB ripple filters having ω_s/ω_c values of 1.5, 2.0, and 3.0 for orders $n = 2$ through $n = 5$.

Elliptic Alignments for $\omega_s/\omega_c = 1.5$

n	$A_{\min}$	i	a_i	b_i	c_i
2	8.3	1	1.2662	1.2275	1.9817
3	21.9	1	1.0720	2.3672	1.6751
		2	0.76695		
4	36.3	1	1.0298	4.0390	1.5923
		2	0.68690	0.74662	3.4784
5	50.6	1	1.0158	6.2145	1.5574
		2	0.75895	1.3310	2.3319
		3	0.42597		

Elliptic Alignments for $\omega_s/\omega_c = 2.0$

n	$A_{\min}$	i	a_i	b_i	c_i
2	13.9	1	1.2617	1.0134	2.7321
3	31.2	1	1.0717	1.9924	2.2701
		2	0.69212		
4	48.6	1	1.0308	3.4228	2.1432
		2	0.64056	0.72416	4.9221
5	66.1	1	1.0169	5.2811	2.0893
		2	0.72457	1.2481	3.2508
		3	0.39261		

Elliptic Alignments for $\omega_s/\omega_c = 3.0$

n	$A_{\min}$	i	a_i	b_i	c_i
2	21.5	1	1.2471	0.91893	4.1815
3	42.8	1	1.0703	1.8159	3.4392
		2	0.65236		
4	64.1	1	1.0312	3.1267	3.2335
		2	0.61471	0.71264	7.6466
5	85.5	1	1.0174	4.8296	3.1457
		2	0.70455	1.2056	5.0077
		3	0.37452		

6.9 The Thompson Phase Approximation

Let $v_i(t)$ be the input voltage to a filter. If the output voltage is given by $v_o(t) = v_i(t - \tau)$, the filter is said to introduce a time delay of τ seconds to the signal. The transfer function of such a filter is given by

$$T(s) = \frac{V_o(s)}{V_i(s)} = e^{-\tau s} \tag{6.69}$$

For $s = j\omega$, the phase of the transfer function in radians is given by

$$\varphi(\omega) = -\omega\tau = -2\pi f \tau \tag{6.70}$$

Thus the phase lag through the filter is directly proportional to the frequency. A filter having such a phase function is called a linear-phase filter. The Thompson approximation is a low-pass filter which has a phase function that approximates a linear-phase function. It can only be realized as a low-pass filter. This is because a time delay cannot be realized with, for example, a high-pass filter. The Thompson filter is also called a Bessel filter. This is because Bessel polynomials are used in obtaining the filter coefficients.

Given the phase function $\varphi(\omega)$ for any filter, there are two delays which can be defined. These are the phase delay τ_φ and the group delay τ_g. These are given by

$$\tau_\varphi = -\frac{\varphi(\omega)}{\omega} = -\frac{\varphi(\omega)}{2\pi f} \tag{6.71}$$

$$\tau_g = -\frac{d\varphi(\omega)}{d\omega} = -\frac{1}{2\pi}\frac{d\varphi(\omega)}{df} \tag{6.72}$$

These are equal for a linear-phase filter.

Let n be the order of the filter. For n even, the transfer function can be put into the form

$$T_{\text{LP}}(s) = \prod_{i=1}^{n/2} \frac{1}{(s/a_i\omega_c)^2 + (1/b_i)(s/a_i\omega_c) + 1} \tag{6.73}$$

For n odd, the form is

$$T_{\text{LP}}(s) = \frac{1}{s/a_{(n+1)/2}\omega_c + 1} \times \prod_{i=1}^{(n-1)/2} \frac{1}{(s/a_i\omega_c)^2 + (1/b_i)(s/a_i\omega_c) + 1} \qquad (6.74)$$

where the time delay of the filter is given by $\tau = 1/\omega_c$. It is beyond the scope of the treatment here to describe how the a_i and b_i are obtained. The tables below give the values for $n = 1$ to $n = 6$. Also given is the ratio of the -3 dB frequency f_3 to the cutoff frequency f_c for each case.

Thompson Alignments for $1 \leq n \leq 4$

n	f_3/f_c	i	a_i	b_i
1	1	1	1.0	
2	1.3617	1	1.7321	0.57735
3	1.7557	1	2.5415	0.69105
		2	2.3222	
4	2.1139	1	3.0233	0.52193
		2	3.3894	0.80554

Thompson Alignments for $5 \leq n \leq 6$

n	f_3/f_c	i	a_i	b_i
5	2.4274	1	3.7779	0.56354
		2	4.2610	0.91648
		3	3.64674	
6	2.8516	1	4.3360	0.51032
		2	4.5665	0.61119
		3	5.1492	1.0233

Figure 6.6 shows the plot of $|T(j2\pi f)|$ as a function of frequency for the 4th-order Thompson filter for the case $\omega_c = 1$ rad/s. This choice gives a delay of 1 second. The -3 dB frequency of the filter is $f_3 = 2.1139\omega_0/2\pi = 0.3364$ Hz. If the -3 dB frequency is multiplied by a constant, the delay time is divided by that constant. Fig. 6.7 shows the phase response of the filter. The phase delay is shown plotted in Fig. 6.8. It can be seen from this figure that the delay is approximately equal to 1 second up to the -3 dB frequency of the filter.

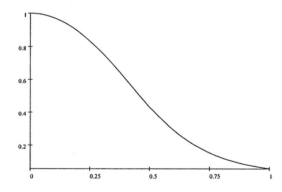

Figure 6.6: Plot of $|T(j2\pi f)|$ versus f.

6.10 First-Order Filter Topologies

6.10.1 First-Order Low-Pass Filter

Figure 6.9 shows the circuit diagrams of two first-order RC low-pass filters, where each is followed by an op-amp buffer. The circuit in Fig. 6.9(a) is a non-inverting filter whereas the one in Fig. 6.9(b) is an inverting filter. The voltage-gain transfer function of each circuit can be written

$$T(s) = K\frac{1}{s/\omega_c + 1} \qquad (6.75)$$

6.10. FIRST-ORDER FILTER TOPOLOGIES

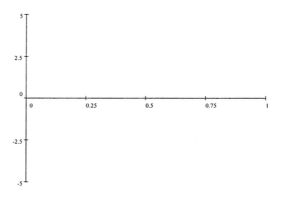

Figure 6.7: Plot of the phase in degrees of $T(j2\pi f)$ versus f.

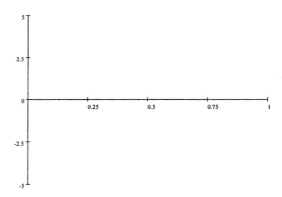

Figure 6.8: Plot of the phase delay in seconds of $T(j2\pi f)$ versus f.

where K is the dc gain constant and ω_c is the radian normalization frequency. The magnitude, phase, and group delay functions are given by

$$A(\omega) = \frac{K}{\sqrt{1 + (s/\omega_c)^2}} \tag{6.76}$$

$$\varphi(\omega) = -\tan^{-1}\left(\frac{\omega}{\omega_c}\right) \tag{6.77}$$

$$\Delta t = \frac{1}{\omega_c} \times \frac{1}{1 + (s/\omega_c)^2} \tag{6.78}$$

For the non-inverting filter of Fig. 6.9a, $K = 1 + R_F/R_2$ and $\omega_c = 1/R_1 C$. For the inverting filter of Fig. 6.9b, $K = -R_F/(R_1 + R_2)$ and $\omega_c = 1/(R_1 \| R_2) C$. For each circuit, the half-power cutoff frequency ω_c is equal to the normalization frequency ω_c. The half-power cutoff frequency in Hertz is given by $f_c = \omega_c/2\pi$.

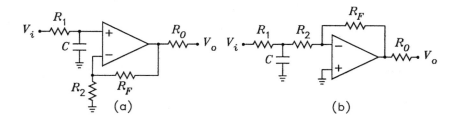

Figure 6.9: Circuit diagrams of (a) a first-order non-inverting low-pass filter and (b) a first-order inverting low-pass filter.

6.10.2 First-Order High-Pass Filter

Figure 6.10 shows the circuit diagrams of two first-order RC high-pass filters, where each is followed by an op-amp buffer. The circuit in Fig. 6.10(a) is a non-inverting filter whereas the one in Fig. 6.10(b) is an inverting filter. The voltage-gain transfer function of each circuit can be written

$$T(s) = K \frac{s/\omega_c}{s/\omega_c + 1} \tag{6.79}$$

where K is the gain constant and ω_c is the radian normalization frequency. For the non-inverting filter of Fig. 6.10(a), $K = 1 + R_F/R_2$ and $\omega_c = 1/R_1 C$. For the inverting filter of Fig. 6.10(b) $K = -R_F/R_1$ and $\omega_c = 1/R_1 C$. For each circuit, the half-power cutoff frequency ω_3 is equal to the normalization frequency ω_c. The half-power cutoff frequency in Hertz is given by $f_c = \omega_c/2\pi$.

6.11 Second-Order Filter Topologies

6.11.1 Sallen-Key Low-Pass Filter

Figure 6.11 shows the circuit diagram of the second-order Sallen-Key low-pass filter. The voltage-gain transfer function is given by

$$\frac{V_o}{V_i} = K \frac{1}{(s/\omega_0)^2 + (1/Q)(s/\omega_0) + 1} \tag{6.80}$$

6.11. SECOND-ORDER FILTER TOPOLOGIES

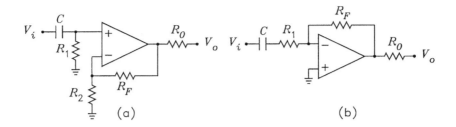

Figure 6.10: (a) First-order non-inverting high-pass filter. (b) First-order inverting high-pass filter.

where K is the dc gain constant, ω_0 is the radian resonance frequency, and Q is the quality factor. These are given by

$$K = 1 + \frac{R_4}{R_3} \qquad \omega_0 = \frac{1}{\sqrt{R_1 R_2 C_1 C_2}} \qquad (6.81)$$

$$Q = \frac{\sqrt{R_1 R_2 C_1 C_2}}{(1-K) R_1 C_1 + (R_1 + R_2) C_2} \qquad (6.82)$$

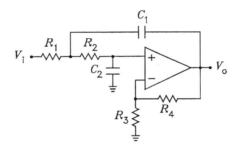

Figure 6.11: Second-order Sallen-Key low-pass filter.

Special Case 1

Let $K = 1$ (R_4 a short and R_3 an open). If R_1, R_2, ω_0, and Q are specified, C_1 and C_2 are given by

$$C_1 = \frac{Q}{\omega_0}\left(\frac{1}{R_1} + \frac{1}{R_2}\right) \qquad C_2 = \frac{1}{Q\omega_0 (R_1 + R_2)} \qquad (6.83)$$

The filter is often realized with $R_1 = R_2$.

Special Case 2

Capacitor values are often difficult to obtain. It is possible to specify C_1 and C_2 and calculate the resistors. Let $K = 1$ (R_4 a short and R_3 an open). If C_1, C_2, ω_0, and Q are specified, R_1 and R_2 are given by

$$R_1, R_2 = \frac{1}{2Q\omega_0 C_2}\left[1 \pm \sqrt{1 - 4Q^2 \frac{C_2}{C_1}}\right] \qquad (6.84)$$

Any value for the ratio C_2/C_1 can be chosen provided $4Q^2 C_2/C_1 \leq 1$. Note that the values for R_1 and R_2 are interchangeable.

Example 6 *Design a unity-gain second-order Sallen-Key low-pass filter with $f_0 = 1$ kHz and $Q = 1/\sqrt{2}$.*

Solution. Let $C_1 = 0.1 \ \mu F$ and $C_2 = 0.022 \ \mu F$. R_1 and R_2 are given by

$$R_1, R_2 = \frac{\sqrt{2}}{2 \times 2\pi 1000 \times 0.022 \times 10^{-6}} \left[1 \pm \sqrt{1 - 4 \times \frac{1}{2} \times 0.22} \right]$$
$$= 1.29 \ k\Omega \text{ and } 8.94 \ k\Omega$$

Either value may be assigned to R_1. The other value is then assigned to R_2.

Special Case 3

Let $R_1 = R_2 = R$ and $C_1 = C_2 = C$. If C, ω_0, and Q are specified, R and K are given by

$$R = \frac{1}{\omega_0 C} \qquad K = 3 - \frac{1}{Q} \tag{6.85}$$

6.11.2 Infinite-Gain Multi-Feedback Low-Pass Filter

Figure 6.12 shows the circuit diagram of the second-order infinite-gain multi-feedback low-pass filter. The voltage-gain transfer function is given by

$$\frac{V_o}{V_i} = -K \frac{1}{(s/\omega_0)^2 + (1/Q)(s/\omega_0) + 1} \tag{6.86}$$

where K is the dc gain constant, ω_0 is the radian resonance frequency, and Q is the quality factor. These are given by

$$K = \frac{R_3}{R_1} \qquad \omega_0 = \frac{1}{\sqrt{R_2 R_3 C_1 C_2}} \tag{6.87}$$

$$Q = \frac{\sqrt{R_2 R_3 C_1 / C_2}}{R_2 + R_3 + R_2 R_3 / R_1} \tag{6.88}$$

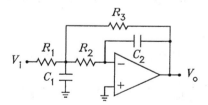

Figure 6.12: Second-order infinite gain multi-feedback low-pass filter.

Special Case 1

Let $R_1 = R_2 = R_3 = R$ so that $K = -1$. If R, ω_0, and Q are specified, C_1 and C_2 are given by

$$C_1 = \frac{Q}{\omega_0} \left(\frac{2}{R_1} + \frac{1}{R_2} \right) \qquad C_2 = \frac{1}{Q\omega_0 (R_1 + 2R_2)} \tag{6.89}$$

The filter is often designed with $R_1 = R_2$.

6.11. SECOND-ORDER FILTER TOPOLOGIES

Special Case 2

Capacitor values are often difficult to obtain. It is possible to specify C_1 and C_2 and calculate the resistors. Let $K = -1$ ($R_3 = R_1$). If C_1, C_2, ω_0, and Q are specified, $R_1/2$ and R_2 are given by

$$\frac{R_1}{2}, R_2 = \frac{1}{4Q\omega_0 C_2}\left[1 \pm \sqrt{1 - 8Q^2\frac{C_2}{C_1}}\right]$$

Any value for the ratio C_2/C_1 can be chosen provided $8Q^2 C_2/C_1 \leq 1$. Note that the values for $R_1/2$ and R_2 are interchangeable.

Example 7 *Design a unity-gain second-order infinite-gain multi-feedback low-pass filter with $f_0 = 1$ kHz and $Q = 1/\sqrt{2}$.*

Solution. Let $C_1 = 0.1$ μF and $C_2 = 0.01$ μF. R'_1 and R_2 are given by

$$\begin{aligned}\frac{R_1}{2}, R_2 &= \frac{\sqrt{2}}{4 \times 2\pi 1000 \times 0.01 \times 10^{-6}}\left[1 \pm \sqrt{1 - 8 \times \frac{1}{2} \times 0.1}\right]\\ &= 1.27 \text{ k}\Omega \text{ and } 9.99 \text{ k}\Omega\end{aligned}$$

The filter can be designed either with $R_1 = 2 \times 1.27$ kΩ = 2.54 kΩ and $R_2 = 9.99$ kΩ or with $R_1 = 2 \times 9.99$ kΩ = 20 kΩ and $R_2 = 1.27$ kΩ.

6.11.3 Sallen-Key High-Pass Filter

Figure 6.13 shows the circuit diagram for the second-order Sallen-Key high-pass filter. The voltage-gain transfer function is given by

$$\frac{V_o}{V_i} = K\frac{(s/\omega_0)^2}{(s/\omega_0)^2 + (1/Q)(s/\omega_0) + 1} \tag{6.90}$$

where K is the asymptotic high-frequency gain, ω_0 is the resonance frequency, and Q is the quality factor. These are given by

$$K = 1 + \frac{R_4}{R_3} \qquad \omega_0 = \frac{1}{\sqrt{R_1 R_2 C_1 C_2}} \tag{6.91}$$

$$Q = \frac{\sqrt{R_1 R_2 C_1 C_2}}{R_1(C_1 + C_2) + (1 - K)R_2 C_2} \tag{6.92}$$

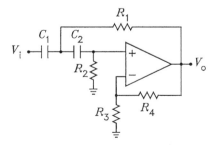

Figure 6.13: Second-order Sallen-Key high-pass filter.

Special Case 1

Let $C_1 = C_2 = C$ and $K = 1$ (R_F a short and R_3 an open). If C, ω_0, and Q are specified, R_1 and R_2 are given by

$$R_1 = \frac{1}{2Q\omega_0 C} \qquad R_2 = \frac{2Q}{\omega_0 C} \tag{6.93}$$

Special Case 2

Let $R_1 = R_2 = R$ and $C_1 = C_2 = C$. If R, ω_0, and Q are specified, C and K are given by

$$C = \frac{1}{\omega_0 R} \qquad K = 3 - \frac{1}{Q} \tag{6.94}$$

6.11.4 Infinite-Gain Multi-Feedback High-Pass Filter

The second-order infinite-gain multi-feedback high-pass filter is not a stable circuit in practice. This is because its input node connects through two series capacitors to the inverting input node of the op amp. At high frequencies, the input node becomes shorted to a virtual ground. This can cause oscillation problems in the source that drives the filter. Therefore, this topology is not recommended.

6.11.5 Sallen-Key Band-Pass Filter

Figure 6.14 shows the circuit diagram of the second-order Sallen-Key band-pass filter. The voltage-gain transfer function is given by

$$\frac{V_o}{V_i} = K \frac{(1/Q)(s/\omega_0)}{(s/\omega_0)^2 + (1/Q)(s/\omega_0) + 1} \tag{6.95}$$

where K is the gain at resonance, ω_0 is the resonance frequency, and Q is the quality factor. These are given by

$$K = \frac{R_2}{R_1 + R_2} \times \frac{K_0 R_3 C_2}{(R_1 \| R_2)(C_1 + C_2) + R_3 C_2 [1 - K_0 R_1/(R_1 + R_2)]} \tag{6.96}$$

$$\omega_0 = \frac{1}{\sqrt{(R_1 \| R_2) R_3 C_1 C_2}} \tag{6.97}$$

$$Q = \frac{\sqrt{(R_1 \| R_2) R_3 C_1 C_2}}{(R_1 \| R_2)(C_1 + C_2) + R_3 C_2 [1 - K_0 R_1/(R_1 + R_2)]} \tag{6.98}$$

where K_0 is the gain from the non-inverting input of the op amp to the output. This is given by

$$K_0 = 1 + \frac{R_5}{R_4} \tag{6.99}$$

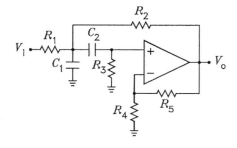

Figure 6.14: Second-order Sallen-Key band-pass filter.

6.11. SECOND-ORDER FILTER TOPOLOGIES

Special Case

Let $R_1 = R_2 = R_3 = R$ and $C_1 = C_2 = C$. If Q, ω_0, and C are specified, R, K_0 and K are given by

$$R = \frac{\sqrt{2}}{\omega_0 C} \qquad K_0 = 4 - \frac{1}{Q^2} \qquad K = 4Q^2 - 1 \tag{6.100}$$

6.11.6 Infinite-Gain Multi-Feedback Band-Pass Filter

Figure 6.15 shows the circuit diagram of the second-order infinite-gain multi-feedback band-pass filter. The voltage-gain transfer function is given by

$$\frac{V_o}{V_i} = -K \frac{(1/Q)(s/\omega_0)}{(s/\omega_0)^2 + (1/Q)(s/\omega_0) + 1} \tag{6.101}$$

where K is the gain at resonance, ω_0 is the resonance frequency, and Q is the quality factor. These are given by

$$K = \frac{R_3 C_1}{R_1 (C_1 + C_2)} \qquad \omega_0 = \frac{1}{\sqrt{(R_1 \| R_2) R_3 C_1 C_2}} \tag{6.102}$$

$$Q = \frac{\sqrt{R_3 C_1 C_2 / (R_1 \| R_2)}}{C_1 + C_2} \tag{6.103}$$

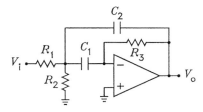

Figure 6.15: Second-order infinite-gain multi-feedback band-pass filter.

6.11.7 Second-Order Band-Pass Filter Bandwidth

The -3 dB bandwidth $\Delta\omega_3$ of a second-order band-pass filter is related to the resonance frequency ω_0 and the quality factor Q by

$$\Delta\omega_3 = \frac{\omega_0}{Q} \tag{6.104}$$

Let the lower and upper -3 dB cutoff frequencies, respectively, be denoted by ω_a and ω_b. These frequencies are related to the bandwidth and the quality factor by

$$\Delta\omega_3 = \omega_b - \omega_a \qquad \omega_0 = \sqrt{\omega_a \omega_b} \tag{6.105}$$

6.11.8 A Biquad Filter

A biquadratic transfer function, or biquad for short, is a transfer function that is the ratio of two second-order polynomials. It has the form

$$T_{\text{BQ}}(s) = K \frac{(s/\omega_n)^2 + (1/Q_n)(s/\omega_n) + 1}{(s/\omega_d)^2 + (1/Q_d)(s/\omega_d) + 1} \tag{6.106}$$

where K, ω_n, Q_n, ω_d, and Q_d are constants. The function can be thought of as the sum of three transfer functions: a low-pass, a band-pass, and a high-pass. For the case $Q_n \to \infty$, the zeros are on the $j\omega$ axis and the transfer function exhibits a notch at $\omega = \omega_c$, i.e. $T_{BQ}(j\omega_n) = 0$. Band-reject filters and elliptic filters have biquad terms of this form.

In general, the simplest biquad circuits to realize are the ones which require more than one op amp. Compared to the single op amp biquads, these circuits can be realized with only one capacitor value and they are usually easier to tweak. Fig. 6.16 shows the circuit diagram of an example three op amp biquad. For this circuit, write

$$V_a = -\left(R_4 \| \frac{1}{Cs}\right)\left(\frac{V_i}{R_3} + \frac{V_b}{R}\right)$$
$$V_o = -R\left(\frac{V_a}{R} + \frac{V_i}{R_2}\right) \qquad V_b = \frac{-1}{Cs}\left(\frac{V_o}{R} + \frac{V_i}{R_1}\right) \qquad (6.107)$$

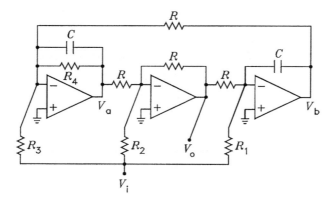

Figure 6.16: Biquad filter.

Although these equations look deceptively simple, a lot of algebra is required to solve for V_o/V_i. The solution is of the form of $T_{BQ}(s)$ given above. For a specified K, ω_d, Q_d, ω_n, Q_n, and C, the design equations for the circuit are

$$R = \frac{1}{\omega_d C} \qquad R_1 = \frac{R}{K} \qquad R_4 = Q_d R$$
$$R_2 = \left(\frac{\omega_n}{\omega_d}\right)^2 R_1 \qquad R_3 = \frac{\omega_d R_2}{\omega_d/Q_d - \omega_n/Q_n} \qquad (6.108)$$

Example 8 *Design a third-order elliptic low-pass filter having 0.5 dB ripple, a notch in its transfer function at $f = 15,734$ Hz, and a stopband to cutoff frequency ratio of 1.5. The gain at dc is to be unity.*

Solution. The third-order elliptic transfer function can be obtained from table given in Section 6.8 for $\omega_s/\omega_c = 1.5$. The minimum attenuation in the stopband for this filter is 21.9 dB. The transfer function is

$$F(s) = \frac{1}{s/0.76695\omega_c + 1} \times \frac{(s/1.6751\omega_c)^2 + 1}{(s/1.0720\omega_c)^2 + (1/2.3672)(s/1.0720\omega_c) + 1}$$

where ω_c is the cutoff frequency. For the null to be at $15,734$ Hz, this requires that $\omega_c = 2\pi 15734/1.6751 = 59,021$. Thus the transfer function is

$$F(s) = \frac{1}{s/45269 + 1} \times \frac{(s/98860)^2 + 1}{(s/63270)^2 + (1/2.367)(s/63270) + 1}$$

6.11. SECOND-ORDER FILTER TOPOLOGIES

The biquadratic term can be realized with the three op amp biquad. Let $C = 1000$ pF. The element values are given by

$$R = \frac{1}{63270C} = 15.81 \text{ k}\Omega \qquad R_1 = \frac{R}{1} = 15.81 \text{ k}\Omega$$

$$R_2 = \left(\frac{98860}{63270}\right)^2 R_1 = 38.59 \text{ k}\Omega \qquad R_4 = 2.367R = 37.41 \text{ k}\Omega$$

$$R_3 = \frac{63270 R_2}{63270/2.367 - 98860/\infty} = 2.367 R_2 = 91.34 \text{ k}\Omega$$

The first-order term can be realized as a passive filter at either the input or the output of the biquad. The preferred realization does not require a buffer stage for isolation. One way of accomplishing this is to divide R_1, R_2, and R_3 each into two series resistors, where the ratio of the resistors in each pair is the same. By voltage division, the voltage at the node where the resistors in each pair connect is the same for the three pairs. Thus these three nodes can be connected together. A capacitor to ground from this node then realizes the first-order term.

Let the ratio of the resistors in each pair be $0.33/0.67$. Divide R_1 into series resistors of values $0.33R_1 = 5.217$ kΩ and $0.67R_1 = 10.59$ kΩ. Similarly, divide R_2 into resistors of values $0.33R_2 = 12.73$ kΩ and $0.67R_2 = 25.86$ kΩ and divide R_3 into resistors of values $0.33R_3 = 30.14$ kΩ and $0.67R_3 = 61.20$ kΩ. Let the smaller of the resistors in each pair connect to the V_i node. These three resistors are in parallel in the circuit and can be replaced with a single resistor of value $5.217 \| 12.73 \| 30.14 = 3.296$ kΩ. The time constant for the capacitor which sets the single-pole term in the transfer function is $R_p C$, where $R_p = 3.296 \| 10.59 \| 25.86 \| 61.20 = 2.208$ kΩ. Thus $C = 1/(45269 \times 2208) = 0.01$ μF. The completed circuit is given in Fig. 6.17. The magnitude response is shown in Fig. 6.18.

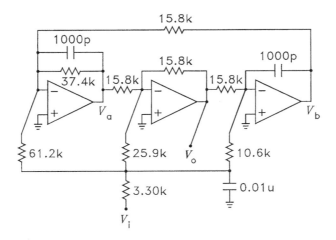

Figure 6.17: Completed elliptic filter.

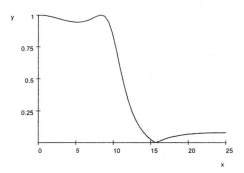

Figure 6.18: Magnitude response versus frequency in kHz.

6.11.9 A Second Biquad Filter

A second biquad filter that has its zeros on the $j\omega$ axis is shown in Fig. 6.19. The following equations can be written for the circuit:

$$V_a = -\frac{V_i}{R_1 C_1 s} - \frac{V_o}{R_3 C_1 s} \qquad V_b = -\frac{R_4}{R_5} V_i$$

$$V_c = \left(\frac{V_a}{R_2} + V_b C_2 s\right)\left(R_2 \| \frac{1}{C_2 s}\right) \qquad V_o = \left(1 + \frac{R_7}{R_6}\right) V_c \qquad (6.109)$$

These equations can be solved for V_o/V_i to obtain

$$\frac{V_o}{V_i} = -K \frac{(s/\omega_n)^2 + 1}{(s/\omega_d)^2 + (1/Q_d)(s/\omega_d) + 1} \qquad (6.110)$$

where

$$K = \frac{R_3}{R_1} \qquad \omega_n = \sqrt{\frac{R_5}{R_1 R_2 R_4 C_1 C_2}} \qquad (6.111)$$

$$\omega_d = \sqrt{\frac{1 + R_7/R_6}{R_2 R_3 C_1 C_2}} \qquad Q_d = \sqrt{\left(1 + \frac{R_7}{R_6}\right) \frac{R_2 C_2}{R_3 C_1}} \qquad (6.112)$$

Because there are more element values than equations that relate them, values must be assigned to some of the elements before the others can be calculated. Let values for C_1, C_2, K, R_5, and $1 + R_7/R_6$ be specified. The other element values are given by

$$R_1 = \frac{1 + R_7/R_6}{K Q_d C_1 \omega_d} \qquad R_2 = \frac{Q_d}{\omega_d C_2} \qquad R_3 = K R_1 \qquad R_4 = K \left(\frac{\omega_d}{\omega_n}\right)^2 \frac{R_5}{1 + R_7/R_6} \qquad (6.113)$$

6.11.10 State-Variable Filter

Fig. 6.20 shows the circuit diagram of a four op-amp state-variable filter. This is a variation of the biquadratic which simultaneously has low-, high-, and band-pass outputs. Although state-variable filters can be realized with three op amps, the one given in Fig. 6.20 has the feature that is simpler to adjust in the laboratory. The circuit has the four outputs V_{hp}, V_{bp1}, V_{bp2}, and V_{lp}. The equations for these are

$$V_{hp} = -\frac{R_2}{R_1} V_i + \frac{R_3}{R_4} V_{bp1} - V_{lp} \qquad (6.114)$$

6.11. SECOND-ORDER FILTER TOPOLOGIES

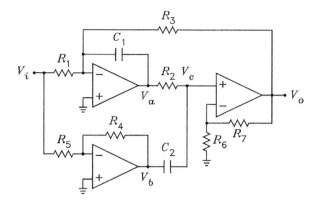

Figure 6.19: A second biquad circuit.

$$V_{bp1} = \frac{-1}{RCs} V_{hp} \tag{6.115}$$

$$V_{bp2} = -\frac{R_3}{R_4} V_{bp1} \tag{6.116}$$

$$V_{lp} = \frac{-1}{RCs} V_{bp1} \tag{6.117}$$

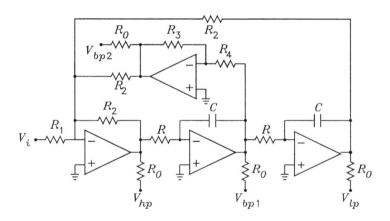

Figure 6.20: A state-variable filter.

These equations can be solved for the following voltage-gain transfer functions:

$$T_{hp}(s) = \frac{V_{hp}}{V_i} = -K \frac{(s/\omega_0)^2}{(s/\omega_0)^2 + (1/Q)(s/\omega_0) + 1} \tag{6.118}$$

$$T_{bp1}(s) = \frac{V_{bp1}}{V_i} = K \frac{s/\omega_0}{(s/\omega_0)^2 + (1/Q)(s/\omega_0) + 1} \tag{6.119}$$

$$T_{bp2}(s) = \frac{V_{bp2}}{V_i} = -K \frac{(1/Q)(s/\omega_0)}{(s/\omega_0)^2 + (1/Q)(s/\omega_0) + 1} \tag{6.120}$$

$$T_{lp}(s) = \frac{V_{lp}}{V_i} = -K \frac{1}{(s/\omega_0)^2 + (1/Q)(s/\omega_0) + 1} \tag{6.121}$$

where K, ω_0, and Q are given by

$$K = \frac{R_2}{R_1} \qquad \omega_0 = \frac{1}{RC} \qquad Q = \frac{R_4}{R_3} \tag{6.122}$$

Because the Q is a function of only R_3 and R_4 it can be adjusted independently of the center frequency, f_o, which is a function of only R and C.

It can be seen that the state-variable filter has simultaneous low-pass, band-pass, and high-pass outputs. There are two band-pass outputs. These differ by the constant factor $(1/Q)$. It is called the state-variable filter because if the differential equation for the circuit were written and expressed in state-variable form, the low-pass and band-pass outputs would be the state-variables.

6.12 Third-Order Sallen-Key Filter Circuits

The filter circuits given in this section consist of a first-order stage in cascade with a second-order stage with no buffer amplifier isolating the two stages. The transfer functions are difficult to derive.

6.12.1 Low-Pass Filters

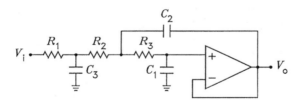

Figure 6.21: Third-order Sallen-Key low-pass filter.

Figure 6.21 shows the circuit diagram of a third-order unity-gain Sallen-Key low-pass filter. For a cutoff frequency ω_c, the element values for Butterworth and Chebyshev filters are given in the following table. The Butterworth filters are 3 dB down at ω_c whereas the Chebyshev filters are down by an amount equal to the dB ripple.

Alignment	R_1	R_2	R_3	C_1	C_2	C_3
Butterworth	R	R	R	$0.20245/R\omega_c$	$3.5465/R\omega_c$	$1.3926/R\omega_c$
0.01 dB Chebyshev	R	R	R	$0.07130/R\omega_c$	$2.5031/R\omega_c$	$0.8404/R\omega_c$
0.03 dB Chebyshev	R	R	R	$0.07736/R\omega_c$	$3.3128/R\omega_c$	$1.0325/R\omega_c$
0.10 dB Chebyshev	R	R	R	$0.09691/R\omega_c$	$4.7921/R\omega_c$	$1.3145/R\omega_c$
0.30 dB Chebyshev	R	R	R	$0.08582/R\omega_c$	$7.4077/R\omega_c$	$1.6827/R\omega_c$
1.0 dB Chebyshev	R	R	R	$0.05872/R\omega_c$	$14.784/R\omega_c$	$2.3444/R\omega_c$

6.12.2 High-Pass Filters

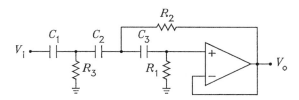

Figure 6.22: Third-order Sallen-Key high-pass filter.

Figure 6.22 shows the circuit diagram of a third-order unity-gain Sallen-Key high-pass filter. For a cutoff frequency ω_c, the element values for Butterworth and Chebyshev filters are given in the following table. The Butterworth filters are 3 dB down at ω_c whereas the Chebyshev filters are down by an amount equal to the dB ripple.

Alignment	R_1	R_2	R_3	C_1	C_2	C_3
Butterworth	$4.93949/\omega_c C$	$0.28194/\omega_c C$	$0.71808/\omega_c C$	C	C	C
0.01 dB Chebyshev	$10.9130/\omega_c C$	$0.39450/\omega_c C$	$1.18991/\omega_c C$	C	C	C
0.03 dB Chebyshev	$10.09736/\omega_c C$	$0.30186/\omega_c C$	$0.96852/\omega_c C$	C	C	C
0.1 dB Chebyshev	$10.3188/\omega_c C$	$0.20868/\omega_c C$	$0.76075/\omega_c C$	C	C	C
0.3 dB Chebyshev	$11.65230/\omega_c C$	$0.13499/\omega_c C$	$0.59428/\omega_c C$	C	C	C
1.0 dB Chebyshev	$17.0299/\omega_c C$	$0.06764/\omega_c C$	$0.42655/\omega_c C$	C	C	C

6.13 Impedance Transfer Functions

6.13.1 RC Network

The impedance transfer function for a two-terminal RC network which contains only one capacitor and is not an open circuit at dc can be written

$$Z = R_{\rm dc} \frac{1 + \tau_z s}{1 + \tau_p s} \tag{6.123}$$

where $R_{\rm dc}$ is the dc resistance of the network, τ_p is the pole time constant, and τ_z is the zero time constant. The pole time constant is the time constant of the network with the terminals open circuited. The zero time constant is the time constant of the network with the terminals short circuited. Figure 6.23(a) shows the circuit diagram of an example two-terminal RC network. The impedance transfer function can be written by inspection to obtain

$$Z = R_1 \frac{1 + R_2 C s}{1 + (R_1 + R_2) C s} \tag{6.124}$$

Figure 6.23: Example RC and RL impedance networks.

6.13.2 RL Network

The impedance transfer function for a two-terminal RL network which contains only one inductor and is not a short circuit at dc can be written

$$Z = R_{\text{dc}} \frac{1 + \tau_z s}{1 + \tau_p s} \qquad (6.125)$$

where R_{dc} is the dc resistance of the network, τ_p is the pole time constant, and τ_z is the zero time constant. The pole time constant is the time constant of the network with the terminals open circuited. The zero time constant is the time constant of the network with the terminals short circuited. Figure 6.23(b) shows the circuit diagram of an example two-terminal RL network. The impedance transfer function can be written by inspection to obtain

$$Z = R_1 \| R_2 \frac{1 + (L/R_2)\,s}{1 + [L/(R_1 + R_2)]\,s} \qquad (6.126)$$

6.14 Voltage Divider Transfer Functions

6.14.1 RC Network

The voltage-gain transfer function of a RC voltage-divider network containing only one capacitor and having a non-zero gain at dc can be written

$$\frac{V_o}{V_i} = k \frac{1 + \tau_z s}{1 + \tau_p s} \qquad (6.127)$$

where k is the dc gain (C an open circuit), τ_p is the pole time constant, and τ_z is the zero time constant. The pole time constant is the time constant of the network with $V_i = 0$ and V_o open circuited. The zero time constant is the time constant of the network with $V_o = 0$ and V_i open circuited. Figure 6.24(a) shows the circuit diagram of an example RC network. The voltage-gain transfer function can be written by inspection to obtain

$$\frac{V_o}{V_i} = \frac{R_2 + R_3}{R_1 + R_2 + R_3} \times \frac{1 + (R_2 \| R_3)\,Cs}{1 + [(R_1 + R_2)\| R_3]\,Cs} \qquad (6.128)$$

Figure 6.24(b) shows the circuit diagram of a second example RC network. The voltage-gain transfer function can be written by inspection to obtain

$$\frac{V_o}{V_i} = \frac{R_3}{R_1 + R_3} \times \frac{1 + (R_1 + R_2)\,Cs}{1 + [(R_1 \| R_3) + R_2]\,Cs} \qquad (6.129)$$

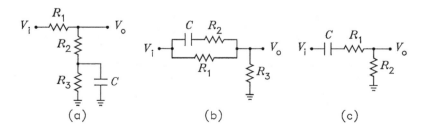

Figure 6.24: Example RC voltage divider networks.

6.14.2 High-Pass RC Network

The voltage-gain transfer function of a high-pass RC voltage-divider network containing only one capacitor can be written

$$\frac{V_o}{V_i} = k\frac{\tau_p s}{1 + \tau_p s} \tag{6.130}$$

where k is the infinite frequency gain (C a short circuit) and τ_p is the pole time constant. The pole time constant is calculated with $V_i = 0$ and V_o open circuited. Figure 6.24(c) shows the circuit diagram of a third example RC network. The voltage-gain transfer function can be written by inspection to obtain

$$\frac{V_o}{V_i} = \frac{R_2}{R_1 + R_2} \times \frac{(R_1 + R_2)\,Cs}{1 + (R_1 + R_2)\,Cs} \tag{6.131}$$

6.14.3 RL Network

The voltage-gain transfer function of a RL voltage-divider network containing only one inductor and having a non-zero gain at dc can be written

$$\frac{V_o}{V_i} = k\frac{1 + \tau_z s}{1 + \tau_p s} \tag{6.132}$$

where k is the zero frequency gain (L a short circuit), τ_p is the pole time constant, and τ_z is the zero time constant. The pole time constant is the time constant of the network with $V_i = 0$ and V_o open circuited. The zero time constant is the time constant of the network with $V_o = 0$ and V_i open circuited. Figure 6.25(a) shows the circuit diagram of an example RL network. The voltage-gain transfer function can be written by inspection to obtain

$$\frac{V_o}{V_i} = \frac{R_2}{R_1 + R_2} \times \frac{1 + [L/\left(R_2 \| R_3\right)]\,s}{1 + \{L/\left[(R_1 + R_2)\,\| R_3\right]\}\,s} \tag{6.133}$$

Figure 6.25(b) shows the circuit diagram of a second example RL network. The voltage-gain transfer function can be written by inspection to obtain

$$\frac{V_o}{V_i} = \frac{R_3}{R_1 + R_3} \times \frac{1 + (L/R_1)\,s}{1 + \{L/\left[R_1 \| \left(R_2 + R_3\right)\right]\}\,s} \tag{6.134}$$

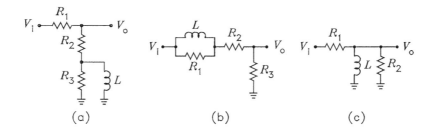

Figure 6.25: Example RL voltage divider circuits.

6.14.4 High-Pass RL Network

The voltage-gain transfer function of a high-pass RL voltage-divider network containing only one inductor can be written

$$\frac{V_o}{V_i} = k\frac{\tau_p s}{1 + \tau_p s} \tag{6.135}$$

where k is the infinite frequency gain (L an open circuit) and τ_p is the pole time constant. The pole time constant is calculated with $V_i = 0$ and V_o open circuited. Figure 6.25(c) shows the circuit diagram of a third example RL network. The voltage-gain transfer function can be written by inspection to obtain

$$\frac{V_o}{V_i} = \frac{R_2}{R_1 + R_2} \times \frac{[L/(R_1\|R_2)]\,s}{1 + [L/(R_1\|R_2)]\,s} \tag{6.136}$$

6.15 Preliminary Derivations

1. Derive the expressions for V_o/V_i for the circuits of Fig. 6.9.

2. Derive the expressions for V_o/V_i for the circuits of Fig. 6.10.

3. Derive the expression for V_o/V_i for the circuit of Fig. 6.11. To do this, write node equations at the node common to R_1, R_2, and C_1 and at the V_+ input to the op amp. Use these equations and the equation $V_o = KV_+$, where $K = 1 + R_4/R_3$, to solve for V_o/V_i.

4. Derive the expression for V_o/V_i for the circuit of Fig. 6.12. To do this, write node equations at the node common to R_1, R_2, R_3, and C_1 and at the V_- input to the op amp. (Because the V_- input is a virtual ground, the voltage at this node is zero.) Use these equations to solve for V_o/V_i.

5. Derive the expression for V_o/V_i for the circuit of Fig. 6.13. To do this, write node equations at the node common to C_1, C_2, and R_1 and at the V_+ input to the op amp. Use these equations and the equation $V_o = KV_+$, where $K = 1 + R_4/R_3$, to solve for V_o/V_i.

6. Derive the expression for V_o/V_i for the circuit of Fig. 6.14. To do this, write node equations at the node common to R_1, R_2, C_1, and C_2 and at the V_+ input to the op amp. Use these equations and the equation $V_o = KV_+$, where $K = 1 + R_5/R_4$, to solve for V_o/V_i.

7. Derive the expression for V_o/V_i for the circuit of Fig. 6.15. To do this, write node equations at the node common to R_1, R_2, C_1, and C_2 and at the V_- input to the op amp. (Because the V_- input is a virtual ground, the voltage at this node is zero.) Use these equations to solve for V_o/V_i.

8. Solve Eqs. 6.114 through 6.117 to verify the transfer functions given in Eqs. 6.118 through 6.121.

6.16 Preliminary Calculations

For the following calculations, the filter characteristics specified by the laboratory instructor are to be used. The recommended range for the cutoff frequencies is from 100 Hz to 10 kHz. (At frequencies below 100 Hz it would be difficult to observe the waveforms with the oscilloscope and at frequencies above 10 kHz the frequency dependence of the op-amp gain may degrade performance.) Capacitor values in the range 0.001 μF to 0.1 μF should be used if possible. All component values should be reasonably close to those available in the laboratory since each circuit will be experimentally examined. Remember that $\omega = 2\pi f$ for all calculations.

1. Design the low-pass filters.

2. Design the high-pass filters.

3. Design the band-pass filters.

4. Design the band-reject filters.

5. Design the state-variable filters.

6.17 Preliminary SPICE Simulations

Write the SPICE code to simulate the circuits designed in the Preliminary Calculations. Perform an ".AC" analysis with at least 50 points per decade for each circuit. Use the PROBE graphics processor of PSpice to display the gain, phase, and group delay as a function of frequency. For the gain plots, use a log scale for the vertical axis. Using a single AC input voltage source, all low-pass circuits can be simulated simultaneously so that the gains versus frequency can be displayed on the same graph, and similarly for the group delay. Similarly, all high-pass circuits can be simulated simultaneously. If more than one band-pass and band-reject filter was specified, they should also be plotted on the same sheet of graph paper. For the state variable filter, display the gain for all four outputs on the same graph, and similarly for the group delay. Use the small-signal AC op-amp macromodel for either the 741 op amp or the TL071 op amp given in Experiment 2.

The group delay is obtained by performing an AC analysis and then specifying that probe plot VG(output node).

6.18 Experimental Procedures

6.18.1 Preparation

(a) Prepare the electronic breadboard to provide buses for the positive and negative power supply rails and the circuit ground. Each power supply rail should be decoupled with a 100 Ω, 1/4 W resistor, and a 100 μF, 25 V (or greater) capacitor. The resistors are connected in series with the external power supply leads and the capacitors are connected from power supply rail to ground on the circuit side of the resistors. The capacitors must be installed with proper polarity to prevent reverse polarity breakdown. (b) The op amps specified for the experiment are either the 741 or the TL071. The TL071 is preferred because of its lower input offset currents, higher gain-bandwidth product, and higher slew rate.

The laboratory instructor will specify which filters are to be measured and the measurements that are to be made. The following instructions are generic.

6.18.2 Low-Pass Filter

(a) Assemble the low-pass filter circuits specified in part 1 of the Preliminary Calculations. Use the element values calculated in that part. (b) Set the function generator to produce a sine wave with a DC level of 0 V (the default value) and a peak-to-peak level of 2 V. Connect the output of the function generator to each filter input (one at a time) and use the oscilloscope to measure the cutoff frequency f_c and gain constant K. The oscilloscope inputs should be DC coupled. If the filters do not meet the specifications, the errors should be found and corrected. (c) Measure the resonance frequency f_0 and quality factor Q of each second-order filter. (First-order filters do not have a f_0 or Q.) To do this, set the oscilloscope for $X - Y$ operation with the filter input connected to the X input and the filter output set to the Y input. Both inputs should be DC coupled. The resonance frequency f_0 can be found by locating the frequency at which the Lissajous figure becomes an ellipse whose major and minor axes are parallel to the vertical and horizontal graticule lines on the screen of the oscilloscope; if the low-frequency gain is unity and the Volts/div for the horizontal and vertical inputs are equal, then this ellipse will become a circle. The gain of the filter at this frequency divided by K is the quality factor Q. (d) Examine the transient response of each filter by applying square-wave input signals with the frequencies $0.2f_0$, f_0, and $5f_0$. Record (sketch or print) sketches of the output waveforms in the time domain for each case. (e) Measure the amplitude response of the filter by measuring the ratio of the output to the input voltage. Measurements should be made from a frequency one decade below to one decade above the filter cut-off frequency. Use the filter analysis program to obtain the amplitude and phase frequency response; several plots should be made so that the rejection capabilities of the filters are display as well as the flatness in the pass-band.

6.18.3 High-Pass Filter

Repeat the preceding part for the high-pass filters specified in part 2 of the Preliminary Calculations. Use the element values calculated in that part.

6.18.4 Band-Pass Filter

Repeat part 6.18.2 for the band-pass filter. The frequency f_0 for the band-pass filter may be found by finding the frequency at which the Lissajous pattern collapses to a straight line. Measure the gain and phase as a function of frequency from one decade below to one decade above the filter center frequency. Use the filter analysis program to obtain the magnitude and phase frequency response.

6.18.5 Band-Reject Filter

Repeat part 6.18.2 for the band-reject filter. Measure the gain and phase as a function of frequency from one decade below to one decade above the filter center frequency. Use the filter analysis program to obtain the magnitude and phase frequency response.

6.18.6 State-Variable Filter

Repeat part 6.18.2 for the state-variable filter. Use the filter analysis program to plot magnitude and phase of each of the three types of outputs as functions of frequency.

6.19 Laboratory Report

The laboratory report should include all appropriate sketches, plots, and table of experimental data as well as all SPICE assignments, derivations, and calculations. The verification sheet should be included. Answer any supplementary questions that may have been posed by the laboratory instructor.

6.20 References

1. C. Chen, *Active Filter Design,* Hayden, 1982.
2. R. F. Coughlin and F. F. Driscoll, *Operational Amplifiers and Linear Integrated Circuits*, Prentice-Hall, 1987.
3. R. A. Gayakwad, *Op-Amps and Linear Integrated Circuits*, 3rd edition, Prentice-Hall, 1993.
4. P. Horowitz and W. Hill, *The Art of Electronics,* 2nd edition, Cambridge University Press, 1989.
5. P. Horowitz and I. Robinson, *Laboratory Manual for The Art of Electronics,* Cambridge University Press, 1981.
6. E. J. Kennedy, *Operational Amplifier Circuits*, Holt Rhinehart and Winston, 1988.
7. J. H. Krenz, *An Introduction to Electrical and Electronic Devices,* Prentice-Hall, 1987.
8. J. M. McMenamin, *Linear Integrated Circuits*, Prentice-Hall, 1985.
9. Motorola, *Linear and Interface Circuits*, Series F, Motorola, 1988.
10. National Semiconductor, *Linear Databook 1,* National Semiconductor, 1988.
11. R. Schaurmann, *Modern Design of Analog Filters*, Oxford, 2000.
12. A. S. Sedra and K. C. Smith, *Microelectronic Circuits,* 4th edition, Oxford, 1998.
13. W. D. Stanley, *Operational Amplifiers with Linear Integrated Circuits*, 2nd edition, Merrill, 1989.
14. Texas Instruments, *Linear Circuits Operational Amplifier Macromodels,* Texas Instruments, 1990.
15. Texas Instruments, *Linear Circuits Amplifiers, Comparators, and Special Functions,* Texas Instruments, 1989.
16. M. E. Van Valkenburg, *Analog Filter Design*, Oxford, 1982.
17. L Weinberg, *Network Analysis and Synthesis,* McGraw-Hill, 1962.

Chapter 7

Characteristics of Active Devices

7.1 Object

The object of this experiment is to study the basic terminal characteristics of the bipolar junction transistor (BJT), the metal oxide semiconductor field effect transistor (MOSFET), and the junction field effect transistor (JFET). The dominant SPICE parameters are obtained from experimental data.

7.2 The Bipolar Junction Transistor

7.2.1 Large-Signal Model

The BJT is fabricated as two back-to-back diodes which share either a common cathode or a common anode. If the anodes are common, it is a NPN type. If the cathodes are common, it is a PNP type. Fig. 7.1 shows the circuit symbols and the large-signal models for the NPN and the PNP BJTs. For each device, the terminal connected to the common diode region is called the base (B). One outer terminal is called the collector (C) and the other the emitter (E). The emitter is designated with an arrow that points in the same direction as the diode in the large-signal model.

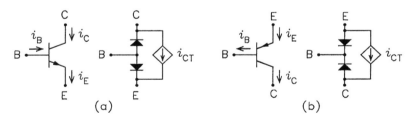

Figure 7.1: (a) NPN BJT symbol. (b) Large-signal model.

There are four modes of operation of the BJT. It is *cutoff* when both diodes are reverse biased. It is *saturated* when both diodes are forward biased. It is in the *active mode* when the base-collector diode is reverse biased and the base-emitter diode is forward biased. It is in the *reverse active mode* when the base-collector diode is forward biased and the base-emitter diode is reverse biased. In the active mode, the current flow between the collector and the emitter is in the direction indicated by the emitter arrow on the circuit symbol. To meet the convention that current flows from top to bottom in a circuit diagram, the NPN circuit symbol is commonly drawn with the emitter at the bottom and the PNP symbol is drawn with the emitter at the top.

With $i_{CT} = 0$, the large-signal circuits in Fig. 7.1 can be used to predict the response of an ohmmeter in measuring the resistance between any two terminals of the BJT. An ohmmeter connected between the collector and the emitter reads an open circuit with both polarities of the test leads. When it is connected between the base and either the collector or the emitter, it reads an open circuit with one polarity and a low resistance with the other polarity. These simple tests can be used to check for a defective transistor in the laboratory. Quite often, a transistor which has failed exhibits a short circuit from the collector to the emitter.

7.2.2 Active-Mode Equations

Because the base-collector diode in the large-signal model is reverse biased in the active mode, the current through it is neglected in the active mode. Fig. 7.2 shows the active-mode approximations for the NPN and PNP BJTs. For the NPN model, the currents i_C and i_B are approximately given by

$$i_C = I_S \exp\left(\frac{v_{BE}}{V_T}\right) \qquad i_B = \frac{i_C}{\beta} \tag{7.1}$$

where $V_T = kT/q$ is the thermal voltage (0.0259 V at $T = 300\ K$), I_S is the saturation current, and β is the base-collector current gain. These are given by

$$I_S = I_{S0}\left(1 + \frac{v_{CE}}{V_A}\right) \qquad \beta = \beta_0\left(1 + \frac{v_{CE}}{V_A}\right) \tag{7.2}$$

where I_{S0} is the zero-bias value of I_S (i.e. the value with $v_{CE} = 0$), β_0 is the zero-bias value of β, and V_A is the Early voltage. For the PNP device, v_{BE} is replaced by v_{EB} and v_{CE} is replaced by v_{EC} in the equations.

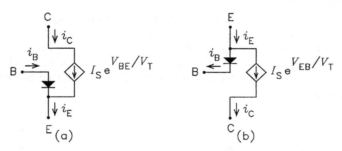

Figure 7.2: Active-mode NPN and PNP models.

Because $i_E = i_C + i_B$ and $i_C = \beta i_B$, it follows that $i_C = i_E \beta/(1+\beta)$. It is common to denote $\alpha = \beta/(1+\beta)$, where α is the emitter-to-collector current gain. Thus a general relation between the currents is

$$i_C = \beta i_B = \alpha i_E \tag{7.3}$$

7.2.3 Output Characteristics

The BJT is connected as a common-emitter (CE) stage when the base is the input and the collector is the output. For the NPN, the CE output characteristics are a plot of i_C versus v_{CE} with i_B constant. For the PNP, i_C is plotted versus v_{EC}. It follows from Eqs. (7.1) and (7.2) that the equation for the NPN characteristics is

$$i_C = \beta_0 \left[1 + \frac{v_{CE}}{V_A}\right] i_B \tag{7.4}$$

7.2. THE BIPOLAR JUNCTION TRANSISTOR

Fig. 7.3 shows the typical output characteristics for a NPN BJT fabricated for use as a small-signal amplifier. The plots are made for $i_B = 10$ μA, 20 μA, $\cdots$, 90 μA. The assumed parameters are $I_{S0} = 1.26 \times 10^{-14}$ A, $\beta_0 = 100$, and $V_A = 75$ V.

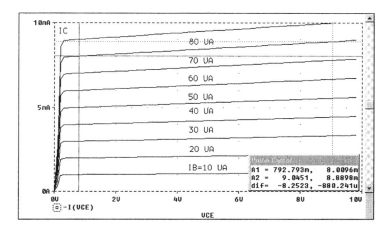

Figure 7.3: Example NPN BJT Output Characteristics for $i_B = 10$ μA, 20 μA, $\cdots$, 90 μA.

Fig. 7.4 shows the typical output characteristics for a PNP BJT fabricated for use as a small-signal amplifier. The plots are made for $i_B = 10$ μA, 20 μA, $\cdots$, 90 μA. The assumed parameters are $I_{S0} = 1 \times 10^{-15}$ A, $\beta_0 = 100$, and $V_A = 36$ V.

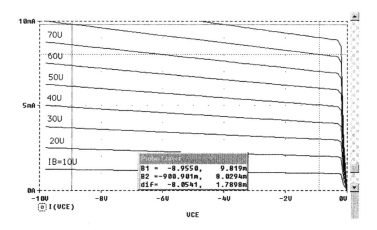

Figure 7.4:

Calculation of V_A and β_0 from the Output Characteristics

From the values for i_C, i_B, v_{CE}, and the slope $m = \Delta i_C / \Delta v_{CE}$ at a point on the curves, it follows from Eq. (7.4) that V_A and β_0 can be calculated from the equations

$$V_A = \frac{i_C}{m} - v_{CE} \tag{7.5}$$

$$\beta_0 = \frac{i_C}{i_B} \times \frac{1}{1 + v_{CE}/V_A} \tag{7.6}$$

7.2.4 Transfer Characteristics

The CE transfer characteristics for the NPN BJT are a plot of i_C versus v_{BE} with v_{CE} constant. For the PNP device, i_C is plotted versus v_{EB}. It follows from Eqs. (7.1) and (7.2) that the equation for the NPN characteristics is

$$i_C = I_{S0}\left[1 + \frac{v_{CE}}{V_A}\right]\exp\left(\frac{v_{BE}}{V_T}\right) \tag{7.7}$$

Figs. 7.5 and 7.6 show the typical transfer characteristics of NPN and PNP BJTs fabricated for use as a small-signal amplifier. The plots are made with the same model parameters used above for the output characteristics. For the plots, $v_{CE} = 5$ V for the NPN device and $v_{EC} = 5$ V for the PNP device.

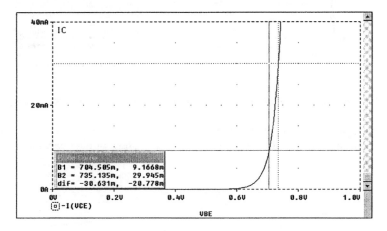

Figure 7.5: Example NPN BJT Transfer Characteristics

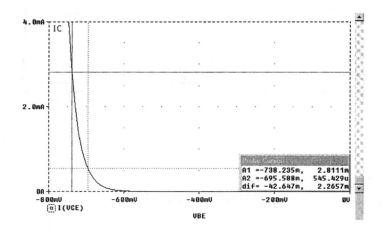

Figure 7.6: Example PNP BJT transfer characteristics.

Calculation of I_{S0} from the Transfer Characteristics

From the values for i_C and v_{BE} at a point on the transfer characteristics curve, it follows from Eq. (7.7) that the parameter I_{S0} can be calculated from the equation

$$I_{S0} = \frac{i_C}{1 + (v_{CE} - v_{BE})/V_A} \exp\left(-\frac{v_{BE}}{V_T}\right) \tag{7.8}$$

7.2.5 SPICE Input Decks

The SPICE input decks for the output and transfer characteristics for the NPN and PNP BJTs are given below.

```
NPN OUTPUT CHARACTERISTICS      NPN TRANSFER CHARACTERISTICS
Q1 2 1 0 QM1                    Q1 2 1 0 QM1
VCE 2 0 DC 0V                   VCE 2 0 DC 5V
IB 0 1 DC 0A                    VBE 1 0 DC 0V
.MODEL QM1 NPN                  .MODEL QM1 NPN
+(IS=1.26E-14,BF=100,VA=75)     +(IS=1.26E-14,BF=100,VA=75)
.DC VCE (0,9.8,0.1)             .DC VBE (0,0.75,0.01)
+IB (0.01M,0.09M,0.01M)         .PROBE
.PROBE                          .END
.END

PNP OUTPUT CHARACTERISTICS      PNP TRANSFER CHARACTERISTICS
Q1 2 1 0 QM1                    Q1 2 1 0 QM1
VCE 2 0 DC 0V                   VCE 2 0 DC -5V
IB 1 0 DC 0A                    VBE 1 0 DC 0V
.MODEL QM1 PNP                  .MODEL QM1 PNP
+(IS=1E-15,BF=100,VA=36)        +(IS=1E-15,BF=100,VA=36)
.DC VCE (-9.8,0,0.1)            .DC VBE (-0.75,0,0.01)
+IB (0.01M,0.09M,0.01M)         .PROBE
.PROBE                          .END
.END
```

7.3 The Junction Field Effect Transistor

The circuit symbols for the n-channel and p-channel JFETs are shown in Fig. 7.7. The three terminals are designated the source (S), the drain (D), and the gate (G). For proper operation, the gate-channel junction must be zero or reverse biased. This requires $v_{GS} \leq 0$ for the n-channel device and $v_{SG} \leq 0$ for the p-type. When the gate-channel junction is reverse biased, the gate current is so small that it can be assumed to be zero. In this case, the drain and source currents are equal. The JFET is fabricated symmetrically with respect to the drain and source leads so that these two leads are interchangeable in a circuit.

7.3.1 Saturation-Mode Equations

The active or amplifying mode for the JFET is the saturation or pinch-off mode. The n-channel JFET is in the saturation mode if $v_{DS} \geq v_{GS} - V_{TO}$ and $V_{TO} \leq v_{GS} \leq 0$, where V_{TO} is the threshold or pinch-off voltage, which is negative. In this mode, the drain current in the n-channel device is given by either of the two equivalent equations

$$i_D = \beta(v_{GS} - V_{TO})^2 = I_{DSS}\left(1 - \frac{v_{GS}}{V_{TO}}\right)^2 \tag{7.9}$$

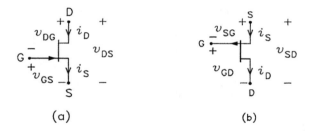

Figure 7.7: JFET circuit symbols. (a) N-channel. (b) P-channel.

where β is the transconductance parameter which has the units A/V^2 and I_{DSS} is the drain-to-source saturation current. These are related by $\beta = I_{DSS}/V_{TO}^2$. Both β and I_{DSS} are functions of the drain-to-source voltage given by

$$\beta = \beta_0(1 + \lambda v_{DS}) \qquad I_{DSS} = I_{DSS0}(1 + \lambda v_{DS}) \tag{7.10}$$

where β_0 and I_{DSS0} are the zero-bias values and λ is the channel-length modulation factor which has the units V^{-1}. The reciprocal of λ in the JFET equations plays the same role as the Early voltage V_A in the BJT equations. For the p-channel device, v_{GS} is replaced with v_{SG} and v_{DS} is replaced with v_{SD} in the equations.

7.3.2 Triode- or Linear-Mode Equations

A JFET biased in the triode or linear mode, is often used as a variable resistor. The n-channel device is in the triode mode if $V_{TO} \leq v_{GS} \leq 0$ and $v_{DS} \leq v_{GS} - V_{TO}$. In this mode, the drain current is given by

$$i_D = 2\beta \left[(v_{GS} - V_{TO})v_{DS} - \frac{v_{DS}^2}{2} \right] \tag{7.11}$$

7.3.3 Saturation-Linear Boundary

The equation for i_D at the boundary between the triode and the saturation modes is obtained by eliminating v_{GS} between Eqs. (7.9) and (7.11). It is given by

$$i_D = \beta v_{DS}^2 = I_{DSS}\left(\frac{v_{DS}}{V_{TO}}\right)^2 \tag{7.12}$$

A JFET can be used as an electronic switch when it is alternately switched between its saturation and triode modes.

7.3.4 Output Characteristics

A set of typical output characteristics for a n-channel JFET are shown in Fig. 7.8. The figure shows a plot of the drain current i_D as a function of the drain-to-source voltage v_{DS} for the gate-to-source voltages $v_{GS} = 0, -0.5, -1, -1.5, -2, -2.5$, and -3 V, where the upper curve is for $v_{GS} = 0$. The parameters used for the curves are $\beta_0 = 0.1875$ mA, $V_{TO} = -4$ V, and $\lambda = 0.01$ V^{-1}.

Figure 7.9 shows the output characteristics for a typical p-channel JFET having the same model parameters as the n-channel device.

7.3. THE JUNCTION FIELD EFFECT TRANSISTOR

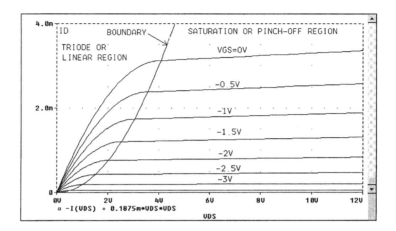

Figure 7.8: Plot i_D in mA versus v_{DS} for the hypothetical n-Channel JFET.

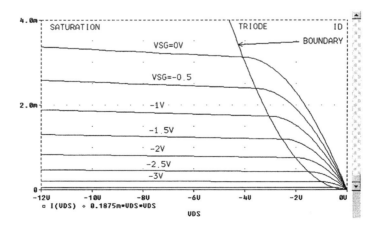

Figure 7.9: Plot i_D in mA versus v_{DS} for the hypothetical p-Channel JFET.

7.3.5 Transfer Characteristics

A plot of the transfer characteristics for the example n-channel JFET are shown in Fig. 7.10. The figure shows plots of the drain current i_D as a function of the gate-to-source voltage v_{GS} for the drain-to-source voltages $v_{DS} = 4$, 8, and 12 V, where the lower curve is for $v_{DS} = 4$ V.

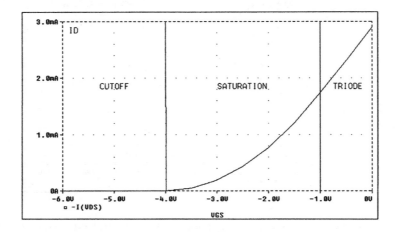

Figure 7.10: Plot of i_D in mA versus v_{GS} for the hypothetical n-channel JFET.

Figure 7.11 shows the transfer characteristics for a typical p-channel JFET having the same model parameters as the n-channel device.

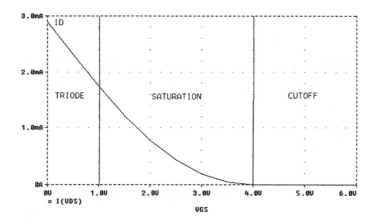

Figure 7.11: Plot of i_D in mA versus v_{GS} for the hypothetical p-channel JFET.

7.3.6 SPICE Input Decks

The SPICE input decks for the output and transfer characteristics for the n-channel and p-channel JFETS are given below.

7.3. THE JUNCTION FIELD EFFECT TRANSISTOR

```
N CH JFET OUTPUT CHARACTERISTICS
VGS 1 0 DC 0
VDS 2 0 DC 0
J 2 1 0 BASIC
.MODEL BASIC NJF
+(VTO=-4,BETA=0.1875M,LAMBDA=0.01)
.DC VDS 0 12 0.1 VGS -6 0 0.5
.PROBE
.END
```

```
N CHANNEL JFET TRANSFER CHARACTERISTICS
VGS 1 0 DC 0
VDS 2 0 DC 3
J 2 1 0 BASIC
.MODEL BASIC NJF
+(VTO=-4,BETA=0.1875M,LAMBDA=0.01)
.DC VGS -6 0 0.05
.PROBE
.END
```

```
P CH JFET OUTPUT CHARACTERISTICS
VGS 1 0 DC 0
VDS 2 0 DC 0
J 2 1 0 BASIC
.MODEL BASIC PJF
+(VTO=-4,BETA=0.1875M,LAMBDA=0.01)
.DC VDS -12 0 0.1 VGS 0 6 0.5
.PROBE
.END
```

```
P CHANNEL JFET TRANSFER CHARACTERISTICS
VGS 1 0 DC 0
VDS 2 0 DC -3
J 2 1 0 BASIC
.MODEL BASIC PJF
+(VTO=-4,BETA=0.1875M,LAMBDA=0.01)
.DC VGS 0 6 0.05
.PROBE
.END
```

7.3.7 Measuring β, I_{DSS}, and V_{TO}

I_{DSS} can be measured by connecting the gate to the source and applying a dc voltage between the drain and the source. The current which flows is I_{DSS}. The drain-to-source voltage should be approximately equal to the value anticipated for the application. Next, insert a potentiometer connected as a variable resistor in series with the source so that $v_{GS} = -i_S R_P$, where i_S is the source current and R_P is the resistance of the potentiometer. With $R_P = 0$, the source current is I_{DSS}. Let the potentiometer be adjusted so that the source current is I_{DSS}/n, where $n > 1$. (A convenient value is $n = 4$). For the n-channel device, measure the gate-to-source voltage and denote it be V_{GS1}. Note that $V_{GS1} < 0$. Eq. (7.7) yields

$$\frac{1}{n} I_{DSS} = I_{DSS} \left(1 - \frac{V_{GS1}}{V_{TO}}\right)^2 \tag{7.13}$$

This can be solved for V_{TO} to obtain

$$V_{TO} = \frac{V_{GS1}}{1 - 1/\sqrt{n}} \tag{7.14}$$

β can then be calculated from

$$\beta = \frac{I_{DSS}}{V_{TO}^2} \tag{7.15}$$

For the p-channel device, V_{GS1} is replaced by V_{SG1}.

An alternate method for solving for V_{TO} is to use a plot of the transfer characteristics to determine the value of v_{GS} for which i_D just reaches zero. However, because the slope of the i_D versus v_{GS} curve is zero at this point, the value of V_{TO} is difficult to read with accuracy.

7.3.8 Bias Considerations

In the design of small-signal amplifiers, the biasing of a JFET is different from the biasing of a BJT. This is because the BJT is turned on by the bias whereas the JFET is turned off, or partially turned off. That is, a gate-to-source bias voltage is applied to cause the drain current to be less than its value with zero bias. Once the BJT is turned on, its base-to-emitter voltage exhibits very little change with collector current so that the base-to-emitter voltage can be approximated by a constant (typically 0.6 to 0.7 V) in the BJT bias equation. In contrast, the JFET gate-to-source voltage exhibits considerable variation with the drain current. For this reason, it is important to know both β and V_{TO} (or I_{DSS} and V_{TO}) for a particular JFET in order to properly bias it as a small-signal amplifier.

7.4 Metal Oxide Semiconductor Field Effect Transistor

The MOSFET is a four terminal active device which has applications in both analog and digital electronics. The four terminals are the source (S), the drain (D), the gate (G), and the bulk (B). In many applications, the bulk is connected to the source. In this case, the MOSFET becomes a three terminal device that is similar to the JFET with the exception that the gate is electrically insulated from the channel. The insulated gate permits the gate-to-source voltage to be either positive or negative without the flow of gate current.

There are two types of MOSFETs – depletion mode and enhancement mode. In each, the current which flows in the drain-source channel is a function of the voltages applied from gate to source and from bulk to source. In the depletion mode device, the channel is conducting when both the gate-to-source voltage and the bulk-to-source voltage are zero. In contrast, the channel is non-conducting in the enhancement mode device when the gate-to-source voltage and the bulk-to-source voltage are zero. For the latter devices, the drain current is zero until the gate-to-source voltage exceeds a value that is known as the threshold voltage. Only enhancement mode devices are considered in this experiment.

7.4.1 Circuit Symbols

Figure 7.12: Enhancement mode MOSFET symbols. (a) N-channel. (b) P-channel.

The circuit symbols for the enhancement mode MOSFETs are shown in Fig. 7.12. In each symbol, the broken line between the drain and source represents the channel. It is broken to indicate that the channel is non-conducting until the applied gate-to-source voltage is greater than the device threshold voltage. The arrow in the bulk lead points from the bulk to the channel, i.e. from p to n. The abbreviated circuit symbols shown in Fig. 7.13 are used when the bulk is connected to the source. In this case, the MOSFET a three terminal device. The arrow in the source lead indicates the direction of normal current flow.

7.4.2 Large-Signal Device Equations

Three model equations define the drain current in the n-channel enhancement-mode MOSFET. These are

$$i_D = 0 \quad \text{for} \quad v_{GS} \leq V_{T0} \tag{7.16}$$

$$i_D = 2K\left[(v_{GS} - V_{TH})v_{DS} - \frac{v_{DS}^2}{2}\right] \quad \text{for} \quad v_{DS} \leq v_{GS} - V_{TH} \tag{7.17}$$

$$i_D = K(v_{GS} - V_{TH})^2 \quad \text{for} \quad v_{DS} \geq v_{GS} - V_{TH} \tag{7.18}$$

7.4. METAL OXIDE SEMICONDUCTOR FIELD EFFECT TRANSISTOR

Figure 7.13: Abbreviated MOSFET symbols for $v_{BS} = 0$. (a) N-channel. (b) P-channel.

where K is the transconductance parameter and V_{TH} is the threshold voltage. Eq. 7.16 defines the cutoff region, Eq. (7.17) defines the linear or triode region, and Eq. (7.18) defines the saturation region. When the MOSFET is used as an amplifier, it is biased in the saturation region. When it is biased in the linear or triode region, it exhibits a low resistance between the drain and source which is a function of the gate-to-source and bulk-to-source voltages.

The model equations for the transconductance parameter K and the threshold voltage V_{TO} are

$$K = K_0(1 + \lambda v_{DS}) \qquad V_{TH} = V_{TO} + \gamma \left[\sqrt{\phi - v_{BS}} - \sqrt{\phi}\right] \tag{7.19}$$

where K_0 is the zero bias value of K (a typical value is 8×10^{-4} A/V^2), λ is the length-modulation factor of the channel (a typical value is 0.01 V^{-1}), V_{TO} is the zero bias value of V_{TH} (a typical value is 2 V), γ is the bulk threshold parameter (a typical value is 0.5 V$^{1/2}$), and ϕ is the surface potential (a typical value is 0.6 V). The parameter K_0 is given by

$$K_0 = \frac{K'}{2}\frac{W}{L} \tag{7.20}$$

where W is the MOSFET channel width, L is the channel length, and K' is given by

$$K' = \mu C_{ox} \tag{7.21}$$

In this equation, μ is the electron mobility in the induced channel and C_{OX} is the capacitance per unit area of the gate-to-channel capacitor. The latter is given by

$$C_{ox} = \frac{\epsilon}{t} \tag{7.22}$$

where ϵ is the permittivity of the insulator and t is its thickness.

7.4.3 Bulk-to-Source Bias Voltage

For proper operation, the bulk-to-channel junction must be either zero biased or reverse biased. For the n-channel device, this requires the bulk-to-source voltage to be either zero or negative. The threshold voltage V_{TH} in the n-channel device is positive. This means that the gate-to-source voltage must be positive and greater than V_{TH} for current to flow in the channel. As shown in Eq. (7.19), the threshold voltage increases as the bulk-to-source voltage is made more negative. Thus the gate-to-source voltage must be increased if the channel current is to be held constant as the bulk-to-source voltage is made more negative.

7.4.4 Linear-Saturation Boundary

Equation (7.17) gives the output characteristics of the MOSFET for what is called the linear or triode region. Eq. (7.18) gives the characteristics for the saturation region. The boundary between the these two regions is obtained by eliminating v_{GS} between the two equations to obtain

$$i_D = K v_{DS}^2 \tag{7.23}$$

7.4.5 Comparison with the JFET

A comparison of the equations for the n-channel enhancement mode MOSFET with the equations for the n-channel JFET discloses that they have essentially the same form. However, the JFET is a depletion-mode device while the MOSFET can be either a depletion-mode or and enhancement-mode device. Unlike the JFET, the threshold voltage of the MOSFET can be controlled by another voltage. Both devices have a high input impedance. The JFET gate current is small because it is the reverse saturation current for a pn junction. The MOSFET gate current is essentially zero because the gate is insulated from the channel.

7.4.6 Output Characteristics

The plots of typical output characteristics for n-channel and p-channel MOSFETs are given in Figs. 7.14 and 7.16. The model parameters used for these plots are $K' = 0.5606$ mA, $V_{TO} = 1.1$ V, $\lambda = 0.02$ V^{-1}, and $\gamma = 0.2$ V$^{1/2}$. The parameter K' is the SPICE parameter KP, V_{TO} is the SPICE parameter VTO, λ is the SPICE parameter LAMBDA, and γ is the SPICE parameter GAMMA.

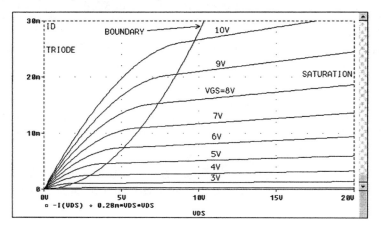

Figure 7.14: Plot of i_D in mA versus v$_{DS}$ for the hypothetical n-channel enhancement mode MOSFET.

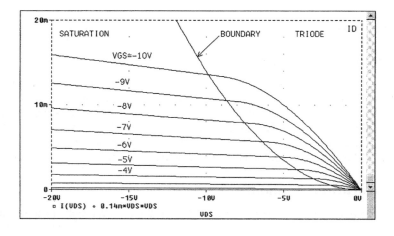

Figure 7.15: Plot i_D in mA versus v_{DS} for the hypothetical p-channel MOSFET.

7.4. METAL OXIDE SEMICONDUCTOR FIELD EFFECT TRANSISTOR

7.4.7 Transfer Characteristics

The plots of typical transfer characteristics for the n-channel and p-channel MOSFETs are given in Figs. 7.16 and 7.17. The model parameters are the same as for the output characteristic curves.

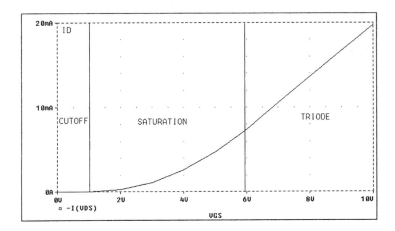

Figure 7.16: Plot of i_D in mA versus v_{GS} for the hypothetical n-channel enhancement mode MOSFET.

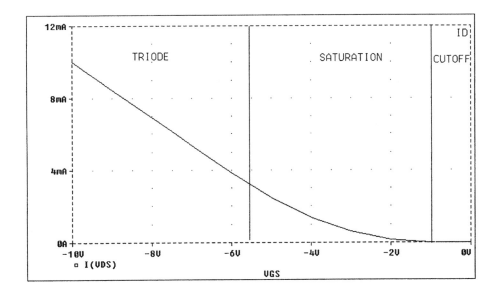

Figure 7.17: Plot of i_D in mA versus v_{GS} for the hypothetical p-channel enhancement mode MOSFET.

7.4.8 SPICE Input Decks

The SPICE input decks used to generate the output and transfer characteristics are given below. The plots are given in Figs. 7.14 and 7.16.

```
OUTPUT CHAR N CH MOSFET          TRANSFER CHAR N CH MOSFET
VGS 1 0 DC 0                      VGS 1 0 DC 0
VBS 3 0 DC 0                      VBS 3 0 DC 0
VDS 2 0 DC 0                      VDS 2 0 DC 5
M 2 1 0 3 MOSEX                   M 2 1 0 3 MOSEX
.MODEL MOSEX NMOS (KP=0.5606M,    .MODEL MOSEX NMOS (KP=0.5606M,
+VTO=1.1,LAMBDA=0.02)             +VTO=1.1,LAMBDA=0.02)
.DC VDS 0 20 0.1 VGS 0 10 1       .DC VGS 0 10 0.1
.PROBE                            .PROBE
.END                              .END

OUTPUT CHAR FOR P CH MOSFET       TRANSFER CHAR P CH MOSFET
VGS 1 0 DC 0                      VGS 1 0 DC 0
VBS 3 0 DC 0                      VBS 3 0 DC 0
VDS 2 0 DC 0                      VDS 2 0 DC -5
M 2 1 0 3 MOSEX                   M 2 1 0 3 MOSEX
.MODEL MOSEX PMOS (KP=0.28M,      .MODEL MOSEX PMOS (KP=0.28M,
+VTO=-1,LAMBDA=0.02)              +VTO=-1,LAMBDA=0.02)
.DC VDS -20 0 0.1 VGS -10 0 1     .DC VGS -10 0 0.1
.PROBE                            .PROBE
.END                              .END
```

7.4.9 P-Channel Device Equations

The equations for the p-channel MOSFET have the same form as those for the n-channel device. The differences are that the reference direction for the drain current is reversed and the subscripts on all voltages are reversed. The drain current is zero unless the source-to-gate voltage v_{SG} exceeds the threshold voltage. In order for the bulk-to-channel junction to be zero biased or reverse biased, the source-to-bulk voltage V_{SB} must be either zero or negative.

7.5 Preliminary Derivations

7.5.1 BJT

Use Eqs. (7.1) and (7.2) to derive Eqs. (7.3) through (7.8).

7.5.2 MOSFET

Use Eqs. (7.17) and (7.18) to derive Eq. (7.23) for the boundary between the MOSFET linear and saturation regions.

7.6 Preliminary Calculations

1. **(a)** Use the numerical data labeled in Fig. 7.3 to calculate the Early voltage V_A for the transistor. **(b)** Use the numerical data labeled in Fig. 7.3 to calculate the zero-bias current gain β_0 of the transistor. **(c)** Use the numerical data labeled in Fig. 7.5 to calculate the zero-bias saturation current I_{S0} for the transistor. For the thermal voltage, use the value $V_T = 0.0259$ V.

2. **(a)** For the drain currents 0.1 mA, 1 mA, and 10 mA, calculate values for the small-signal transconductance g_m and the source intrinsic resistance $r_s = 1/g_m$ for a MOSFET for which $K = 4 \times 10^{-4}$ A/V^2. **(b)** For BJT emitter currents equal to the MOSFET drain currents, compare the values obtained in the preceding part to the values of g_m and r_e for a BJT for which the thermal voltage is 25.9 mV.

3. Experimental measurements on a n-channel enhancement mode MOSFET give the following data: $V_{TO} = 2$ V, $I_D = 2$ mA when $V_{GS} = 4$ V, and $\Delta I_D/\Delta V_{DS} = 10^{-5}$ S for $V_{DS} = 10$ V and $I_D = 2$ mA. Calculate values of K_0 and λ for the MOSFET.

7.7 Preliminary SPICE Simulations

7.7.1 BJT

(a) Use SPICE to calculate the output characteristics of a PNP BJT having the parameters $I_{S0} = 1.83 \times 10^{-15}$ A, $\beta_0 = 200$, and $V_A = 50$ V. Sweep the collector-to-emitter voltage from 0 V to 9.8 V in steps of 0.1 V and the base current from 10 μA to 90 μA in steps of 10 μA. Use the graphics post processor of SPICE to display the characteristic curves. **(b)** Use SPICE to calculate the transfer characteristics for the transistor of the preceding part for the collector-to-emitter voltage $V_{CE} = 5$ V. Sweep the base-to-emitter voltage from 0 V to 0.75 V in steps of 0.01 V. Use the graphics post processor of SPICE to display the transfer characteristics for the range $0 \leq I_C \leq 10$ mA.

7.7.2 JFET

One of the JFETs used in this experiment is the 2N5457 n-channel JFET. Typical SPICE parameters for this device are as follows: BETA=0.333E-3 (β_0), VTO=-3 (V_{TO}), LAMBDA=0.01 (λ), CGD=6E-12 (c_{gdo}), CGS=6E-12 (c_{gso}). Although these parameters are representative, the parameters of individual 2N5457 JFETs can be different. **(a).** Write the SPICE input deck to generate the plot of drain current versus drain-to-source voltage for stepped values of gate-to-source voltage from 0 to -3 V in steps of 0.5 V.

7.7.3 MOSFET

Write the SPICE input deck to obtain a plot of the output characteristics for a n-channel enhancement mode MOSFET with the bulk connected to the source. The MOSFET has the model parameters $K = 2 \times 10^{-4}$ A/V^2, $V_{TO} = 2$ V, and $\lambda = 0.01$ V^{-1}. Vary v_{DS} from 0 to 15 V in steps of 0.5 V, and vary v_{GS} from 3 to 7 V in steps of 1 V.

7.8 Procedure

This procedure assumes that a **Tektronix 370B** transistor curve tracer is available to obtain the output and transfer characteristics of the semiconductor devices. This instrument is self contained and must not be used with external dc power supplies, function generators, or oscilloscopes. Indeed, the function generator and dc power supply are not used in experiment. Do not turn them on or connect them to anything. The active devices will be placed into a test module in the curve tracer. The laboratory instructor will specify whether the manual or computer automated method will be used to obtain the characteristics. The computer automated method uses a LabVIEW program to control the instrument.

PROCEDURE for Computer Acquisition of Transistor Characteristics

NPN BJT

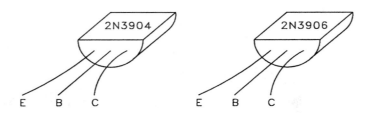

Figure 7.18: Lead placement for 2N3904 NPN and 2N3906 PNP BJTs.

1. Junction Resistance Measurement

The BJT transistors specified for the experiment are the 2N3904 (NPN) and the 2N3906 (PNP). The physical arrangement of the leads for both transistors is shown in Fig. 7.18. For each transistor, use the digital multimeter to measure the resistance between two leads at a time for all permutations of the leads. (The digital multimeter should be set to the "diode test" setting.) From the measurements, verify that the transistors are fabricated as back-to-back diodes. Whilst using the multimeter to measure the resistance of a reverse biased junction, the resistance of the human body can affect the measurements. Therefore, the transistor leads and the metal terminals of the multimeter leads should not be touched with the hands when making the resistance measurements.)

7.8.1 Safety Precautions

The **Tektronix 370B Curve Tracer** is an industrial instrument used to measure the characteristics of semiconductor devices which can produce large voltages and currents. Certain obvious precautions should be observed:

- When inserting the transistor or other solid state device to be tested the **OUTPUTS** switch should be set to **DISABLED**. The is a toggle switch located under the CRT display for the instrument.

- Never change the **MAX PEAK VOLTS** or **MAX PEAK POWER** from their default settings of 16 V and 0.08 Watts.

- Never open the cover for the breadboard section when the **OUTPUTS** switch on the **Tektronix 370B Curve Tracer** is the **ENABLED** position.

If these precautions are not strictly adhered to a painful, horrific, excruciating, and, possibly, debilitating electrical

SHOCK!!!!

may be experienced. Therefore, when it is turned on none of the electrical terminals should be touched by the user. Changes should be made with this instrument only when the **OUTPUTS** are disabled.

7.8.2 NPN BJT Output Characteristic

- Turn the **OUTPUTS** switch on the **Tektronix 370B Curve Tracer** to the **DISABLED** position (down).

7.8. PROCEDURE

- Turn the plastic cover to the up position and insert the 2N3904 onto the left side of the breadboard section.

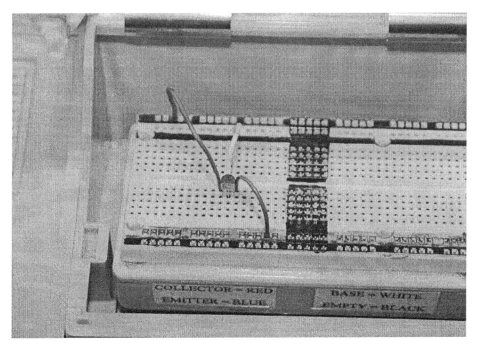

Connect the base to the white colored bus on the breadboard, the collector to the red, and the emitter to the blue. Close the plastic cover. The base current will be stepped through positive values while the collector-to-emitter voltage is swept through positive values. From this a plot of the collector current versus collector-to-emitter voltage is made for stepped values of base current.

- Log onto the pc attached to the **Tektronix 370B Curve Tracer**.

- Double click the icon for the Curve Tracer on the desktop of the pc attached to the curve tracer or on the curve tracer program on the network server. Double click curvetracer. Press Run on the LabVIEW front panel (upper left on toolbar). Select NPN for transistor type and Output Characteristic. On the LabVIEW front panel select **Lift** or **Right** depending on whether the transistor was placed in the left or right socket. Press **Read Curve Tracer** on the LabVIEW front panel. Check on the curve tracer to see if the **POLARITY** is set to the picture of the positive full wave rectified sine wave, **STEPS** to current, positive polarity steps, and the display invert button is off. If not, press **LOCAL** on the Curve Tracer and press Run Curve Trace again and verify.

- Turn the **OUTPUTS** switch on the **Tektronix 370B Curve Tracer** to the **ENABLED** position. Use the pointer on the pc to rotate the **% Collector Supply** knob on the LabVIEW front panel to 100%. Press **Read Curve Tracer** again on the LabVIEW Front Panel. Use the pointer on the pc to position the two cursors until the display appear asThe intercept is the negative of the Early voltage; it should be a voltage in the range of 100 to 300 volts.

- Position the cursor so that a value for I_C and I_B can be measured.

- Print the display and obtain a verification signature from the laboratory instructor.

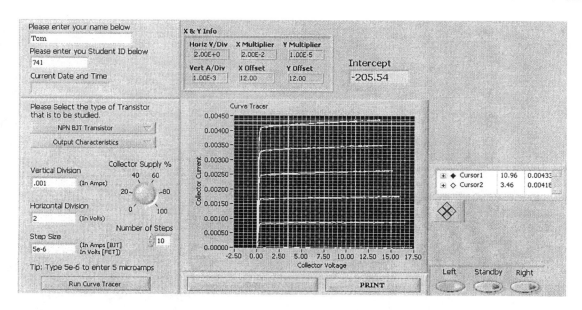

Figure 7.19:

7.8.3 NPN BJT Transfer Characteristic

- On the LabVIEW front panel switch the characteristic from **Output** to **Transfer**. Press **Read Curve Tracer**. This should result in an exponential plot of collector current versus base-to-emitter voltage.

- Plot the display using the above procedure.

- Turn the **OUTPUTS** switch on the Tektronix 370 Curve Tracer to the **DISABLED** position. Lift the cover and remove the transistor.

- Obtain a verification signature from the laboratory instructor.

7.8.4 PNP BJT Output Characteristic

- Turn the **OUTPUTS** switch on the Tektronix 370B Curve Tracer to the **DISABLED** position.

- Turn the plastic cover to the up position and insert the 2N3906 onto the left side of the breadboard section. The lead placements are identical to those of the 2N3904. Connect the base to the white colored bus on the breadboard, the collector to the red, and the emitter to the blue. Close the plastic cover. The base current will be stepped through negative values while the collector-to-emitter voltage is swept through negative values. The display will be inverted and a plot of collector current versus emitter-to-collector voltage will be obtained for stepped values of the base current.

- Select PNP transistor on the LabVIEW Front Panel and the Output Characteristic.

- Press Read Curve Tracer on the LabVIEW Front Panel. Check to see if **POLARITY** is set to the picture of the negative full wave rectified signal on the Curve Tracer, the polarity of the steps is negative, and the **DISPLAY INVERT** is on. If so, switch the **OUTPUTS** switch on the Curve Tracer to **ENABLED**. Press Run Curve Tracer again on the LabVIEW front panel. Use the pointer on the pc to rotate the % Collector Supply knob on the LabVIEW panel from 100% to 0% and then back to 100%. Press **Read Curve Tracer** again if necessary.

7.8. PROCEDURE

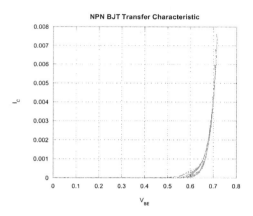

Figure 7.20: NPN BJT Transfer Characterictic

- Position the cursors as shown. Print the output and obtain a verification signature from the laboratory instructor.

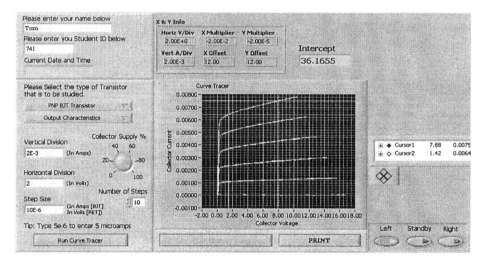

7.8.5 PNP BJT Transfer Characteristic

- Select PNP transfer characteristic on the LabVIEW front panel. Press **Read Curve Tracer**. This should yield a plot illustrating the exponential dependence of the collector current on the base-to-emitter voltage.

- Print the output and transfer characteristics and obtain a verification signature from the laboratory instructor. Switch the **OUTPUTS** switch on the **Tektronix 370B Curve Tracer** to **DISABLED** and remove the transistor.

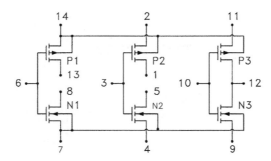

Figure 7.21: Pinouts for 4007 MOSFET IC

7.8.6 N Channel Enhancement Mode MOSFET

The major difference in obtaining the terminal characteristics of FETs rather than BJTs is that the input or stepping variable is gate voltage rather than base current. Gate voltage is stepped as the drain-to-source voltage is swept yielding a plot of drain current versus drain-to-source voltage for stepped values of gate-to-source voltage.

The MOSFETs to be examined in this experiment are contained in a 14 pin mini dip package known as the CD4007 or MC14007. Because the output characteristics for the N Channel Enhancement Mode MOSFET are being obtained in this section, transistor N1 in Fig. 7.21 will be used. For this transistor, pin 6 is the gate (analogous to base), pin 7 is the source (analogous to emitter), and pin 8 is the drain (analogous to collector). The drain of N1 is connected internally to the body whereas it floats for N2 and N3.

- Turn the **OUTPUTS** switch on the **Tektronix 370B Curve Tracer** to the **DISABLED** position.

- Turn the plastic cover to the up position and insert the CD4007 integrated circuit onto the left side of the breadboard section. Connect gate, source, and drain pins to the appropriate buses on the breadboard. Namely connect gate pin 6 to the white bus (second from top), source pin 7 to the blue bus (top bus), and drain pin 8 to the red bus (bottom bus). Close the plastic cover.

- Select N Channel Enhancement Mode MOSFET Output Characteristic on the LabVIEW front panel. Press **Read Curve Tracer**. Check to verify that the **STEP GENERATOR** is set to **VOLTAGE**, the **POLARITY** on the **COLLECTOR SUPPLY** is set to the picture of the positive full wave rectified waveform, the **POLARITY** set to positive steps.

- Check to see if the side is set to **LEFT**.

- Change the **OUTPUTS** selector on the **Tektronix 370B Curve Tracer** to **ENABLED**. Press **Read Curve Tracer** on the LabVIEW Front Panel. Vary the % Collector Supply knob on the LabVIEW front panel from 100% to 0 % and then back to 100%. Increase the number of steps from the default value of 5 to 10. Press **Read Curve Tracer** again.

- Position the cursors as shown. The Intersect on the LabVIEW Front Panel is the reciprocal of the channel length modulation factor, λ.

- Position the cursors to obtain data of drain current as a function of gate-to-source voltage for a constant drain-to-source voltage. Pick a specific value of the drain-to-source voltage for which the curves are reasonably straight and record the value of drain current for each value of the gate-to-source voltage.

- Print the display and obtain a verification signature from the laboratory instructor.

- Turn the **OUTPUTS** switch to the **DISABLED** position.

7.8. PROCEDURE

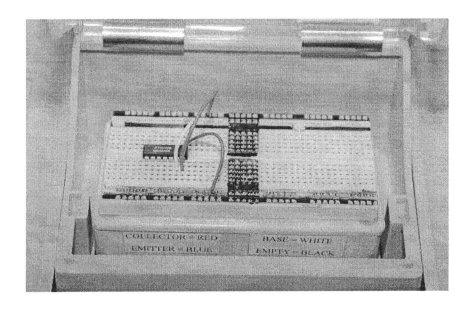

Figure 7.22:

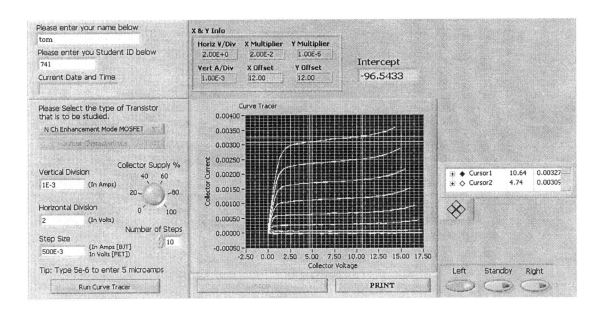

Figure 7.23:

7.8.7 P Channel Enhancement Mode MOSFET

The MOSFETs to be examined in this experiment are contained on an integrated circuit known as the CD4007 or MC14007. Because the output characteristics for the P Channel Enhancement Mode MOSFET are being obtained in this section, transistor P1 in Fig. 7.21 will be used. For this transistor, pin 6 is the gate (analogous to base), pin 14 is the source (analogous to emitter), and pin 13 is the drain (analogous to collector). The drain of P1 is connected internally to the body whereas it floats for P2 and P3.

- Turn the **OUTPUTS** switch on the **Tektronix 370B Curve Tracer** to the **DISABLED** position.

- Turn the plastic cover to the up position and insert the CD4007 onto the left side of the breadboard section. Connect gate, source, and drain pins to the appropriate buses on the breadboard. Namely connect gate pin 6 to the white bus (second from top), source pin 14 to the blue bus (top bus), and drain pin 13 to the red bus (bottom bus). Close the plastic cover.

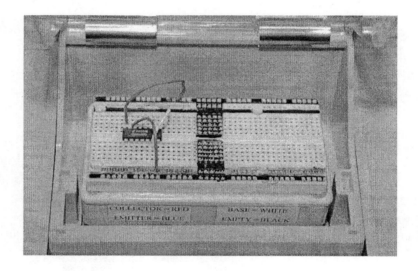

- Press P Channel Enhancement Mode MOSFET Output Characteristic on the LabVIEW Front Panel. Press **Read Curve Tracer**.

- Check to verify **POLARITY** on the Curve Tracer is set to the picture of the negative full wave rectified sine wave, the **STEP** polarity to negative, the **STEP GENERATOR** to voltage, and the **DISPLAY** to **INVERT**. If so, switch the **OUTPUTS** switch on the **Tektronix 370B Curve Tracer** to **ENABLED** and press **Read Curve Tracer** on the LabVIEW Front Panel. Rotate the % Collector Supply button on the LabVIEW front panel from 100% to 0 % and back to 100% and press **Read Curve Tracer** again.

7.8. PROCEDURE

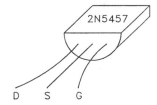

Figure 7.24: Lead Placements for 2N5457 N Channel JFET.

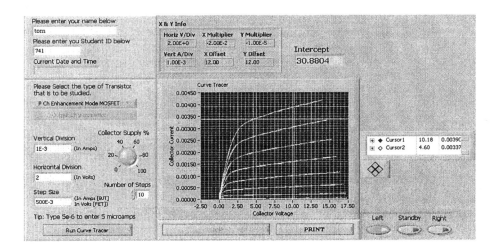

- Position the cursors as shown. The intercept is the reciprocal of the channel length modulation factor, λ.

- Position the cursors to obtain data of drain current as a function of the gate-to-source voltage for constant values of the drain-to-source voltage. Pick a value for the drain-to-source voltage for which the curves are reasonably straight and record the value of the drain current for each value of the gate-to-source voltage.

- Flip the Output Switch on the **Tektronix 370B Curve Tracer** to **DISABLED** and remove the CD 4007.

7.8.8 N Channel JFET

The output characteristics for the N Channel JFET will be obtained. The JFET that will be used is the 2N5457. This is a depletion mode device. Maximum current occurs when the gate-to-source voltage is zero and lower currents are obtained by making the gate-to-source voltage more negative. Gate voltage will be stepped through negative values as the drain-to-source voltage is swept through positive values. This yields a plot of drain current versus drain-to-source voltage for stepped values of gate-to-source voltage.

The lead placements for the 2N5457 N Channel JFET is shown in Fig. 7.24. The drain and source terminals are interchangeable. The gate is the lead on the right. So it is left from right drain, source, gate or source, drain, gate.

- Turn the **OUTPUTS** switch on the **Tektronix 370B Curve Tracer** to the **DISABLED** position.

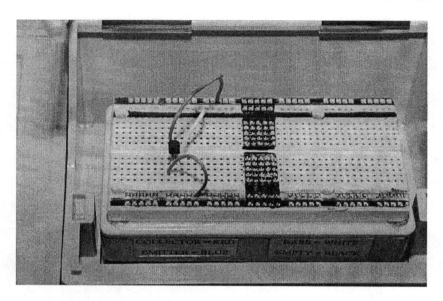

- Turn the plastic cover to the up position and insert the 2N5457 onto the left side of the breadboard section. Connect gate, source, and drain leads to the appropriate buses on the breadboard. Namely connect gate to the white bus (second from top), source to the blue bus (top bus), and drain to the red bus (bottom bus). Close the plastic cover.

- Select JFET on the LabVIEW Front Panel. Press **Read Curve Tracer** on the front panel.

- Verify that the **COLLECTOR SUPPLY POLARITY** is set to the picture of the positive full wave rectified sine wave, the **DISPLAY INVERT** is off, and the step **POLARITY** is set to negative. If not, press **LOCAL** on the Curve Tracer and set them.

- Switch the **OUTPUTS** switch to **ENABLED** on the **Tektronix 370B Curve Tracer** and press **Read Curve Tracer** on the Front Panel. If necessary rotate the % Collector Supply from 100% to 0% and then back to 100 % and press **Read Curve Tracer** again.

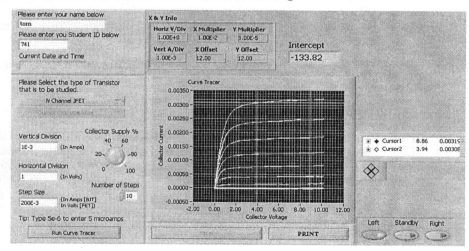

7.8. PROCEDURE

- Position the cursors as shown. The intercept is the reciprocal of the channel length modulation factor, λ.

- Move the cursors to obtain data of drain current as a function of gate-to-source voltage for constant values of drain-to-source voltage. Pick a value for the drain-to-source voltage for which the curves are reasonably straight and record the value of the drain current for each value of the gate-to-source voltage. The top curve corresponds to a gate-to-source voltage of zero and the ones beneath it are for negative values of the gate to source voltage given the per step setting.

- Switch the **OUTPUTS** switch on the **Tektronix 370 Curve Tracer** to **DISABLED** and remove the transistor.

- Print the display and obtain the verification signature of the laboratory instructor..

- Log off the computer.

PROCEDURE for Manual Acquisition of transistor characteristics

NPN BJT

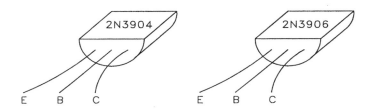

Figure 7.25: Lead placement for 2N3904 NPN and 2N3906 PNP BJTs.

1. Junction Resistance Measurement

The transistors specified for the experiment are the 2N3904 (NPN) and the 2N3906 (PNP). The physical arrangement of the leads for both transistors is shown in Fig. 7.25. For each transistor, use the digital multimeter to measure the resistance between two leads at a time for all permutations of the leads. (The digital multimeter should be set to the "diode test" setting.) From the measurements, verify that the transistors are fabricated as back-to-back diodes. Whilst using the multimeter to measure the resistance of a reverse biased junction, the resistance of the human body can affect the measurements. Therefore, the transistor leads and the metal terminals of the multimeter leads should not be touched with the hands when making the resistance measurements.)

7.8.9 Safety Precautions

The Tektronix 370B is an industrial curve tracer which can produce large voltages and currents. Certain obvious precautions should be observed:

- When inserting the transistor or other solid state device to be tested the **OUTPUTS** switch should be set to **DISABLED**. The is a toggle switch located under the CRT display for the instrument.

- Never change the **MAX PEAK VOLTS** or **MAX PEAK POWER** from their default settings of 16 V and 0.08 Watts.

- Never open the cover for the breadboard section when the **OUTPUTS** switch is the **ENABLED** position.

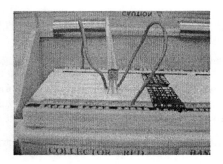

Figure 7.26:

If these precautions are not strictly adhered to a painful, horrific, excruciating, and, possibly, debilitating electrical

SHOCK!!!!

may be experienced. Therefore, when it is turned on none of the electrical terminals should be touched by the user. Changes should be made with this instrument only when the **OUTPUTS** are disabled..

7.8.10 NPN BJT Output Characteristic

- Turn the **OUTPUTS** switch to the **DISABLED** position.

- Turn the plastic cover to the up position and insert the 2N3904 onto the left side of the breadboard section. Connect the base to the white colored bus on the breadboard, the collector to the red, and the emitter to the blue. Close the plastic cover. The base current will be stepped through positive values while the collector-to-emitter voltage is swept through positive values. From this a plot of the collector current versus collector-to-emitter voltage is made for stepped values of base current.

- Under **DISPLAY** the **INVERT** button should not be illuminated.

- Set the **STEP GENERATOR** to **CURRENT** (push button) and the **STEP AMPLITUDE** (rotary knob) to 5 μA. The step amplitude is indicated on the display graticule as **PER STEP**. Don't confuse this with the number of steps.

- Set the **POLARITY** on the **COLLECTOR SUPPLY** to the picture of the positive full wave rectified waveform.

- Set the **POLARITY** to positive up steps.

- Set the **VERTICAL CURRENT/DIV** to 1 mA. This is indicated on the graticule.

- Set the **HORIZONTAL VOLTS/DIV** to 2 V. This is indicated on the graticule.

- Change the side selector to **LEFT** (push button lower right).

- Change the **OUTPUTS** selector to **ENABLED**.

- Rotate the **VARIABLE COLLECTOR SUPPLY** knob clockwise (to the right) until the output characteristics are seen. If necessary adjust the **VERTICAL CURRENT/DIV** so that the entire display can be seen. This is a plot of the collector current versus the collector-to-emitter voltage for stepped values of the base current.

7.8. PROCEDURE

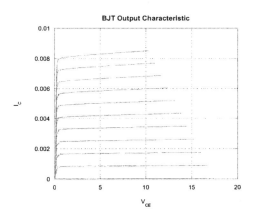

Figure 7.27: NPN BJT Output Characteristic

- Turn the **CURSOR** from **OFF** to **DOT**.

- On the **CURSOR POSITION** push the **UP** button until the dot ends up on the top curve. This may be speeded up by holding down the **FAST** button while holding down the **UP** button.

- Change the **CURSOR** setting from **DOT** to **LINE**. Press the **UP** button until the line cursor lines up with the top curve. The value indicated on the graticule as intercept is the negative of the Early Voltage.

- Insert a formatted, high quality, 3.5 inch floppy disk into the floppy disk drive of the curve tracer. Press the blue **SHIFT** button and hold it down and then press the blue **BMP** button. Wait until the instrument has written the bit map of the screen to the floppy disk.

- (If a csv file is needed to insert the numerical values of the characteristic, it can be obtained by pressing and holding the blue **ADDR** button and then the blue **BMP** button.)

- Turn the **CURSOR** to the **OFF** position.

7.8.11 NPN BJT Transfer Characteristic

- Rotate the **HORIZONTAL VOLTS/DIV** knob counter clockwise (to the left) until it passes 500 V and then until it reaches 100 mV. This is a plot of collector current versus base-to-emitter voltage. Note that the **HORIZONTAL VOLTS/DIV** green LED annunciator changes from collector to base which indicates that the horizontal axis is now base to emitter voltage instead of collector-to-emitter voltage.

- Plot the display using the above procedure.

- Turn the **OUTPUTS** switch to the **DISABLED** position. Lift the cover and remove the transistor and the floppy disk.

- Use any Windows PC to print the output and transfer characteristics and obtain a verification signature from the laboratory instructor.

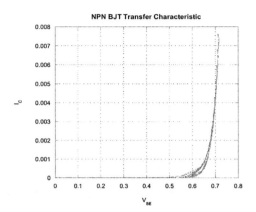

Figure 7.28: NPN BJT Transfer Characterictic

7.8.12 PNP BJT Output Characteristic

- Turn the **OUTPUTS** switch to the **DISABLED** position.

- Turn the plastic cover to the up position and insert the 2N3906 onto the left side of the breadboard section. The lead placements are identical to those of the 2N3904. Connect the base to the white colored bus on the breadboard, the collector to the red, and the emitter to the blue. Close the plastic cover. The base current will be stepped through negative values while the collector-to-emitter voltage is swept through negative values. The display will be inverted and a plot of collector current versus emitter-to-collector voltage will be obtained for stepped values of the base current.

- Set the **STEP GENERATOR** to **CURRENT** (push button) and the **STEP AMPLITUDE** (rotary knob) to 5 μA. The step amplitude is indicated on the display graticule as **PER STEP**.

- Set the **POLARITY** on the **COLLECTOR SUPPLY** to the picture of the negative full wave rectified waveform.

- Note that the **POLARITY** switched to negative down steps. Do not change it.

- Set the **VERTICAL CURRENT/DIV** to 1 mA. This is indicated on the graticule.

- Set the **HORIZONTAL VOLTS/DIV** to 2 V. This is indicated on the graticule.

- Under the **DISPLAY** section press **INVERT**.

- Change the side selector to **LEFT** (push button lower right).

- Change the **OUTPUTS** selector to **ENABLED**.

- Rotate the **VARIABLE COLLECTOR SUPPLY** knob clockwise (to the right) until the output characteristics are seen. If necessary adjust the **VERTICAL CURRENT/DIV** so that the entire display can be seen. This is a plot of the collector current versus the collector-to-emitter voltage for stepped values of base current.

- Turn the **CURSOR** from **OFF** to **DOT**.

7.8. PROCEDURE

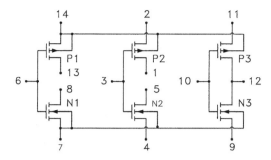

Figure 7.29: Pinouts for 4007 MOSFET IC

- On the **CURSOR POSITION** push the **UP** button until the dot ends up on the top curve. This may be speeded up by holding down the **FAST** button while holding down the **UP** button.

- Change the **CURSOR** setting from **DOT** to **LINE**. Press the **UP** button until the line cursor lines up with the top curve. The value indicated on the graticule as intercept is the negative of the Early Voltage.

- Insert a formatted, high quality, 3.5 inch floppy disk into the floppy disk drive of the curve tracer. Press the blue **SHIFT** button and hold it down and then press the blue **BMP** button. Wait until the instrument has written the bit map of the screen to the floppy disk.

- (If a csv file is needed to insert the numerical values of the characteristic, it can be obtained by pressing and holding the blue **ADDR** button and then the blue **BMP** button.)

- Turn the **CURSOR** to the **OFF** position.

7.8.13 PNP BJT Transfer Characteristic

- Rotate the **HORIZONTAL VOLTS/DIV** know counter clockwise (to the left) until it passes 500 V and then until it reaches 100 mV. This is a plot of collector current versus base-to-emitter voltage.

- Plot the display using the above procedure.

- Turn the **OUTPUTS** switch to the **DISABLED** position. Lift the cover and remove the transistor and the floppy disk.

- Use any Windows PC to print the output and transfer characteristics and obtain a verification signature from the laboratory instructor.

7.8.14 N Channel Enhancement Mode MOSFET

The major difference in obtaining the terminal characteristics of FETs rather than BJTs is that the input or stepping variable is gate voltage rather than base current. Gate voltage is stepped as the drain-to-source voltage is swept yielding a plot of drain current versus drain-to-source voltage for stepped values of gate-to-source voltage.

The MOSFETs to be examined in this experiment are contained on an integrated circuit known as the CD4007 or MC14007. Because the output characteristics for the N Channel Enhancement Mode MOSFET are being obtained in this section, transistor N1 in Fig. 7.29 will be used. For this transistor, pin 6 is the gate (analogous to base), pin 7 is the source (analogous to emitter), and pin 8 is the drain (analogous to collector). The drain of N1 is connected internally to the body whereas it floats for N2 and N3.

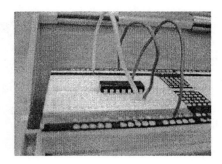

Figure 7.30:

- Turn the **OUTPUTS** switch to the **DISABLED** position.

- Turn the plastic cover to the up position and insert the CD4007 onto the left side of the breadboard section. Connect gate, source, and drain pins to the appropriate buses on the breadboard. Namely connect gate pin 6 to the white bus (second from top), source pin 7 to the blue bus (top bus), and drain pin 8 to the red bus (bottom bus). Close the plastic cover.

- Under **DISPLAY** the **INVERT** button should not be illuminated.

- Set the **STEP GENERATOR** to **VOLTAGE** (push button) and the **STEP AMPLITUDE** (rotary knob) to 1 V. The step amplitude is indicated on the display graticule as **PER STEP**.

- Set the **POLARITY** on the **COLLECTOR SUPPLY** to the picture of the positive full wave rectified waveform.

- Set the **POLARITY** to positive up steps.

- Set the **VERTICAL CURRENT/DIV** to 2 mA. This is indicated on the graticule.

- Set the **HORIZONTAL VOLTS/DIV** to 2 V. This is indicated on the graticule.

- Change the side selector to **LEFT** (push button lower right).

- Change the **OUTPUTS** selector to **ENABLED**.

- Rotate the **VARIABLE COLLECTOR SUPPLY** knob clockwise (to the right) until the output characteristics are seen. If necessary adjust the **VERTICAL CURRENT/DIV** so that the entire display can be seen. This is a plot of the drain current versus the drain-to-source voltage for stepped values of the gate-to-source voltage.

- Change the **NUMBER OF STEPS** to 7.

- Turn the **CURSOR** from **OFF** to **DOT**.

- On the **CURSOR POSITION** push the **UP** button until the dot ends up on the top curve. This may be speeded up by holding down the **FAST** button while holding down the **UP** button.

- Change the **CURSOR** setting from **DOT** to **LINE**. Press the **UP** button until the line cursor lines up with the top curve. The value indicated on the graticule as intercept is the negative of the reciprocal of the channel length modulation factor.

- Insert a formatted, high quality, 3.5 inch floppy disk into the floppy disk drive of the curve tracer. Press the blue **SHIFT** button and hold it down and then press the blue **BMP** button. Wait until the instrument has written the bit map of the screen to the floppy disk.

7.8. PROCEDURE

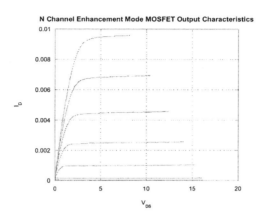

Figure 7.31: N Channel Enhancement Mode MOSFET Output Characteristics

- (If a csv file is needed to insert the numerical values of the characteristic, it can be obtained by pressing and holding the blue **ADDR** button and then the blue **BMP** button.)
- Turn the **CURSOR** to the **OFF** position.

7.8.15 P Channel Enhancement Mode MOSFET

The MOSFETs to be examined in this experiment are contained on an integrated circuit known as the CD4007 or MC14007. Because the output characteristics for the P Channel Enhancement Mode MOSFET are being obtained in this section, transistor P1 in Fig. 7.29 will be used. For this transistor, pin 6 is the gate (analogous to base), pin 14 is the source (analogous to emitter), and pin 13 is the drain (analogous to collector). The drain of P1 is connected internally to the body whereas it floats for P2 and P3.

- Turn the **OUTPUTS** switch to the **DISABLED** position.
- Turn the plastic cover to the up position and insert the CD4007 onto the left side of the breadboard section. Connect gate, source, and drain pins to the appropriate buses on the breadboard. Namely connect gate pin 6 to the white bus (second from top), source pin 14 to the blue bus (top bus), and drain pin 13 to the red bus (bottom bus). Close the plastic cover.
- Under **DISPLAY** the **INVERT** button should selected or illuminated.
- Set the **STEP GENERATOR** to **VOLTAGE** (push button) and the **STEP AMPLITUDE** (rotary knob) to 1 V. The step amplitude is indicated on the display graticule as **PER STEP**.
- Set the **POLARITY** on the **COLLECTOR SUPPLY** to the picture of the negative full wave rectified waveform.
- Set the **POLARITY** to negative down steps (**INVERT** illuminated).
- Set the **VERTICAL CURRENT/DIV** to 2 mA. This is indicated on the graticule.
- Set the **HORIZONTAL VOLTS/DIV** to 2 V. This is indicated on the graticule.
- Change the side selector to **LEFT** (push button lower right).

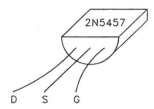

Figure 7.32: Lead Placements for 2N5457 N Channel JFET.

- Change the **OUTPUTS** selector to **ENABLED**.

- Rotate the **VARIABLE COLLECTOR SUPPLY** knob clockwise (to the right) until the output characteristics are seen. If necessary adjust the **VERTICAL CURRENT/DIV** so that the entire display can be seen. This is a plot of the drain current versus the drain-to-source voltage for stepped values of the gate-to-source voltage.

- Change the **NUMBER OF STEPS** to 7.

- Turn the **CURSOR** from **OFF** to **DOT**.

- On the **CURSOR POSITION** push the **UP** button until the dot ends up on the top curve. This may be speeded up by holding down the **FAST** button while holding down the **UP** button.

- Change the **CURSOR** setting from **DOT** to **LINE**. Press the **UP** button until the line cursor lines up with the top curve. The value indicated on the graticule as intercept is the negative of the reciprocal of the channel length modulation factor.

- Insert a formatted, high quality, 3.5 inch floppy disk into the floppy disk drive of the curve tracer. Press the blue **SHIFT** button and hold it down and then press the blue **BMP** button. Wait until the instrument has written the bit map of the screen to the floppy disk.

- (If a csv file is needed to insert the numerical values of the characteristic, it can be obtained by pressing and holding the blue **ADDR** button and then the blue **BMP** button.)

- Turn the **CURSOR** to the **OFF** position.

- Turn the **OUTPUTS** switch to the **DISABLED** position and remove the IC and floppy disk.

7.8.16 N Channel JFET

The output characteristics for the N Channel JFET will be obtained. The JFET that will be used is the 2N5457. This is a depletion mode device. Maximum current occurs when the gate-to-source voltage is zero and lower currents are obtained by making the gate-to-source voltage more negative. Gate voltage will be stepped through negative values as the drain-to-source voltage is swept through positive values. This yields a plot of drain current versus drain-to-source voltage for stepped values of gate-to-source voltage.

The lead placements for the 2N5457 N Channel JFET is shown in Fig. 7.32. The drain and source terminals are interchangeable. The gate is the lead on the right. So it is left from right drain, source, gate or source, drain, gate.

- Turn the **OUTPUTS** switch to the **DISABLED** position.

7.8. PROCEDURE

Figure 7.33:

- Turn the plastic cover to the up position and insert the 2N5457 onto the left side of the breadboard section. Connect gate, source, and drain leads to the appropriate buses on the breadboard. Namely connect gate to the white bus (second from top), source to the blue bus (top bus), and drain to the red bus (bottom bus). Close the plastic cover.

- Under **DISPLAY** the **INVERT** button should not be illuminated.

- Set the **STEP GENERATOR** to **VOLTAGE** (push button) and the **STEP AMPLITUDE** (rotary knob) to 100 mV. The step amplitude is indicated on the display graticule as **PER STEP**.

- Set the **POLARITY** on the **COLLECTOR SUPPLY** to the picture of the positive full wave rectified waveform.

- Set the **POLARITY** to negative down steps (**INVERT** illuminated).

- Set the **VERTICAL CURRENT/DIV** to 1 mA. This is indicated on the graticule.

- Set the **HORIZONTAL VOLTS/DIV** to 2 V. This is indicated on the graticule.

- Change the side selector to **LEFT** (push button lower right).

- Change the **OUTPUTS** selector to **ENABLED**.

- Rotate the **VARIABLE COLLECTOR SUPPLY** knob clockwise (to the right) until the output characteristics are seen. If necessary adjust the **VERTICAL CURRENT/DIV** so that the entire display can be seen. This is a plot of the drain current versus the drain-to-source voltage for stepped values of the gate-to-source voltage.

- Change the **NUMBER OF STEPS** to 7.

- Turn the **CURSOR** from **OFF** to **DOT**.

- On the **CURSOR POSITION** push the **UP** button until the dot ends up on the top curve. This may be speeded up by holding down the **FAST** button while holding down the **UP** button.

- Change the **CURSOR** setting from **DOT** to **LINE**. Press the **UP** button until the line cursor lines up with the top curve. The value indicated on the graticule as intercept is the negative of the reciprocal of the channel length modulation factor.

- Insert a formatted, high quality, 3.5 inch floppy disk into the floppy disk drive of the curve tracer. Press the blue **SHIFT** button and hold it down and then press the blue **BMP** button. Wait until the instrument has written the bit map of the screen to the floppy disk.

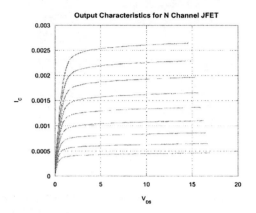

Figure 7.34: Output Characteristic for N Channel JFET.

- (If a csv file is needed to insert the numerical values of the characteristic, it can be obtained by pressing and holding the blue **ADDR** button and then the blue **BMP** button.)
- Turn the **CURSOR** to the **OFF** position.
- Turn the **OUTPUTS** switch to the **DISABLED** position and remove the floppy disk and transistor.

7.9 Laboratory Report

7.9.1 NPN BJT

The data taken for the output characteristic of the NPN BJT will indicate the X axis intercept. The magnitude of this voltage is essentially the Early voltage. The Early voltage V_A is the SPICE parameter **VA**. The Early voltages for PNP transistors are normally lower than those for NPN transistors.

The parameter β_0 may now be determined using the output characteristic. Using the same curve for which the Early voltage was measured

$$\beta_0 = \frac{I_C/I_B}{1 + V_{CE}/V_A} \tag{7.24}$$

Pick the curve the cursors were placed on and determine I_C, I_B, and V_{CE} in a region where the curve is essentially straight. The parameter β_0 is the SPICE parameter **BF**.

The data taken for the transfer characteristic may be used to obtain the saturation current I_{S0} The zero bias saturation current may then be determined as

$$I_{S0} = \frac{I_C}{1 + V_{CE}/V_A} \exp\left(-\frac{V_{BE}}{V_T}\right) \tag{7.25}$$

where V_T is the thermal voltage (25.852 mV for $T = 300$ K). Pick a point on the transfer characteristic and measure the actual value of V_{BE} for a particular I_C. The zero bias saturation current is the SPICE parameter **IS**.

Use the SPICE parameters (**IS**, **BF**, and **VA** all three of which are positive numbers) obtained above to obtain a simulation of the output and transfer characteristics of the NPN BJT. Use the default values for the other SPICE parameters, i.e. don't specify them in the model statement for the device. The required SPICE input deck is found in the theory section.

7.9. LABORATORY REPORT

Compare the experimental and theoretical output and transfer characteristics for the NPN BJT. Provide reasons for any appreciable differences.

7.9.2 PNP BJT

Use the same procedure for the PNP BJT curves to obtain the SPICE parameters for these devices. It is normal for the Early voltage to be smaller for the PNP device.

Use the SPICE parameters (IS, BF, and VA all three of which are positive numbers) obtained above to obtain a simulation of the output and transfer characteristics of the PNP BJT. Use the default values for the other SPICE parameters, i.e. don't specify them in the model statement for the device. The required SPICE input deck is found in the theory section.

Compare the experimental and theoretical output and transfer characteristics for the PNP BJT. Provide reasons for any appreciable differences.

7.9.3 N Channel Enhancement Mode MOSFET

Use the output characteristic of the N Channel Enhancement Mode MOSFET to obtain the channel length modulation factor λ. The magnitude of the reciprocal of the intercept is essentially λ. The channel length modulation factor λ is the SPICE parameter LAMBDA.

The output characteristic may be used to obtain the transconductance K and threshold voltage V_{TO} for the MOSFET. If the square root of each side of Eq. (7.18) is taken, the following relationship is obtained:

$$\sqrt{I_D} = \sqrt{K} \times (V_{GS} - V_{TO}) \tag{7.26}$$

It follows that a plot of $y = \sqrt{I_D}$ versus $x = V_{GS}$ should be a straight line. The intercept of this line with the x or V_{GS} axis gives V_{TO}. The square of the slope of the line gives K. Data should be taken at the drain-to-source voltage at which the channel length modulation factor λ was measured (the average drain-to-source voltage). A more accurate method for estimating V_{TO} and K from experimental data is to use a least squares curve fit of a straight line to the data. Most scientific calculators have built in programs for doing this. If the experimental data does not give a reasonably straight line, the MOSFET may not be biased in its saturation region. Simply pick a value for V_{DS} for which the curves were reasonably straight and record the value of the drain current for the different values of gate-to-source voltage. This provides a set of data points which may be used in a least squares curve fitting routine.

Once K is known, Eqs. (7.19) and (7.20) may be used to calculate K_0 as

$$K_0 = \frac{K}{1 + \lambda V_{DS}} \tag{7.27}$$

where K is the SPICE parameter KP and it has been assumed that $W = L$ in the equation for K. The parameters V_{TO} and λ are the SPICE parameters VTO and LAMBDA. Use the default values for the other SPICE parameters, i.e. don't specify them in the model statement for the device.

Use the SPICE parameters (KP, VTO, and LAMBDA all three of which are positive numbers) obtained above to obtain a simulation of the output and transfer characteristics of the N Channel Enhancement Mode MOSFET. The required SPICE input deck is found in the theory section.

Compare the experimental and theoretical output and transfer characteristics for the N Channel Enhancement Mode MOSFET. Provide reasons for any discrepancies.

7.9.4 P Channel Enhancement Mode MOSFET

Use the same procedure to obtain the SPICE parameters for the P Channel device. It is normal for λ to be larger for the P Channel device.

Use the SPICE parameters (KP and LAMBDA are positive numbers while VTO is negative) obtained above to obtain a simulation of the output and transfer characteristics of the P Channel Enhancement Mode MOSFET. The required SPICE input deck is found in the theory section.

Compare the experimental and theoretical output and transfer characteristics for the P Channel Enhancement Mode MOSFET. Provide reasons for any discrepancies.

7.9.5 N Channel JFET

Use the data for the output characteristics for the N Channel JFET to obtain the channel length modulation factor λ The channel length modulation factor λ is the SPICE parameter LAMBDA.

The output characteristic may be used to obtain the transconductance parameter β and threshold voltage V_{TO} for the JFET. If the square root of each side of Eq. (7.9) is taken, the following relationship is obtained:

$$\sqrt{I_D} = \sqrt{\beta} \times (V_{GS} - V_{TO}) \tag{7.28}$$

It follows that a plot of $y = \sqrt{I_D}$ versus $x = V_{GS}$ should be a straight line. The gate-to-source voltage V_{GS} and the threshold voltage V_{TO} are negative for this device. The intercept of this line with the x or V_{GS} axis gives V_{T0}. The square of the slope of the line gives β. Data should be taken at the drain-to-source voltage at which the channel length modulation factor λ was measured (the average drain-to-source voltage). A more accurate method for estimating V_{TO} and β from experimental data is to use a least squares curve fit of a straight line to the data. Most scientific calculators have built in programs for doing this. If the experimental data does not give a reasonably straight line, the JFET may not be biased in its saturation region. This procedure is identical for that for the MOSFET.

Once β is known, β_0 can be calculated from

$$\beta_0 = \frac{\beta}{1 + \lambda V_{DS}} \tag{7.29}$$

where β_0 is the SPICE parameter BETA. The parameters V_{TO} and λ are the SPICE parameters VTO and LAMBDA. Use the default values for the other SPICE parameters, i.e. don't specify them in the model statement for the device.

Use the SPICE parameters (BETA, VTO, and LAMBDA) obtained above to obtain a simulation of the output and transfer characteristics of the N Channel JFET. (BETA and LAMBDA are positive while VTO is negative for this device.) The required SPICE input deck is found in the theory section.

Compare the experimental and theoretical output and transfer characteristics for the N Channel JFET. Provide reasons for any discrepancies.

7.10 References

1. E. J. Angelo, *Electronics: BJTs, FETs, and Microcircuits*, McGraw-Hill, 1969.
2. W. Banzhaf, *Computer-Aided Circuit Analysis Using SPICE,* Prentice Hall, 1989.
3. Burns, S. G. and Bond, P. R., *Principles of Electronic Circuits*, 2nd ed., PWS, 1997.
4. Foty, D., *MOSFET Modeling with SPICE*, Prentice Hall, 1997.
5. Hodges, D. A. and Jackson, H. G., *Analysis and Design of Digital Integrated Circuits*, 2nd ed., McGraw-Hill, 1988.
6. M. N. Horenstein, *Microelectronics Circuits and Devices,* 2nd ed., Prentice Hall, 1996.
7. P. Horwitz and W. Hill, *The Art of Electronics,* 2nd edition, Cambridge University Press, 1989.
8. P. Horowitz and I. Robinson, *Laboratory Manual for The Art of Electronics,* Cambridge University Press, 1981.
9. Howe, R. T. and Sodini, C. G., *Microelectronics,* Prentice Hall, 1997.
10. Jaeger, R. C., *Microelectronic Circuit Design,* McGraw-Hill, 1997.
11. J. H. Krenz, *An Introduction to Electrical and Electronic Devices,* Prentice Hall, 1987.
12. Massobrio, G. and Antognetti, P., *Semiconductor Device Modeling with SPICE,* 2nd ed., McGraw-Hill, 1993.
13. R. Mauro, *Engineering Electronics,* Prentice Hall, 1989.

7.10. REFERENCES

14. F. H. Mitchell and F. H. Mitchell, *Introduction to Electronic Design,* Prentice Hall, 1988.
15. Motorola, *Small-Signal Semiconductors,* DL 126, Motorola, 1987.
16. C. J. Savant, M. S. Roden, and G. L. Carpenter, *Electronic Design,* 3rd ed., Discovery Press, 1997.
17. A. S. Sedra and K. C. Smith, *Microelectronic Circuits,* 4th edition, Oxford, 1998.
18. D. L. Schilling and C. Belove, *Electronic Circuits: Discrete and Integrated,* McGraw-Hill, 1968.
19. P. W. Tuinenga, *SPICE,* Prentice Hall, 1988.

Chapter 8

Digital Electronic Circuits

8.1 Object

This experiment examines some elementary electronic circuits that can be used to implement digital logic. Two logic families are examined. These are the BJT TTL family and the CMOS 4000 series family. Transfer characteristics, noise margins, propagation delays, and power dissipation are measured.

8.2 Theory

Data of interest to engineers and scientists often occur naturally in discrete or digital form. Some common examples are whether a switch is open or closed, a seat belt is buckled, a can is on a conveyor belt, a missile is to be launched, a key on a keyboard has been pressed, an elevator door is closed, etc. Any data which can be placed into a one-to-one correspondence with integers in a number system are discrete data.

Analog data can often be processed and/or stored more efficiently in digital form. Circuits known as analog-to-digital converters are used to convert analog data into digital data. After it is processed in digital form, if required, it can then be converted back into analog form with a circuit known as a digital-to-analog converter. For instance, music compact discs systems sample the analog music signal from a microphone every $23\,\mu$s, quantize and convert the analog voltage into a 16 bit binary word, and encode and store this data on a 5 inch metal disc. The disk stores the binary digits in the form of pits and spaces using a recording laser. The user then retrieves this music from the stored digital form by illuminating the spinning disk with another laser to obtain the binary digits stored on the disk. These binary words are passed through a digital-to-analog converter, the output of which is filtered to remove digital "glitches" and then amplified before being applied to a loudspeaker.

Digital electronic circuits use voltage levels to represent discrete data which represents either the state of a system or integers in a number system. Conceptually, there is no reason why electronic circuits with ten voltage levels could not be used to represent the familiar decimal number system. However, modern electronic technology limits digital electronics to the binary number system because the circuits are implemented as switches which have only two states. These two states correspond to the "1" and "0" of the binary number system. Moreover, the binary number system is the natural number system for decisions or events which can be formulated as either true or false propositions, such as whether a switch is open or closed. Therefore, on the circuit level the binary number system is the one universally employed.

Digital logic is implemented with circuits called logic gates. There are five basic logic gates. These are the inverter, the AND gate, the NAND gate, the OR gate, and the NOR gate. The logic implemented by these gates is described with tables called truth tables. The logic symbols and the truth tables for the five basic gates are given in Fig. 8.1. These tables show the logical state of the output of each gate as a function of the logical states of the input variables.

CHAPTER 8. DIGITAL ELECTRONIC CIRCUITS

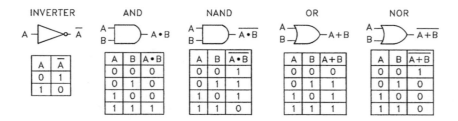

Figure 8.1: Truth tables for the inverter and the AND, NAND, OR, and NOR gates.

8.2.1 Logic Variables and Voltage Levels

Digital electronics uses voltage levels to represent logic variables. The logic variables have two states, which are denoted by 0 and 1. Ideally, only two distinct voltages are required to represent two logic states. Practical considerations dictate that a range of voltage be used for each state as shown in Fig. 8.2(a). A voltage v_H that lies in the range $V_{H1} \leq v_H \leq V_{H2}$ is interpreted as a high and a voltage v_L that lies in the range $V_{L1} \leq v_L \leq V_{L2}$ is interpreted as a low. For example, suppose that the output voltage of a logic gate is constrained to be in the range from 0 to 5 V. A logic 1 might correspond to the range $4.75\,\text{V} \leq v_H \leq 5.0\,\text{V}$ and a logic 0 might correspond to the range $0\,\text{V} \leq v_L \leq 0.25\,\text{V}$. Systems which use the voltage v_H in the high range to represent a logic 1 and the voltage v_L in the low range to represent a logic 0 are known as positive logic systems. Conversely, systems which use the alternative assignment of the high voltage range for a logical 0 and the low voltage range for a logical 1 are known as negative logic systems. This is summarized in Fig. 8.2(b) where v_H is denoted H and v_L is denoted L.

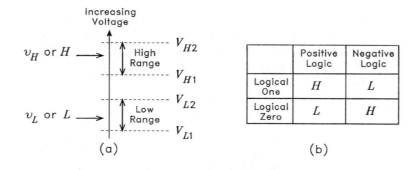

Figure 8.2: (a) Illustration of voltage levels for binary signals. (b) Voltage levels for positive and negative logic.

8.2.2 Ideal inverter

The inverter is a fundamental building block of digital electronics. Fig. 8.3(a) shows the diagram of an ideal inverter. The input and output voltages, respectively, are v_I and v_O. The circuit is implemented with a dc voltage V^+ that is connected in series with two voltage controlled switches S_1 and S_2. When one switch is closed, the other is open, and vice versa. When S_1 is open and S_2 is closed, the output is connected to the power supply and $v_O = V^+$. When S_2 is open and S_1 is closed, the output is grounded and $v_O = 0$. The input voltage v_I controls the switches. This is illustrated in Fig. 8.3(b). When the input is low (L), S_1 is open, S_2 is closed, and the output is high (H). Similarly, when the input is high (H), S_1 is closed, S_2 is

8.2. THEORY

open, and the output is low (L). The value of the input voltage at which the output switches is one-half the power supply voltage, i.e. $V^+/2$.

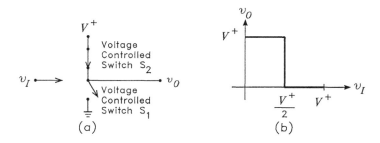

Figure 8.3: f03(a) Ideal inverter. (b) Output voltage versus input voltage.

The ideal inverter has an infinite input impedance so that no current flows into the v_I terminal. This means that it produces no loading on the circuit which drives it. In digital circuits, the number of inputs is called the fan-in. The fan-in of the inverter is one.

The ideal inverter has a low output impedance. When the switch is closed, the output is grounded and the output impedance is zero. When the switch is open, the output impedance is set by the pull-up load. This means that it can drive a large number of loads. The number of similar logic circuit that a digital electronic circuit can drive is known as the fan-out. For a physical inverter the resistance of the switch when it is closed cannot be zero and thus the output impedance of a real inverter will not be zero even when the switch is closed.

The ideal inverter switches instantaneously. When the input crosses the switching threshold of $V_{PP}/2$ the output instantly changes to the state opposite the input. Therefore the rise time and fall time of the ideal inverter are zero. If the input is a rectangular square wave with a dc level of $V_{PP}/2$ and a peak-to-peak value of V_{PP} the output is also a perfect rectangular square wave that is $180°$ out of phase with the input. Therefore there is no propagation delay associated with this circuit. Any changes at the input are instantly felt at the output.

The ideal inverter is unidirectional. This means that any type of load connected to the output node will have no effect on the input node of this circuit. Changes in the output are not reflected back to the input.

Power is dissipated in the ideal inverter. For simplicity it will be assumed that the pull-up load is simply a resistor R. When the switch is closed a current V_{PP}/R flows through the pull-up load which causes a power of V_{PP}^2/R to be dissipated in this resistor. When the switch is open, no current flows and no power is dissipated. If this circuit were driven by a rectangular square wave with a frequency of f in Hz, such as the clock signal of a digital system, the average power dissipated per cycle would be $V_{PP}^2/2R$ because the switch would be open for one half a cycle of the input and closed for the other.

8.2.3 Physical Inverter

Transfer Characteristic

Physical inverters have characteristics that differ from those of the ideal. A typical transfer characteristic for a physical inverter is shown in Fig. 8.4. The output is not constrained to the two values, V_{PP} and 0 as is the case for the ideal inverter. The output may be an value between these two voltage levels. Voltage ranges must be selected corresponding to high (H) and low (L). The transition between these two ranges will not be $V_{PP}/2$ as was the case for the ideal inverter.

The points on the transfer characteristic that are employed as the transition points are those for which the slope of the transfer characteristic is one. The voltage V_{OH} is the smallest output voltage that will unambiguously be interpreted as a high (H). The voltage V_{OL} is the largest output voltage that will be

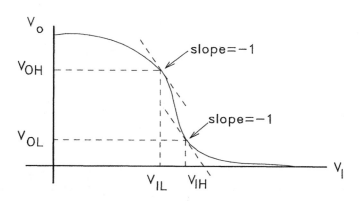

Figure 8.4: Physical INVERTER.

unambiguously interpreted as a logic low (L). The voltage V_{IL} is the largest input voltage that will be unambiguously interpreted as a logic low (L). The voltage V_{IH} is the smallest input voltage that will unambiguously be interpreted as a high (H). The region between the two points for which the slope is unity is the forbidden region which is used in analog circuits for inverting amplifiers but is normally avoided by digital electronic circuits.

The permissible ranges for operation of a physical inverter are illustrated in Fig. 8.5. If two or more of these circuits were cascaded, then in order to have a workable system, the output voltage ranges must be smaller than the corresponding inputs. This digital circuit element will still function properly as long as the inputs lie in the regions shown.

Noise Margins

In an actual circuit external unwanted signals may be inductively or capacitively coupled into the input circuitry. These unwanted signals are known as noise. As long as the inputs still lie in the ranges shown no harm is done because the output will still lie in the ranges for a high (H) or low (L). This is one of the principle advantages of digital circuits over analog ones. Namely, they have the ability to reject noise. If two analog circuits are cascaded, the effect of noise becomes cumulative, i.e. the noise in the first stage is amplified by the second. For digital circuits, as long as the noise does not take the input outside the range corresponding to a high (H) or low (L), the output will be correctly interpreted. Thus the noise has, in essence, been rejected. This noise rejecting property of digital circuits means that they can reproduce or regenerate their inputs.

The largest value the input noise voltage can be before the circuit malfunctions (does not produce the correct logic output) is known as the noise margin. It is, in general, different depending on whether the input is a logic high (H) or low (L). When the input is a low (L) the largest value that the input noise can have is

$$NM_L = V_{IL} - V_{OL} \tag{8.1}$$

where NM_L is known as the low noise margin. If the input noise is positive and greater than this value, the input voltage to the inverter will lie in either the forbidden region or the high (H) region. Both conditions will cause the output to be erroneous. Similarly, when the input is in the high state (H) the noise margin,

8.2. THEORY

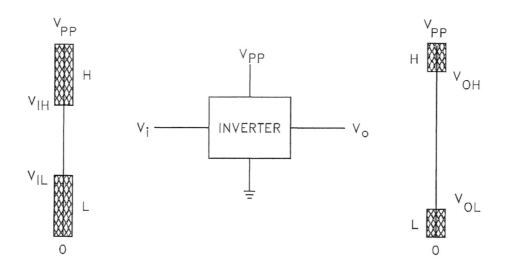

Figure 8.5: Permissible input-output voltage ranges.

NM_H, is given by

$$NM_H = V_{OH} - V_{IH} \tag{8.2}$$

because a negative voltage with a magnitude larger than this will cause the voltage at the input node to lie in either the forbidden region or the low (L) region. As long as the noise voltages lie within the appropriate noise margins, their effect on the output will be the same as if the noise were zero.

Propagation Delay

Any real digital electronic circuit has delays associated with it due to the internal capacitances of the circuit and any load capacitances that the circuit may be driving. Typical input and output waveforms for a digital electronic inverter are shown in Fig. 8.6 where it is assumed that it is being driven by a similar device. These waveforms are square waves with rounded edges, finite rise and fall times, and a phase shift or propagation delay between the input and the output.

The rise time of a waveform is defined to be the amount of time that it takes for the waveform to go from 10% to 90% of the transition from the lower level of V_{OL} to the upper level of V_{OH}. Similarly, the fall time is the amount of time required for the waveform to go from 90% to 10% of the transition from the upper to the lower level.

The propagation delay is the time difference between the 50% points on the input and output waveforms. The propagation time for a high-to-low transition at the output, t_{PHL}, is, in general, different from the low-to-high transition, t_{PLH}. The propagation time, t_P, is defined to be

$$t_P = \frac{t_{PHL} + t_{PLH}}{2} \tag{8.3}$$

The propagation time is one of the parameters that sets the maximum speed that a digital system may be operated.

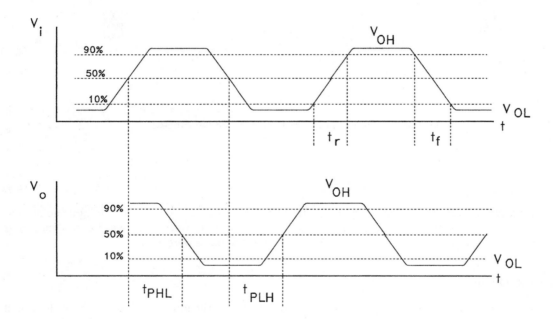

Figure 8.6: Terminal waveforms for physical INVERTER.

8.2. THEORY

Power Dissipation

Any physical inverter has a nonzero power dissipation. This power dissipation has both a static and dynamic component. The static component is the power dissipated while the device is not changing states and the dynamic component is associated with the transition between states. Both are important design parameter since they set the maximum current that must be produced by the dc power supply for the digital electronic circuit and whether or not these devices must be cooled by heat sinks or fans. The maximum allowable power dissipation is one of the parameters that sets the maximum speed at which a digital system may operate.

Physical inverters must be connected to other digital electronic circuit elements that will have a finite input impedance. The load impedance is primarily capacitive. This capacitive load impedance sets the dynamic power dissipation.

The dynamic power dissipation may be obtained for the simple ideal inverter shown in Fig. 8.3 connected to a capacitive load C. The pull-up load will be modeled as a simple resistor R. When the switch is closed the capacitor is discharged and the output voltage, which is the capacitor voltage, is zero. When the switch opens the capacitor begins to charge and the current flowing from the power supply through the pull-up resistor into the capacitor is given by

$$i(t) = \frac{V_{PP}}{R} e^{-t/\tau} \tag{8.4}$$

where $\tau = RC$ is the time constant. If the switch is open for a period of time that is long compared to τ, then the energy dissipated in R is given by

$$E_D = \int_0^\infty i^2(t) R \, dt = C \frac{V_{PP}^2}{2} \tag{8.5}$$

This is also equal to the energy stored in the load capacitor after it has charged up to the voltage V_{PP}. When the switch closes the energy stored in the load capacitor is dissipated in the switch which will have a nonzero resistance for a physical device. Thus the energy lost in one cycle of opening and closing the switch, which corresponds to one cycle of the output voltage if the inverter is being driven by a square wave, is CV_{PP}^2. If the inverter is being drive by a square wave with a frequency, f, then the dynamic component of the power dissipation is

$$P_d = fCV_{PP}^2 \tag{8.6}$$

For some digital electronic circuits, it is the dynamic component of the power dissipation that is dominant.

Delay-Power Product

It is desirable for both the power dissipation and the propagation delay to be small. These goals are in conflict since high currents are required to switch levels rapidly and these high currents are associated with high power dissipations. A figure of merit for any digital electronic circuit element is the delay-power product, DP, which is defined as

$$DP = t_P P \tag{8.7}$$

where t_P is the propagation delay and P is the total power dissipated per cycle. This parameter is dimensionally an energy and has units of joules. It is desirable for this delay-power product to be as small as possible.

8.2.4 BJT Inverter

The circuit shown in Fig. 8.7 can be used as a logic inverter. It consists of a single NPN BJT with the emitter grounded and resistor R_B to limit the base current and collector load resistor R_C. If no load is connected to the output, the dc transfer characteristic is shown next to circuit diagram. This type of inverter would be rarely used but it is instructive to investigate its performance since it forms the basis of several logic families that use BJTs as the active device.

This circuit can also be used as a linear amplifier is the signal swing and operating are properly chosen. The input signal has a dc bias added to it so that the quiescent operating point lies in the linear portion

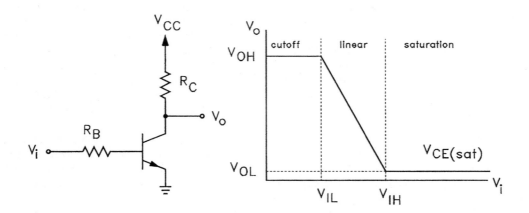

Figure 8.7: BJT INVERTER.

of the output characteristic. The cutoff and saturation regions are avoided because these regions are highly nonlinear. With digital circuits the linear region is to be eschewed and the two nonlinear regions used for the digital signals.

The inputs to the simple BJT inverter could be taken as either 0 which would correspond to low (L) and V_{CC} which would correspond to high (H). However, this voltage swing could not be produced by a similar gate. Therefore, the inputs will be taken as V_{OL} and V_{OH} since these are the outputs levels that would be produced by another such circuit.

The resistor R_B is chosen to be small enough so that the high (H) input will drive the transistor into saturation and the output will then be low (L). If R_B is made too small, excessive minority charge will be stored in the base and it will add to the time required to switch the output of the transistor from low to high.

If the BJT inverter is not loaded and the input is low, the transistor will be turned off and the output will be V_{CC}. Conversely, if the input is high, the transistor will driven into saturation and the output will be $V_{CC(sat)}$ which is normally a few tenths of a volt. In making a transition from low to high or vice versa the transistor circuit must pass through its linear region where the output is given by

$$v_o = V_{CC} - \beta_F (v_i - V_\gamma)\frac{R_C}{R_B} \tag{8.8}$$

where V_γ is the turn-on or cut-in voltage of the transistor ($\cong 0.5$ V) and $\beta_F = i_C/i_B$ is the forward beta of the transistor. Thus the slope of the transfer characteristic in the linear region is given by $-\beta_F R_C/R_B$.

Since it is desirable to make this slope as high as possible so as to make the transition between the high to the low state sharp, this would imply that R_C should be large. However, if R_C is made large, then any load that could be placed on this circuit with a resistive component to the input will result in significantly loading the output of the BJT inverter. This loading will cause the high output level V_{OH} to be lower than V_{CC} and the low output level V_{IH} to be higher than $V_{CE(sat)}$. Even one similar stage connected as the load significantly reduces the output logic swing, i.e. $V_{OH} - V_{OL}$. This means that this type of circuit would not be capable of driving a large number of similar circuits or would have a low fan-out. This is, obviously,

8.2. THEORY

undesirable.

An additional problem with a large value for the collector load resistor R_C deals with the switching properties of this circuit. If a capacitor C were used as the load for this circuit, the capacitor would be essentially uncharged or have a voltage of zero when the output was in the low state (the output is $V_{CE(sat)}$ which is quite small). When the input to the BJT inverter switches from high (H) to low (L) the capacitor prevents the output from changes instantaneously from the low to high state. The transistor is cutoff but the capacitor must charge to the high level output voltage through R_C. The time constant for this charging operation is given by the product of the capacitor C and the collector load resistor R_C. Therefore making R_C large, which is desirable to make the slope of the transfer characteristic in the linear region large, increases the time required to switch the output from the low to the high state as well as decreasing the available logic swing when the circuit is used to drive loads with resistive inputs.

An additional problem relating to the switching characteristics of this simple BJT inverter circuit is the removal of excess minority charges from the base region of the NPN transistor when the transistor is in the saturated state. When the input is high (H) and the output is low (L) the transistor is saturated and the collector-to-emitter voltage is $V_{CE(sat)}$. Both the base-to-emitter and the collector-to-emitter junctions are forward biased and both inject electrons into the base where they are minority carriers. When the input makes a transition from the high to the low state this charge must be withdrawn from the base before the transistor can cutoff. Since this current must flow out of the base through R_B, a large R_B will add to the time required to remove the excess minority charge from the base. However, the smaller R_B is made, the more charge will be forced into the base when the input is high. Therefore, R_B is often picked to be small enough so that the transistor is driven into saturation when the input is high and large enough so that the charge storage is not excessive. To speed up the switching operation, a capacitor known as a speed-up capacitor is often connected in parallel with R_B so that when the input switches from high to low the capacitor acts as a momentary short circuit bypass path around R_B.

Shown in Fig. 8.8 are typical input and output waveforms for a BJT inverter driving and being driven by a similar circuit. Note the finite rise and fall times of the output as well as the current being drawn by the base. The rise time is larger than the fall time and is the limiting factor on the switching speed of this circuit. When the input drops from high (H) to low (L) the excess charge stored in the base must be removed so the base current becomes negative for a period of time. The amount of time required to remove the excess charge from the base region of the transistor is knows as the storage time, t_S. The collector current does not respond to the change in the input until this excess charge is removed from the base.

The problems associated with switching and loading this simple BJT inverter prevent it from being used as a practical digital circuit element. However, variations of this basic circuit are used to implement several integrated circuit families of digital logic. These will be investigated in subsequent sections.

8.2.5 IC Digital Electronic Circuits

The complexity of most modern digital systems requires that the digital electronic circuits be fabricated on integrated circuits. This reduces not only the physical dimensions of the system but the smaller components have concomitant smaller parasitic capacitances, power consumption, and propagation delays.

Levels of Integration

The number of digital electronic circuits or gates on a single integrated circuit may range from less than ten to over one million. The scales of integration for a digital electronic integrated circuit are defined as: SSI (small-scale integration) for 10 or less gates, MSI (medium-scale integration) for 11 to 100 gates, LSI (large-scale integration) for 101 to 1,000 gates, VLSI (very-large scale integration) 1,101 to 1,000,000 gates, and ULSI (ultra-large scale integration) for circuit with over one million gates. Theses rules or definitions are not totally autocratic and somewhat flexible.

Digital systems may be implemented with either custom-made, programmable, or modular integrated circuits. Modular or standard integrated circuits are known as "off-the-shelf" circuits which perform standard logic functions; systems may be implemented by interconnecting a number of these. Custom-made integrated

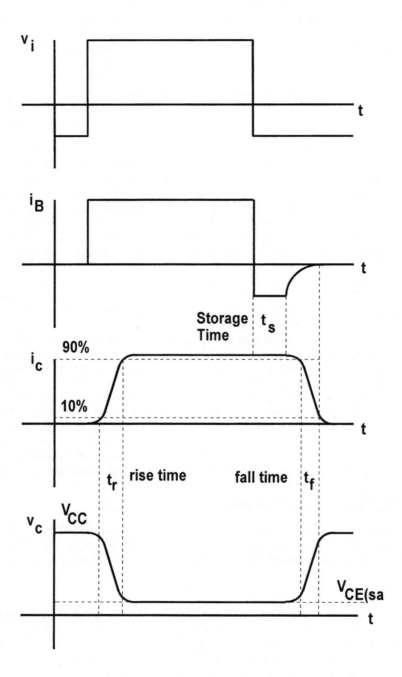

Figure 8.8: Input and output waveforms for loaded BJT INVERTER.

8.2. THEORY

circuits, such as microprocessors or memory circuits, are justifiable only when the production level is high and implementing the same function by interconnecting modular integrated circuits would be prohibitively expensive. Programmable logic offers designers a medium between customized and standard integrated circuits; these devices consists of thousands of unconnected gates which may be configured or programmed by the user.

Logic Families

Most digital integrated circuits use only one type of active device as the switching element and as active loads. The active devices currently used are NPN BJTs and MOS transistors. The type of active device used defines the logic family. The families that employ BJTs are the TTL (transistor-transistor-logic) and ECL (emitter-coupled logic). Devices that use the MOS transistors are the NMOS and CMOS families. There are numerous subdivision among these families as well as hybrids.

The TTL logic family is the oldest and one of the most reliable. It is moderately fast and has good fanout properties. It is impervious to static electricity damage that plagues MOS devices which makes it ideal for testing or "breadboarding" digital circuits in a laboratory setting. It is available in many standard "off-the-shelf" families, e.g. Standard TTL (Series 74), Schottky TTL (Series 74S), Low-Power Schottky TTL (Series 74LS), Advanced Schottky TTL (Series 74AS), Advanced Low-Power Schottky TTL (Series 74ALS), and Fast TTL (Series 74F). The "74" is a prefix that describes a numbering system for the part number that was developed for these off-the-shelf integrated circuits in the 1970s by Texas Instruments. Two or three additional numbers are appended to the prefix to specify the part number, e.g. the 7404 is a HEX inverter (six logic inverters) contained in this integrated circuit.

The other family of BJT digital circuits is emitter coupled logic (ECL). With the exception of state-of-the-art gallium-arsenide circuits, ECL is the fastest of the logic families and is used in high speed communications systems and supercomputers. This family uses NPN BJTs in a differential amplifier configuration to achieve high switching speeds. The differential amplifier topology also provides dual outputs which are the logic function and its complement. The ECL family is available in "off-the-shelf" variety as 10K, 10KH, and 100K families.

The NMOS logic family is not available as an "off-the-shelf" part. They are used exclusively in customized integrated circuits such as microprocessors and memories. They may be easily fabricated and densely packed in a small area which makes them ideal for such circuits. No resistors are used in these circuits; NMOS transistors function as both the switch element and the active load. Enhancement mode NMOS transistors are always used as the switching element while the load may be either an enhancement or depletion mode NMOS transistor.

Digital circuits using CMOS are rapidly becoming the most popular logic families. They have zero static power dissipation; power is dissipated only when the device is switching between states. This low power dissipation makes them ideal for portable applications such as wrist watches and calculators for which the dc power must be supplied by a battery. These circuits use both a NMOS and a complementary PMOS transistor. No resistors or other passive components are used. These circuits are available as either customized integrated circuits or standard "off-the-shelf" families.

The standard series for CMOS logic are: 4000B, 74C, 74HC, 74HCT, 74AC, and 74ACT. The 4000B series is an refinement of the 4000 series which was introduced by RCA in the 1960s and was the first available standard CMOS family. The families that use the "74" prefix indicate that these devices have a pin-to-pin compatibility with the corresponding TTL family. The letter "C" stands for CMOS while the additional letter "H" stands for high speed and "A" indicated advanced. The letter "T" means that these logic families employ voltage levels that are compatible with TTL and can, therefore, be interfaced with the BJT TTL family. Characteristics for some of these families are summarized in Table 1.

Logic Family	Propagation Delay t_P (ns)	Power Dissipation @1MHz (mW)	Delay Power Product (pJ)
74 (Standard TTL)	10	10	100
74S	4.5	20	90
74LS	10	2	20
74ALS	4	1.3	5.2
74AS	2	8	16
74F	3.5	5.4	18.9
ECL 10K	2	25	50
ECL 100K	0.75	40	30
ECL 100KH	1	25	25
4000B/74C			
@ $V_{DD} = 10$ V	30	1.2	36
@ $V_{DD} = 5$ V	50	0.3	15
74AC/ACT	3	0.5	1.5
74HC/HCT	9	0.5	4.5
10G (GaAs)	0.3	125	37.5

Table 1. Characteristics of Digital Logic Families

The logic levels employed by the digital logic families used in this experiment are shown in Fig. 8.9. The TTL families employ a power supply voltage that must lie within five percent of five volts. The CMOS families may employ more than one power supply voltage and its characteristics are functions of the power supply voltage. Note that with the CMOS logic family that increasing the power supply voltages increases the noise margins and that the noise margin for the HIGH and LOW states are the same. With TTL, the power supply is fixed and the noise margins are different.

8.2.6 Standard TTL Logic Family

The circuit shown in Fig. 8.10 is an inverter in the standard TTL logic family. Specifically, this is one of the inverters that is contained in the 7404 *HEX* inverter. The "hex" means that there are six such circuits in each of the 7404 integrated circuits. It is designed to work with a single power supply having a value of +5 V. As the figure illustrates, the logical operation is implemented with NPN BJT transistors, resistors, and diodes.

The analysis of the operation of the TTL inverter necessitates the review of the four modes of operation of the BJT. The four modes of operation of the BJT are: forward active mode, cutoff, saturation, and the reverse active mode. When the BJT is used as the amplifying element in an amplifier, only the forward active mode is employed. All four modes are used in the transistors used in the TTL family of logic gates.

8.2.7 Modes of Operation of the BJT

Cutoff

When the BJT is in the cutoff mode both the emitter-base junction and collector-base junction are reverse biased. No current flows in any terminal of the transistor. The voltages across the terminals of the transistor are set by the external circuit.

Saturation

When the BJT is saturated both the base-emitter and collector-base junctions are forward biased. The currents flowing into the terminals of the transistor are set by the external circuit and the parameters of the transistor. The voltages across the terminals of the transistor are functions of the current levels, the charge

8.2. THEORY

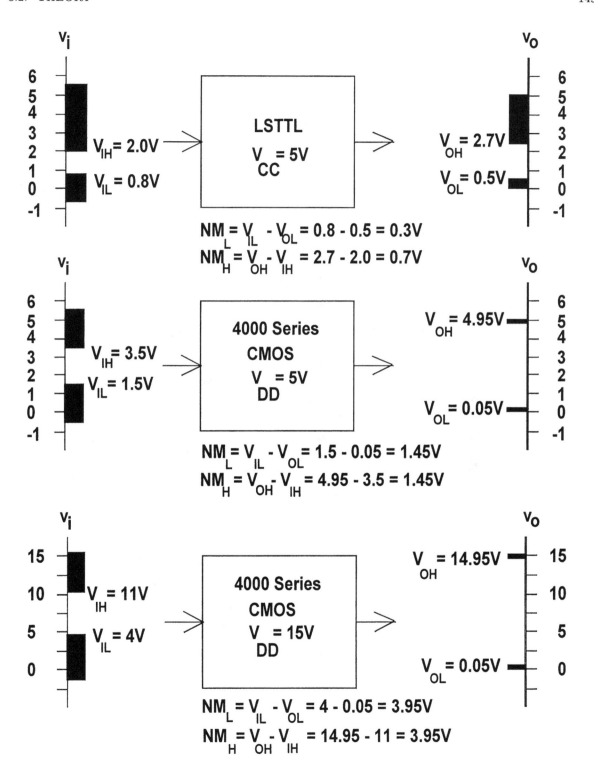

Figure 8.9: Logic levels for various logic families.

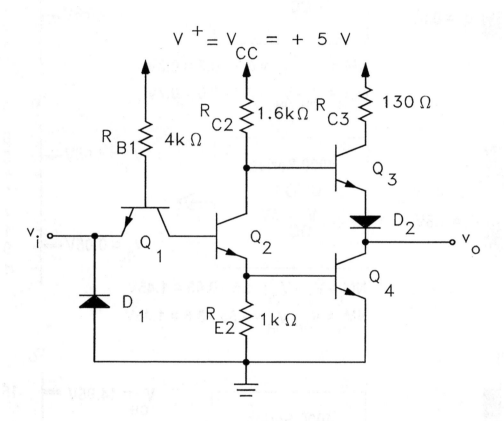

Figure 8.10: 7404 TTL INVERTER.

8.2. THEORY

stored in the base, and the external circuit. For simplicity's sake, it will be assumed that when the transistor is saturated that the base-to-emitter voltage is 0.7 V and that the collector-to-emitter voltage is 0.2 V; these voltages vary but the added precision in using different values for the voltage drops is outweighed by the complexity introduced by using more precise values.

Forward Active Mode

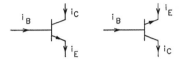

Figure 8.11: (a) Forward active mode (b) reverse active mode.

When the transistor is in the forward active mode, the base-emitter junction is forward biased and the base-collector junction is reverse biased. This mode of operation is shown in Fig. 8.11(a).

The relationship between the collector current and the base current is given by

$$i_C = \beta_F i_B \tag{8.9}$$

where β_F is the forward beta of the BJT. The emitter and collector currents are related by

$$i_C = \alpha_F i_E \tag{8.10}$$

where α_F is the forward alpha of the BJT. As shown in the figure all three terminal currents are positive. The parameters α_F and β_F are related by

$$\alpha_F = \frac{\beta_F}{\beta_F + 1} \tag{8.11}$$

Typical values of the forward beta and alpha for the BJTs used in TTL are 50 and 0.98 respectively.

Reverse Active Mode

. When the transistor is in the reverse active mode the roles of the emitter and collector are exchanged as shown in Fig. 8.11 (b). The emitter and base currents are related by

$$i_E = \beta_R i_B \tag{8.12}$$

where β_R is the reverse beta of the transistor. The collector and emitter currents are related by

$$i_E = \alpha_R i_C \tag{8.13}$$

where α_R is the reverse alpha of the transistor. The reverse alpha and beta are related by

$$\alpha_R = \frac{\beta_R}{\beta_R + 1} \tag{8.14}$$

Because the collector region of a BJT is physically much larger than the emitter, the forward beta is much larger than the reverse beta. Typical values for the reverse beta and alpha are 1 and 0.5 respectively.

Saturation also occurs in the reverse active mode. Both the emitter-base and collector-base junctions are forward biased. The difference between saturation in the forward and reverse active modes is the direction of the flow of current through the transistor. The direction of the flow of current for the forward active mode is indicated by the arrow on the emitter and for the reverse active mode the direction is reversed. The collector-to-emitter voltage when the transistor is in the reverse active mode and saturated may be as small as a fraction of the millivolt; however, the removable of excess minority charge from the base requires even longer than the forward active mode. Hence, the saturation in the reverse active mode is rarely used.

In the development that follows it will be assumed that $\beta_F = 100$, $\beta_R = 0.2$, $V_{CE(sat)} = 0.2\,\text{V}$, and $V_{BE(on)} = 0.7\,\text{V}$. These parameters are variable from transistor to transistor and are functions of the current levels and temperatures of the transistor. However, these values are representative and reasonable accurate answers can be obtained by assuming that they are applicable.

8.2.8 Standard 7404 TTL Inverter

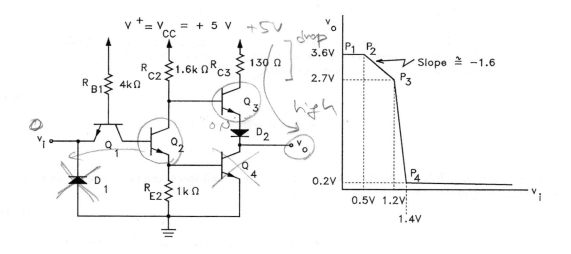

Figure 8.12: 7404 Standard TTL INVERTER.

The circuit shown in Fig. 8.12 is the standard 7404 TTL inverter along with its static transfer characteristic. It can be divided into three sections: input, phase splitter, and output stages. The input stage consists of diode D_1, transistor Q_1, and resistor R_{B1}. The phase splitter consists of transistor Q_2, resistor R_{C2}, and resistor R_{E2}. The output stage consists of transistor Q_3 and Q_4, diode D_2, and resistor R_{C3}. The arrangement of the transistors and the diode in the output stage is knows as a totem pole topology due to its slight resemblance to the totem poles used by certain primitive cultures.

Transistor Q_1 is used to control the direction of current flowing into the base of transistor Q_2; it is used to rapidly saturate or turn off transistor Q_2 by either flooding the base of Q_2 with excess charge carriers or sucking them out. Diode D_1 is used to prevent the input from going below one diode drop below ground. The phase splitter transistor Q_2 is used to turn one transistor in the output stage on and the other off. Thus in the output stage Q_4 is either on or off and transistor Q_3 is in the complementary state. This analysis of this circuit requires an examination of it when it is in the two states corresponding to logic highs and lows and the transition between these states.

8.2. THEORY

TTL Inverter with Input Low

The input is low when it is grounded which makes $V_{E1} = 0\,\text{V}$ or when it is being driven by a similar low circuit in the low state which makes $V_{E1} = 0.2\,\text{V}$. In either case the base-emitter PN junction of transistor Q_1 is forward biased and excess charge carriers are rapidly removed from the base of transistor Q_2 by transistor action which turns it off and reduces the base current in transistor Q_2 to zero. Transistor Q_1 then enters the saturation mode with $V_{CE1} \cong 0.2\,\text{V}$ and $I_{C1} \cong 0$. This makes the voltage at the base of transistor Q_2, $V_{B2} = v_i + V_{CE2} = 0.2 + 0.2 = 0.4\,\text{V}$, which is insufficient to turn transistor Q_2 on. When the input is low a current equal to

$$I_{E1} = (V^+ - V_{BE1} - v_i)/R_{B1} \cong 1\,mA \tag{8.15}$$

flows out of the input to ground.

When transistor Q_2 is off both the emitter and collector currents are zero which means that the base current of Q_4 is also zero. Therefore, when the input is low both transistors Q_2 and Q_4 are off. If transistor Q_3 is in the active mode, the output voltage is then given by

$$v_o = V^+ - \frac{i_o}{\beta_F + 1}R_{C2} - V_{BE3} - V_\gamma \tag{8.16}$$

where i_o is the output current and V_γ is the voltage drop across diode D_2. When the inverter is unloaded, i.e. $i_o = 0$, the output voltage is given by

$$v_o = 5 - 0.7 - 0.7 = 3.6\,\text{V} \tag{8.17}$$

As the load on the inverter increases the output current increases and the output voltage decreases according to Eq. 8.16. Eventually, transistor Q_3 becomes saturated and the output voltage is given by

$$v_o = V^+ - i_o R_{C3} - V_{CE3(\text{sat})} - V_\gamma \tag{8.18}$$

TTL Inverter when Input is High

When the input is high transistor Q_1 is operating in its reverse active mode. Current flows through R_{B1} and into the base of transistor Q_2 which causes both it and transistor Q_4 to saturate. In the reverse active mode the voltage drop from base-to-collector is approximately 0.7 V which makes the voltage at the base of Q_1

$$V_{B1} = V_{BC1} + V_{BE2} + V_{BE4} \cong 0.7 + 0.7 + 0.7 = 2.1\,\text{V} \tag{8.19}$$

and the currents

$$I_{B1} = (V^+ - V_{B1})/R_{B1} \cong 0.73\,\text{mA} \tag{8.20}$$

$$I_{E1} = \beta_R I_{B1} \cong 150\,\mu\text{A} \tag{8.21}$$

$$I_{C1} = (\beta_R + 1)I_{B1} \cong 0.87\,\text{mA} \tag{8.22}$$

where the assumed positive direction of the currents in the reverse active mode are shown in Fig. 8.11b.

Since both transistor Q_2 and Q_4 are saturated, the voltage at the base of transistor Q_3 is

$$V_{B3} = V_{BE4} + V_{CE2(\text{sat})} \cong 0.7 + 0.2 = 0.9\,\text{V} \tag{8.23}$$

which is insufficient to turn transistor Q_3 on. The purpose of diode D_2 is to prevent transistor Q_3 from being turned on when transistor Q_4 is on. Had diode D_2 been omitted transistor Q_3 would be on whenever transistor Q_4 is on.

Whenever Q_4 is saturated the output voltage $v_o = V_{CE4(sat)} \cong 0.2\,\text{V}$. This will be the case as long as the output current is less than $\beta_F I_{B4}$. The current I_{B4} is given by

$$\begin{aligned} I_{B4} &= I_{E2} - V_{BE4}/R_{E2} = I_{C2} + I_{B2} - V_{BE4}/R_{E2} \\ &= (V^+ - V_{C2})/R_{C2} + I_{B2} - V_{BE4}/R_{E2} \cong 2.73\,\text{mA} \end{aligned} \tag{8.24}$$

which sets the upper limit on the load current as $i_o \cong \beta_F I_{B4} = 273$ mA. For currents larger than this the output voltage will rise above $V_{CE4(sat)} \cong 0.2$ V as Q_4 leaves saturation.

When the output is low the output of the 7404 inverter sinks current (the load current flows from the external circuit into the inverter). Current flows from the external circuit through transistor Q_4 to ground since both the diode D_2 and transistor Q_3 are off. When the output is high the output of this inverter sources current (the load current flows through transistor Q_3 and diode D_2 and out through the external circuit). The situation is reversed at the input, i.e. when the input is high current flows into the inverter from the external circuit and when the input is low current flows from the inverter to the external circuit. Were this not the case circuits such as these could not be cascaded.

Standard TTL Inverter Transfer Characteristic

The static transfer characteristic for the standard TTL inverter is shown in Fig. 8.12b. As the input varies from 0 to 0.5 V the characteristic varies from point P_1 to point P_2. At point P_2 transistor Q_2 begin to turn on and the output starts to drop. The slope m of the line from point P_2 to point P_3 is given by

$$m = -\frac{R_{C2}}{R_{E2+}} r_{e2} \cong -1.6 \qquad (8.25)$$

if the intrinsic emitter resistance r_{e2} is neglected. At point P_3 transistor Q_4 starts to turn on and the output voltage is rapidly reduced to $V_{CE4(sat)} \cong 0.2$ V at point P_4. It should be borne in mind that the actual transfer characteristic does not consist of straight line segments but is a smooth curve.

When the output switches from low to high transistor Q_4 remains saturated until the excess charge stored in the base of Q_4 can be removed through the resistor R_{E2}. While this is happening transistor Q_3 could pass from the off state to the saturation state which means that the voltage across the resistor R_{C3} is approximately the power supply voltage minus one diode drop and two $V_{CE9(sat)}$. This case cause huge current to be drawn from the power supply which this circuit changes from the low to the high state. Indeed, the sole purpose of resistor R_{C3} is to limit the current spikes when the output is changing states. These current spikes can wreck havoc on other digital circuits such as sequential logic. Therefore, it is a common practice to liberally place power supply decoupling capacitors near the power supply pins of the 7400 logic family to shunt these current spikes to ground.

From the transfer characteristic shown in Fig. 8.12b it would appear that

$$V_{IL} \cong 0.5\,\text{V} \qquad (8.26)$$

$$V_{IH} \cong 1.4\,\text{V} \qquad (8.27)$$

$$V_{OL} \cong 0.2\,\text{V} \qquad (8.28)$$

$$V_{OH} \cong 3.6\,\text{V} \qquad (8.29)$$

which would result in noise margins of

$$NM_L = V_{IL} - V_{OL} = 0.5 - 0.2 = 0.3\,\text{V} \qquad (8.30)$$

$$NM_H = V_{OH} - V_{IH} = 3.6 - 1.4 = 2.2\,\text{V} \qquad (8.31)$$

which is realistic for the NM_L but NM_H was computed for no load and is larger than found on manufacturers' data sheets.

8.2. THEORY

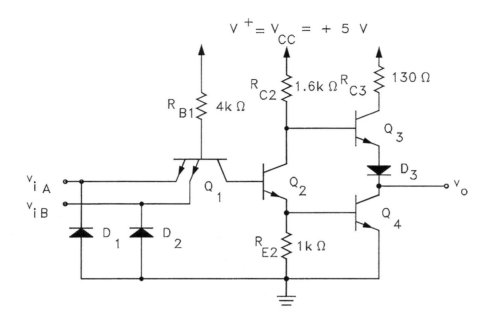

Figure 8.13: Standard TTL NAND gate.

8.2.9 Standard TTL NAND Gate

The circuit shown in Fig. 8.13 is the 7400 standard TTL NAND gate. The input stage consists of a transistor with two emitters which is equivalent to taking two transistors and placing their bases and collectors in parallel. If either input is low, current is shunted to ground through transistor Q_1 and both transistors Q_2 and Q_4 are off which makes the output high. If both inputs are high, current flows into the base of Q_2 and Q_4 which saturates both transistors and the output is low. This circuit, therefore, implements a logical NAND operation. If either input is connected to V^+ this circuit may be used as an inverter.

The standard TTL family is relatively slow because logical lows are obtained by saturating BJTs. When switching to the high state the excess charge must be removed from the base of the transistor. Faster logic can be obtained using BJTs by employing circuits which do not saturate. The fastest of all such logical families is emitter coupled logic which employs differential amplifiers in which the tail current is switched between the transistors. Another fast logic family using BJTs is the LSTTL which uses Schottky transistors which will be examined next.

8.2.10 Low-Power Schottky TTL

The standard TTL logic family was introduced in 1965. Although still available in the 1990s, it has been almost totally supplanted by variants using Schottky-clamped transistors. These logic family exhibit improved performance because Schottky-clamped transistors cannot saturate. One of the most popular of these families is the low-power Schottky-clamped TTL logic which has propagation times identical to that of standard TTL but with a power dissipation of only about one fifth which is examined in this experiment.

A Schottky diode consists of a metal semiconductor junction. It has a rectifying characteristic similar to that of a semiconductor junction diode but there is no mobile carrier storage in the junction. This permits these diodes to switch from the forward biased to reverse biased states rapidly. The voltage drop across

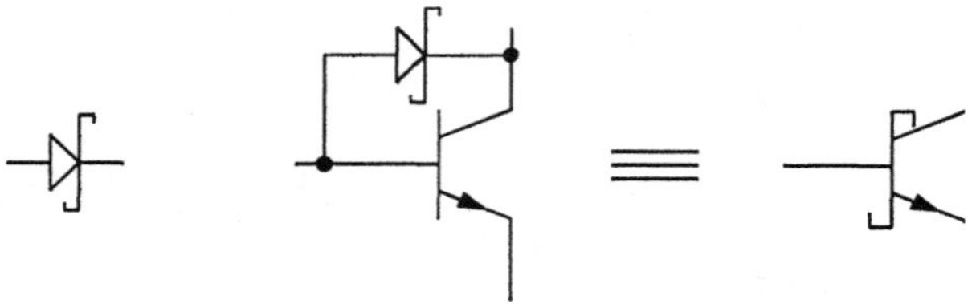

Figure 8.14: (a) Schottky diode (b) Schottky-clamped transistor.

a forward biased Schottky diode is approximately is 0.5 V rather than the 0.7 V found with semiconductor diodes. Thus, for the Schottky diode the on-voltage, V_γ, is 0.5 V. The circuit symbol for the Schottky diode is shown in Fig. 8.14a.

A Schottky-clamped transistor is formed by placing a Schottky diode across the base-collector PN junction of a BJT. The diode's anode is connected to the base and the cathode to the collector. This arrangement is shown in Fig. 8.14b along with the circuit symbol for the Schottky-clamped transistor. For large collector-to-emitter voltages the Schottky diode is reverse biased and is, therefore, an open circuit and plays no role in the performance of the transistor. If the collector-to-emitter voltage drops to a sufficiently small voltage, the Schottky diode becomes forward biased and shunts current from the base terminal to the collector; it, therefore, prevents the transistor from entering saturation. When the collector-to-emitter voltage drops to a small enough value the Schottky diode clamps the collector-to-emitter voltage to a value equal to the base-to-emitter voltage minus the on voltage of the Schottky diode which is approximately $0.7 - 0.5 \cong 0.2$ V. Since the Schottky-clamped transistor cannot saturate, there is no excess charge to be removed from the base when the transistor switches from the low to the high state which means that this transistor can switch much more rapidly than the saturating BJT.

The circuit shown in Fig. 8.15 is the 74LS00 low-power Schottky-clamped NAND gate. This circuit has two inputs and one output. When both inputs are high the output is low and if one or both inputs are low the output is high. Thus this circuit implements the logical NAND operation in positive logic. If either of the input is removed, this circuit becomes the 74LS04 inverter.

The transistor Q_1 if Fig. 8.15 is a phase splitter which serves the same purpose as the phase splitter in the standard TTL circuit. Since transistor Q_1 is a Schottky-clamped transistor and cannot saturate, an input transistor is not required to remove excess charge carriers from its base. Simple diodes, albeit Schottky diodes, can then be used as the input devices.

The transistors Q_2 and Q_3 form a Darlington transistor or configuration. This compound transistor can produce much larger currents than a single transistor. It replaces the single pull-up transistor in the standard TTL circuit. Because the collector-to-emitter voltage of transistor Q_3 is given by the sum of its base-to-emitter voltage and the collector-to-emitter voltage of Q_2, Q_3 can never saturate and it not required to be Schottky-clamped. The diode in the totem pole output of the standard TTL circuit is no longer required because the base-to-emitter voltage of Q_2 assures that Q_3 will never be on when Q_4 is on.

The emitter resistor for the phase splitter transistor in the standard TTL family is replaced by the transistor Q_3 with its associated resistors R_{E1} and R_{C3}. This is knows as a squaring circuit because it "squares" the voltage transfer characteristic of the device. The use of this circuit assures that Q_3 and Q_4 come on simultaneously. This circuit also provides for a more rapid discharge of transistor Q_4 when the circuit is in the low state.

When the device is in either the high or low state the diodes D_5 and D_6 are reverse biased and play no role in the operation of the circuit. These diodes come into play when the output is switching from the high to the low state. The discharge of the base of Q_3 is primarily through diode D_5 and, if the circuit is loaded,

8.2. THEORY

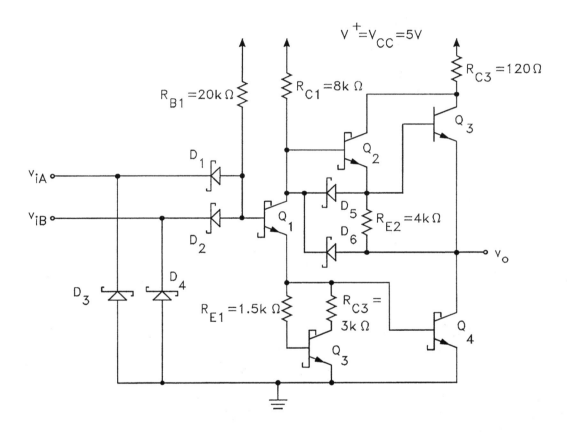

Figure 8.15: Low-power Schottky-clamped NAND gate (74LS00).

the load discharges through diode D_6. These discharge current flow into the collector of Q_1 and enhance this current which increases its emitter current which then assists in turning on transistors Q_4 and Q_3. Diodes D_3 and D_4 serve the same purpose as with the standard TTL circuit, i.e. they prevent the input from going below 0.5 V which could have a disastrous effect on the IC if this forward biased the isolation diodes to the substrata.

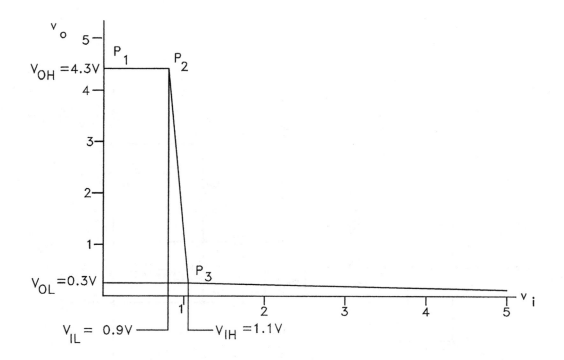

Figure 8.16: Transfer characteristic for TTL LS INVERTER.

The transfer characteristic for the TTL LS inverter is shown in Fig. 8.16 (this is the circuit shown in Fig. 8.15 with input B connected to V^+). When the input is low diode D_1 is forward biased which places the base of Q_1 at 0.5 V, the on-voltage of the Schottky diode. This turns transistors Q_1, Q_3, and Q_4 off. The output voltage is then

$$v_o = V^+ - I_{C1}R_{C1} - V_{BE2} - (I_{E2} - I_{B3})R_{E2} \tag{8.32}$$

which means if the device is unloaded and the currents are negligible $v_o \cong 5 - 0.7 = 4.3$ V. This is point P_1 on the characteristic.

As the input voltage to the inverter increases eventually both transistors Q_3 and Q_4 turn on. This is point P_2 on the characteristic. The output voltage is still $v_o = V_{OH} = 4.3$ V but the input is

$$v_i = V_{IL} = V_{BE4} + V_{BE1} - V_\gamma \cong 0.7 + 0.7 - 0.5 = 0.9 \text{ V} \tag{8.33}$$

As the input continues to increase transistors Q_4 and Q_1 enter a state close to saturation for which the base-to-emitter voltages is somewhat larger than 0.7 V which will be assumed to be 0.8 V. The output voltage is then given by

$$v_o = V_{OL} = V_{BE4} - V_\gamma \cong 0.8 - 0.5 = 0.3 \text{ V} \tag{8.34}$$

8.2. THEORY

and the corresponding input

$$v_i = V_{BE4} + V_{BE1} - V_\gamma \cong 0.8 + 0.8 - 0.5 = 1.1\,\text{V} \tag{8.35}$$

which is point P_3 on the characteristic.

8.2.11 CMOS Inverter

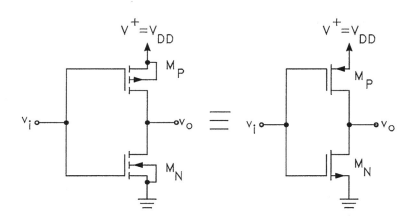

Figure 8.17: CMOS INVERTER.

The CMOS inverter is shown in Fig. 8.17. All MOSFETs are enhancement mode devices with the bulk connected to the source. An n-channel device is used as the output transistor and a p-channel device is used as an active pull-up load. The inputs are constrained to be the power supply rails, e.g. 0 or ground and $V^+ = V_{DD}$. Unlike TTL logic where the power supply voltage is constrained to be 5 V any number of power supply voltages may be used with most CMOS families (there are CMOS families designed to be compatible with TTL and, therefore, are constrained to a power supply voltage of +5 V).

When the input is zero or low the gate of both the n-channel and p-channel devices are zero. This makes gate-to-source voltage of the n-channel device 0 and the gate-to-source voltage of the p-channel device V^+ which means that the n-channel device is off and the p-channel device is on. The output (drain-to-source voltage) of the n-channel device is an open circuit and the output of the p-channel device is a short circuit which makes the output V^+. Therefore, when the input is low the output is high.

When the input is V^+ the gate voltage of both the n-channel and p-channel devices are V^+. This makes the gate-to-source voltage of the n-channel device V^+ which turns it on and makes its output (the drain-to-source voltage) zero. The gate-to-source voltage of the p-channel device is zero which makes its output an open circuit. Therefore, when the input is high the output is low. Therefore, this circuit implements logical inversion which makes its an inverter.

The static voltage transfer characteristic is almost identical to that of the ideal inverter. The transition point between the low and high output states occurs when the input is approximately on half of the power supply voltage (assuming that the n-channel and p-channel devices have matched characteristics). The parameters V_{OH} and V_{0L} are approximately

$$V_{OH} \cong V^+ = V_{DD} \tag{8.36}$$

$$V_{OL} \cong V^- = 0 \tag{8.37}$$

To determine the parameters V_{IL} and V_{IH} the drain currents are equated and, then, making use of the fact that the output voltage is the drain-to-source voltage of the n-channel device and also the power supply voltage minus the drain-to-source of the p-channel device the slope dv_o/dv_i is found and equated to -1. After a fusillade of algebra

$$V_{IL} = \frac{3V^+ + 2V_{TO}}{8} \qquad (8.38)$$

$$V_{IH} = \frac{5V^+ - 2V_{TO}}{8} \qquad (8.39)$$

where V_{TO} is the threshold voltage of the n-channel device and V^+ is the power supply voltage. This makes the noise margins

$$NM_L = V_{IL} - V_{OL} = \frac{3V^+ + 2V_{TO}}{8} \qquad (8.40)$$

$$NM_H = V_{OH} - V_{IH} = \frac{3V^+ + 2V_{TO}}{8} \qquad (8.41)$$

which are equal due to the symmetry of the circuit and the matched characteristics of the n-channel and p-channel devices.

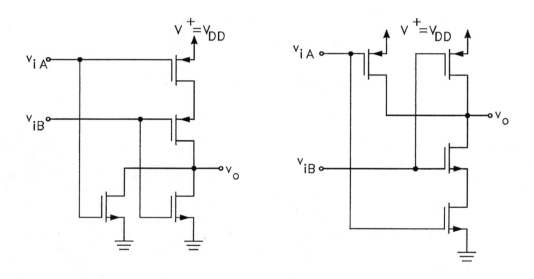

Figure 8.18: (a) CMOS NOR (b) CMOS NAND.

A CMOS NOR gate is shown in Fig. 8.18 (a). When either input is high the n-channel device is on which makes the drain-to-source voltage zero and, therefore the output zero. The p-channel device connected to a high input is off and an open circuit exists between the power supply and the output. If both inputs are high, the same situation exists. When both inputs are low both n-channel devices are off or open circuits and the p-channels devices are on and a short circuit exists between the power supply and the output which makes the output high. Therefore, when either or both inputs are high the output is low and if both inputs are low the output is high. This makes this circuit a logical NOR.

A CMOS NAND gate is shown in Fig. 8.18(b). If both inputs are high the two n-channel transistors are on and there is a short from the output to ground and both p-channel transistors are off and there is no path from the output to the power supply. When either input is low the n-channel device connected to it is

8.2. THEORY

off and the p-channel device is on which means that there is no path from ground to the output but there is a short from the power supply to the output which makes the output high. Therefore, the output is high unless both inputs are high which makes this a NAND gate.

The gates shown in Fig. 8.18 have two inputs. The extension of these circuits to obtain NOR and NAND gates with more than two inputs is obvious. Note that with the NOR gate the pull-up active loads, the p-channel devices, are in series and the switching transistors, the n-channel devices, are in parallel. To obtain NAND gates the series parallel configuration is reversed.

The CMOS device that will be examined in this experiment is the CD4007 or MC14007 CMOS inverter from the 4000 series of the off-the-shelf digital ICs. This is a metal gate device that was first introduced in the mid 1960s. Although still in use, there are many CMOS families available with superior characteristics. Indeed, one day CMOS may totally replace the other logic families.

8.2.12 Ring Oscillator

Modern logic gates have propagation times that are so small that the direct measurement of them with common laboratory instruments is difficult if not impossible. Therefore, this parameter of digital electronic circuits is usually measured indirectly by constructing a ring oscillator which has a frequency of oscillation that is a function of the propagation delay. Once the oscillation frequency is measured, the propagation delay may be calculated.

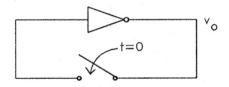

Figure 8.19: Ring oscillator.

Shown in Fig. 8.19 is a ring oscillator. Prior to the closing of the switch at time $t = 0$ it will be assumed that the input is high and the output is low. Once the switch is closed a low is connected to the input of the inverter which means that the output will switch to high t_{PLH} second later. Once the output is high the input is also high which means that the output will switch to low t_{PHL} later. The inverter is now in the same state as when the switch was closed at $t = 0$ which means that this process will continue ad infinitum. Thus this circuit is an oscillator with a period, T, of

$$T = t_{PLH} + t_{PHL} \tag{8.42}$$

If N inverters are placed in the ring where N is an odd integer, the period becomes

$$T = N(t_{PLH} + t_{PHL}) \tag{8.43}$$

But, since the propagation time, t_P, is given by

$$t_P = \frac{t_{PLH} + t_{PHL}}{2} \tag{8.44}$$

this makes the period

$$T = 2N t_P \tag{8.45}$$

and since the frequency of oscillation, f_o, is the reciprocal of the period

$$t_P = \frac{1}{2Nf_o} \tag{8.46}$$

For practical reasons N is usually picked to be large so that the frequency of oscillation can be easily measured.

8.3 Preliminary Derivations

1. Derive Eq. (8.8).

8.4 Preliminary Calculations

1. Calculate the noise margins for a CMOS INVERTER with a power supply voltage of 5 V and a threshold voltage of 1 V for the n-channel device and -1 V for the p-channel device.

2. Repeat Prob. 1 if the power supply voltage is changed to 12 V.

8.5 Preliminary SPICE Simulations

1. Perform a dc sweep of the circuit shown in Fig. 8.12. In the SPICE MODEL statement for the NPN BJTs use the parameters: $IS = 1F$, $BF = 100$, $BR = 0.2$, $CJE = 1P$, $CJC = 0.5P$, $TF = 100P$, and $TR = 100U$.

2. Perform a transient analysis of the BJT inverter shown in Fig. 8.7. The input is to be a 100 kHz square wave with a dc level of 2.5 V, a peak-to-peak value of 5 V, and a frequency of 100 kHz. Use $R_C = 2.2\,\text{k}\Omega$ and $R_B = 270\,\text{k}\Omega$. Plot the collector and base voltage as well as the base and collector currents as functions of time for two cycles of the input.

3. Repeat Prob. 2 if the frequency of the input is changed to 10 kHz.

4. Repeat Prob. 2 if the frequency of the input is changed to 1 MHz.

8.6 Experimental Procedures

8.6.1 Preparation

(a) Prepare the electronic breadboard to provide buses for the positive and negative power supply rails and the circuit ground. Each power supply rail should be decoupled with a $100\,\Omega$ 1/4 W resistor, and a $100\,\mu\text{F}$ 25 V (or greater) capacitor. The resistors are connected in series with the external power supply leads and the capacitors are connected from power supply rail to ground on the circuit side of the resistors. The capacitors must be installed with proper polarity to prevent reverse polarity breakdown.

8.6.2 BJT Inverter

(a) Transfer Characteristic

Assemble the circuit shown in Fig. 8.20 with $Z_L = \infty$ (open circuit) and the dc power supply OFF. Use the fixed 5 V output on the dc power supply. The lead layouts for the 2N3904 NPN BJT are given in Exp. 7. The diode is a 1N4148 small signal diode whose sole purpose in this circuit is to protect the transistor from an inadvertent application of a negative voltage to the base of the transistor which could destroy it. Set the function generator to produce the waveform shown in the (a) portion of Fig. 8.20, i.e. a triangular wave with a peak-to-peak of 5 V, a dc level of 2.5 V, and a frequency of 30 Hz. Turn the dc power supply ON.

8.6. EXPERIMENTAL PROCEDURES

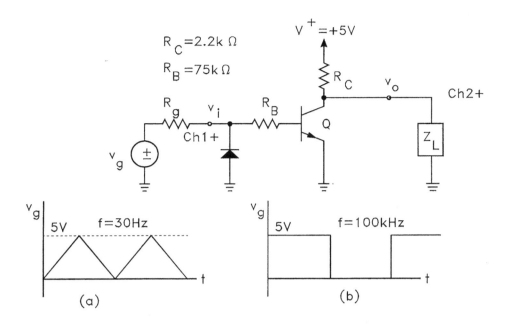

Figure 8.20: BJT INVERTER.

Set the oscilloscope for x-y operation. Use the x and y position controls to set the origin for the plot near the lower left portion of the CRT screen. Connect Ch. 1 to v_i and Ch. 2 to v_o. Both channels should be dc coupled. Set the Volts/Div for each channel to a value that adequately displays v_o versus v_i. Print or sketch, to scale, this plot. From this plot determine V_{IL}, V_{IH}, V_{OL}, V_{OH}, NM_L, and NM_H. (Reference page no. 139.)

Change Z_L to a 6.8 kΩ resistor to ground and repeat the above.

(b) Terminal Waveforms

Set the function generator to produce the waveform shown in the (b) portion of Fig. 8.20, i.e. a 100 kHz square wave with a peak-to-peak value of 5 V and a dc level of 2.5 V. Use $Z_L = \infty$ for this step. Press Auto-scale or set the time per division to display one or two cycles of v_i and v_o versus t. Print or sketch the waveforms to scale. From the sketch determine the rise time, t_r; the fall time, t_f; and the storage time, t_s for the transistor. (Reference page no. 145.)

Place a 300 pF capacitor across the resistor R_B and note the effect on the sketch. Remove the capacitor and place a diode from the collector to the base of the transistor and note the effect on the plot. (The anode of the diode is connected to the base and the cathode to the collector. This simulates a Schottky diode.)

8.6.3 Standard TTL Inverter

Assemble the circuit shown in Fig. 8.21 with the dc power supply turned OFF. (For the 4 kΩ resistor place a 3.9 kΩ resistor in series with a 100 Ω resistor.) Both diodes are 1N4148 small signal switching diodes and all the NPN BJTs are 2N3904s.

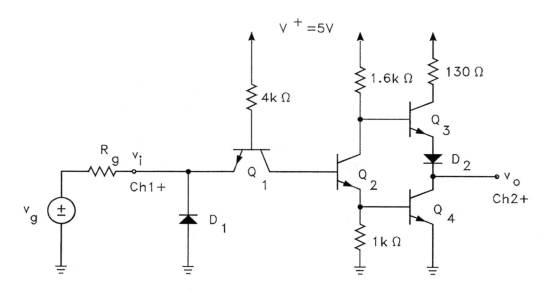

Figure 8.21: Standard TTL INVERTER.

(a) Transfer Characteristic

Set the function generator to produce the waveform shown in the (a) portion of Fig. 8.20, i.e. a triangular wave with a peak-to-peak of 5 V, a dc level of 2.5 V, and a frequency of 30 Hz. Turn the dc power supply ON.

Set the oscilloscope for x-y operation. Use the x and y position controls to set the origin for the plot near the lower left portion of the CRT screen. Connect Ch. 1 to v_i and Ch. 2 to v_o. Both channels should be dc coupled. Set the Volts/Div for each channel to a value that adequately displays v_o versus v_i. Print or sketch, to scale, this plot. If oscillations occur (fuzzy portion on display), place a 100 pF capacitor from the base of Q_2 to ground.) From this plot determine V_{IL}, V_{IH}, V_{OL}, V_{OH}, NM_L, and NM_H.

(b) Terminal Waveforms

Set the function generator to produce the waveform shown in the (b) portion of Fig. 8.20, i.e. a 100 kHz square wave with a peak-to-peak value of 5 V and a dc level of 2.5 V. Turn both Chs. 1 and 2 ON and press Autoset or set the time per division to display one or two cycles of v_i and v_o versus t. Print or sketch the waveforms to scale. Also sketch the collector voltage of Q_3 versus t (explain the spikes in this sketch in the laboratory report). Increase the frequency of the square wave to 1 MHz and repeat.

8.6.4 74LS TTL Inverter

Assemble the circuit shown in Fig. 8.22 with the dc power supply turned OFF. The inverter is one of six contained on the 74LS04 IC. Connect pin 14 of this IC to the +5 V supply and pin 7 to ground. Turn on the dc power supply.

(a) Transfer Characteristic

Set the function generator to produce the waveform shown in the (a) portion of Fig. 8.20, i.e. a triangular wave with a peak-to-peak of 5 V, a dc level of 2.5 V, and a frequency of 30 Hz. Turn the dc power supply

8.6. EXPERIMENTAL PROCEDURES

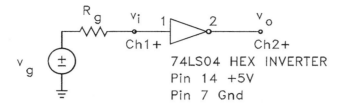

Figure 8.22: 74LS TTL INVERTER.

ON.

Set the oscilloscope for x-y operation. Use the x and y position controls to set the origin for the plot near the lower left portion of the CRT screen. Connect Ch. 1 to v_i and Ch. 2 to v_o. Both channels should be dc coupled. Set the Volts/Div for each channel to a value that adequately displays v_o versus v_i. Print or sketch, to scale, this plot. From this plot determine V_{IL}, V_{IH}, V_{OL}, V_{OH}, NM_L, and NM_H.

(b) Terminal Waveforms

Set the function generator to produce the waveform shown in the (b) portion of Fig. 8.20, i.e. a 100 kHz square wave with a peak-to-peak value of 5 V and a dc level of 2.5 V. Press Autoset or set the time per division to display one or two cycles of v_i and v_o versus t. Print or sketch the waveforms to scale. Increase the frequency of the square wave to 1 MHz and repeat.

(c) Power Dissipation

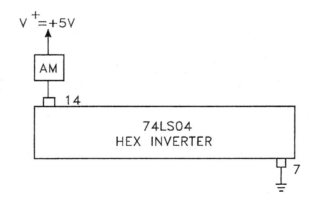

Figure 8.23: Power dissipation measurement.

With the function generator disconnected from the circuit, insert an ammeter in series with the power supply as shown in Fig. 8.23. Compute the total power dissipated in the IC as the product of the power supply voltage (+5 V) and the current flowing into the IC. Since there are six inverters on this IC, the average power dissipation per gate, P_D, is this power divided by 6; this is the static power dissipation. Set the function generator to produce the waveform shown in Fig. 8.20b with a frequency of 100 kHz. Connect the output of the function generator to pin 1 of the IC and measure and record the power dissipation. Change the frequency to 1 MHz and repeat. (Since only one inverter is being switched, the increase in the power dissipation is due to the dynamic power dissipation per gate.)

(d) Propagation Delay

The propagation delay could be measured by observing the delay between the input and output terminal waveforms. This would be difficult to do with the equipment available for this experiment. Instead, the propagation delay will be determined indirectly by constructing a ring oscillator with an frequency of oscillation that is a function of the propagation delay. The ring oscillator is a relaxation oscillator which contains an odd number of inverters in the feedback path. The frequency of oscillation is given by

$$f_o = \frac{1}{2Nt_P} \qquad (8.47)$$

where N is the number of inverters inside the feedback loop and t_P is the propagation delay.

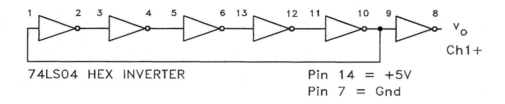

Figure 8.24: 74LS TTL ring oscillator.

Assemble the circuit shown in Fig. 8.24 with the power OFF. Note that since this is an oscillator, there is no input; it only has an output. Note that pin 14 must be connected to the +5 V supply and pin 7 must be grounded. Turn on the power and measure the frequency of oscillation, f_o, in Hertz with the oscilloscope. Use Eq. 8.47 to compute the propagation delay, t_P. Compute the power-delay product, $t_P P_D$, and express it in picojoules.

Measure the power dissipation for the ring oscillator. Why is it much higher than the static dissipation?

8.6.5 CMOS Inverter

(a) Transfer Characteristic

The CMOS inverter that will be used for this experiment is the MC14007 or CD4007. Assemble the circuit shown in Fig. 8.25 with the power OFF. The sole purpose of the 330 kΩ resistor is to protect the device from the possible application of a negative voltage to the gate. Note that the power supply voltage is +10 V.

Set the function generator to produce the same type of waveform as shown in the (a) portion of Fig. 8.20, i.e. a triangular wave but with a peak-to-peak of 10 V, a dc level of 5 V, and a frequency of 30 Hz. Turn the dc power supply ON.

Set the oscilloscope for x-y operation. Use the x and y position controls to set the origin for the plot near the lower left portion of the CRT screen. Connect Ch. 1 to v_i and Ch. 2 to v_o. Both channels should be dc coupled. Set the Volts/Div for each channel to a value that adequately displays v_o versus v_i. Print or sketch, to scale, this plot. From this plot determine V_{IL}, V_{IH}, V_{OL}, V_{OH}, NM_L, and NM_H.

8.7. LABORATORY REPORT

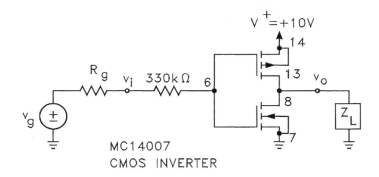

Figure 8.25: CMOS INVERTER.

(b) Terminal Waveforms

Set the function generator to produce the same type of waveform as shown in the (b) portion of Fig. 8.20, i.e. a 100 kHz square wave but with a peak-to-peak value of 10 V and a dc level of 5 V. Use $Z_L = \infty$ for this step. Press Autoset or set the time per division to display one or two cycles of v_i and v_o versus t. Sketch the waveforms to scale. Change Z_L to a 1 nF capacitor and repeat.

(c) Power Dissipation

Place an ammeter in series with the power supply and pin 14 of the IC. Set the function generator to produce a square wave with a lower level of 0 V, an upper level of 10 V, and a frequency of 100 kHz. Measure and record the power dissipation. Change the frequency to 15 MHz and repeat. Since there are 3 inverters on this IC, the average power dissipation per inverter is the total power dissipation divided by three.

(d) Propagation Delay

Assemble the circuit shown in Fig. 8.26. (Pins 14, 2, and 11 are connected to the positive power supply. Pins 7, 4, and 9 are grounded. Pins 13, 8, and 3 are connected together. Pins 1, 5, and 10 are connected together. Finally, pins 6 and 12 are connected together.) This is a ring oscillator with three inverters inside the feedback loop. Measure the frequency of oscillation and use Eq. 8.47 to compute the propagation delay. The output can be taken as the drain of any of the transistors. Compute the power delay product and express it in picojoules.

8.7 Laboratory Report

Turn in all the tables, sketches, printouts, and/or plots of experimental data taken. Turn in all the computed quantities requested in the procedure steps. Answer all questions posed in the procedure steps as well as those posed by the laboratory instructor. Include the verification sheet.

Why does TTL logic float high? Namely, with nothing connected to the input why does the circuit act as if the input were high? Hint: Review the circuit for the standard TTL inverter.

8.8 References

1. E. J. Angelo, *Electronics: BJTs, FETs, and Microcircuits*, McGraw-Hill, 1969.
2. W. Banzhaf, *Computer-Aided Circuit Analysis Using SPICE*, Prentice-Hall, 1989.

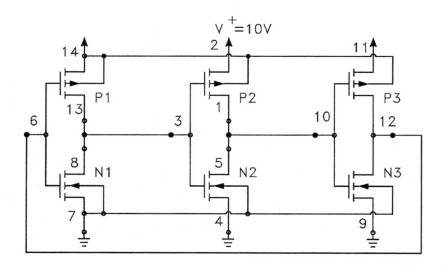

Figure 8.26: CMOS ring oscillator.

3. J. D. Greenfield, *Practical Digital Design Using IC's*, Wiley, 1977.
4. M. N. Horenstein, *Microelectronics Circuits and Devices*, Prentice-Hall, 1990.
5. P. Horwitz and W. Hill, *The Art of Electronics*, 2nd edition, Cambridge University Press, 1989.
6. P. Horowitz and I. Robinson, *Laboratory Manual for The Art of Electronics*, Cambridge University Press, 1981.
7. J. H. Krenz, *An Introduction to Electrical and Electronic Devices*, Prentice-Hall, 1987.
8. R. Mauro, *Engineering Electronics*, Prentice-Hall, 1989.
9. F. H. Mitchell and F. H. Mitchell, *Introduction to Electronic Design*, Prentice-Hall, 1988.
10. Motorola, *Small-Signal Semiconductors*, DL 126, Motorola, 1987.
11. C. J. Savant, M. S. Roden, and G. L. Carpenter, *Electronic Circuit Design*, Benjamin Cummings, 1987.
12. A. S. Sedra and K. C. Smith, *Microelectronics Circuits*, 4th edition, Oxford, 1998.
13. D. L. Schilling and C. Belove, *Electronic Circuits: Discrete and Integrated*, McGraw-Hill, 1968.
14. P. W. Tuinenga, *SPICE*, Prentice-Hall, 1988.
15. J. F. Wakerly, *Digital Design*, Prentice Hall, 1994.

Chapter 9

Analog to Digital and Digital to Analog Conversion Systems

9.1 Object

The object of the experiment is to examine analog-to-digital and digital-to-analog conversion. The specific analog-to-digital converter that is the subject of this experiment is the tracking A/D converter.

9.2 Theory

9.2.1 Introduction

Oftentimes it is more convenient, economical, and efficient to store, process, or transmit signals in digital rather than analog form. The plethora of cheap and powerful microprocessors makes it possible to perform complex signal processing on digitized signals which would be almost impossible if they were in analog form. Additionally, when signals are in digital form they are less susceptible to the deleterious effects of additive noise which makes it desirable to transmit and store them in digital form.

The devices that transform a signal from analog to digital form are known as analog-to-digital converters (acronyms in use are A/D and ADC). When the analog signal is to be recovered from its digital representation a device that performs the reciprocal of analog-to-digital conversion is required which is simply known as the digital-to-analog converter (acronyms in use and D/A and DAC).

This experiment will examine a simple analog-to-digital conversion system which uses a digital-to-analog converter as an integral part of the conversion process. It will be constructed in component form so that each component of the system can be examined individually.

9.2.2 Analog-to-Digital Converter Fundamentals

An analog quantity may assume any of an infinite number of values between its maximum and minimum values. (For simplicity it will be assumed that the analog quantity being discussed is a voltage. Any analog quantity can be converted into a voltage by the choice of an appropriate transducer). The difference between the maximum and minimum value of the voltage is known as the range voltage, $V_r = V_{\max} - V_{\min}$. In order to perform an analog-to-digital conversion, this range must be subdivided into intervals and a digital code word or data word assigned to each interval. Obviously, if these intervals have equal length, the accuracy of the conversion will be a monotonically increasing function of the number of the intervals and consequently the number of code words used to represent the digitized signal.

The number of code words used to represent a signal is dependent on the length of the code word. If the natural binary number system is used, then the length of the code word is equal to the number of bits of information that it contains. For an N bit binary code, the number of available code words is 2^N. If the analog signal is to be divided into equal intervals for digitization, then the length of each interval corresponding to a code word is $V_{\text{ref}}/2^N$ where V_{ref} is known as the reference voltage of the A/D converter (The reference is the largest change in the input voltage that can be encoded. It will be assumed that the range voltage of the input signal and reference voltage of the A/D converter are the same). The length of each of the intervals determines the resolution of the A/D converter (the resolution is the smallest change in the input voltage that guarantees that the output code word will change) and this resolution is given by

$$\Delta V = \frac{V_r}{2^N} \qquad (9.1)$$

where it is assumed that the range and reference voltage are the same (If it weren't a simple op amp circuit could be used to shift the level and make the variation of the input voltage to the A/D equal to the reference voltage.) The resolution represents the quantization error inherent in the conversion of the signal to digital form, i.e. it is impossible to reconstruct the analog signal from the digitized version with a precision greater than ΔV. The resolution is sometimes simply referred to as the LSB (least significant bit) in that it represents the voltage range corresponding to a change in the input code of one bit.

The dynamic range of an A/D converter is the ratio of the maximum to minimum voltage change (usually expressed in decibels) that the converter can encode and is given by $20\log(2^N)$ in decibels. This also provides a measure of the resolution of an A/D converter.

Another important parameter associated with A/D systems is the conversion time, sets a limit to the accuracy of the conversion process since the input signal must not change significantly while the conversion is in progress. If the signal changes by more than ΔV in a period of time equal to t_{conv}, then an error is guaranteed. This sets a limit to the ΔV in a period of time equal to t_{conv}, then an error is guaranteed. This sets a limit to the maximum frequency of a sinusoid that can be successfully digitized.

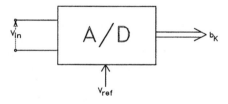

Figure 9.1: Analog to digital converter.

Fig. 9.1 illustrates a general A/D converter. As the input voltage varies from 0 to V_r the output code, b_k, varies from all zeros to all ones. Each code word b_k is of the form $b_k = (B_1 B_2 B_3 \ldots B_N)$ where each B_i is either 0 or 1 and B_1 is the most significant bit (MSB) and B_N is the least significant bit (LSB). This type of binary code is called simple, straight, or natural binary. As V_{in} varies from 0 to V_{ref} the output code increases from 0 to $2^N - 1$ in binary increments of 1.

Fig. 9.2 illustrates the transfer characteristic of a 2 bit A/D converter. For simplicity it is assumed that $V_{\max} = V_{\text{ref}} = V_r$ and that $V_{\min} = 0$. The input analog voltage is converted into a digital code word as

V_{in}	b_k
$0 - \Delta V$	00
$\Delta V - 2\,\Delta V$	01
$2\,\Delta V - 3\,\Delta V$	10
$3\,\Delta V - V_r$	11

9.2. THEORY

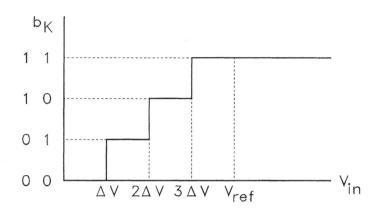

Figure 9.2: Transfer characteristic for 2 bit A/D.

the resolution is $\Delta V = 0.25 V_r$, and the dynamic range is 12 dB. A voltage less than zero would be encoded as all 0's and a voltage greater than V_r would be encoded as all 1's. Sophisticated A/D converters would also have indications of when the input voltage is above V_{ref} (over-range) and below 0 (under-range). Generalization to higher order (more bits) A/D converters is obvious and tedious. This type of A/D converter is known as an equal interval converter since the voltage intervals corresponding to each code word have equal lengths.

A variation of the above encoding process is also used in which the voltage levels are shifted by $\Delta V/2$. This would correspond to for the two bit A/D

V_{in}	b_k
$0 - .5\,\Delta V$	00
$0.5\,\Delta V - 1.5\,\Delta V$	01
$1.5\,\Delta V - 2.5\,\Delta V$	10
$2.5\,\Delta V - V_r$	11

which has the same resolution and dynamic range as the previous conversion scheme. Each interval for V_{in} has a length of ΔV except all 0's (length $\Delta V/2$) and all 1's (length $1.5\Delta V$). This conversion scheme has the advantage of being slightly more linear than the previous one. This is the encoding scheme that is normally used in industrial applications. For A/D converters with a large number of bits the difference between these two coding schemes is slight.

The number of bits used in practical systems is rarely less than 8. That is because many of the cheaper microprocessors employ 8 bits and the cheapest commercially available A/D converters therefore also employ 8 bits. This results in a resolution of $\Delta V = V_r/256$ and a dynamic range of 48 dB.

Compact disc recording systems employ 16 bit A/D converters to encode music. The results in a dynamic range of 96 dB which for sound corresponds to the difference between the largest and smallest sound that can be reproduced.

9.2.3 Digital-to-Analog Converters

Digital-to-analog converters (D/A) are required when it is desired to convert a digital signal to analog form. They are often integral parts of an A/D converter so it is logical to consider them first.

Code words are applied to the input of the D/A and an analog voltage is produced at the output. An N bit D/A converter produces one of 2^N analog output voltages for each of the possible 2^N input binary code words. Many of the terms, such as resolution and reference voltage, used to describe A/D converters are equally applicable to D/A converters.

The transfer characteristic of a D/A converter is normally the complement of the transfer function of the corresponding A/D converter. The analog output voltage is a monotonically increasing function of the input binary code word. Normally an input code word of all 0's would produce an output voltage of zero and an input of all 1's would produce an output voltage of $V_{\text{ref}}(1 - 2^{-N})$ which is the reference voltage minus the resolution. (If a bipolar output were desired, it could be obtained by using an op-amp as a level shifter.)

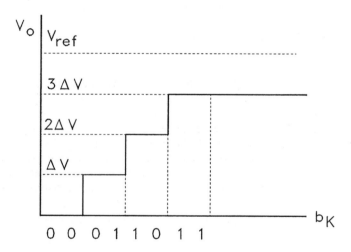

Figure 9.3: Transfer characteristic for 2 bit D/A converter.

Shown in Fig. 9.3 is the transfer characteristic of a 4 bit D/A. The output voltage is given as a function of the input code by

b_k	V_o
00	0
01	ΔV
10	$2\,\Delta V$
11	$3\,\Delta V$

where ΔV is the resolution of the D/A converter given by $\Delta V = V_{\text{ref}}/2^N = V_{\text{ref}}/4$. It is important to note that an input code of all 1's does not produce the reference voltage; it produces the reference voltage minus the resolution voltage. For D/A converters employing a large number of bits, the input code word of all 1's would have an output voltage that would be approximately the reference voltage.

Digital-to-analog converters are normally implemented using a circuit known as the $R - 2R$ ladder. Such a circuit is shown in Fig. 9.4 for an N bit D/A converter. Only two resistor values are needed to implement this circuit and they can be precisely made. The input binary code word $b_k = (B_1 B_2 B_3 ... B_N)$ determines

9.2. THEORY

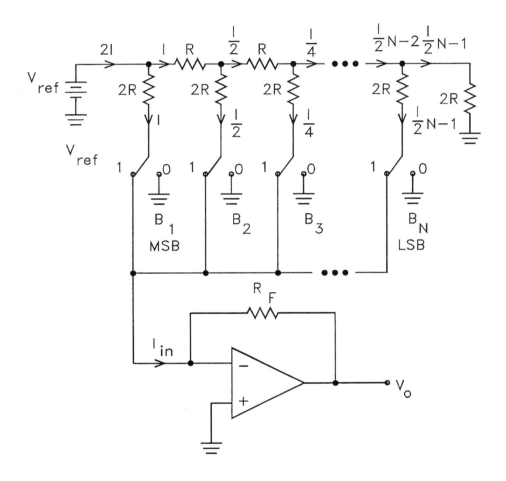

Figure 9.4: An N bit R-2R ladder D/A converter.

the setting of the N switches which determines whether the current flowing through each $2R$ resistor flows to ground or the lead connected to the inverting input of the op-amp. This circuit has the property that the resistance seen looking to the right from just to the left of the top N nodes is always R and the current splits in half at each node. (The currents in the $2R$ branches are not functions of the switch positions since the switch is either set to ground [0 position] or virtual ground [1 position]) The current I_{in} is then given by

$$I_{\text{in}} = I \sum_{i=1}^{N} \frac{B_i}{2^{i-1}} \tag{9.2}$$

where the current I is given by $I = V_{\text{ref}}/(2R)$. Since the current I_{in} flows through the feedback resistor of the op-amp (ideal op-amp assumption invoked), the output voltage of the D/A converter is

$$V_o = V_{\text{ref}} \frac{R_F}{2R} \sum_{i=1}^{N} \frac{B_i}{2^{i-1}} \tag{9.3}$$

where the input code word b_k determines the $B_i's$ which are either 0 or 1. If $R_f = R$, then the output voltage varies from 0 volts when the input code is all 0's to $V_{\text{ref}}(1 - 2^{-N})$ when the input code word is all 1's. This circuit would normally be contained in a single monolithic integrated circuit with inputs for the N bit code word and the reference voltage; the switches are electronic rather than mechanical. The accuracy of the D/A converters (or A/D) can never be more accurate than the precision of the reference voltage. For this reason, a special circuit known as a voltage reference would normally be used with a D/A converter.

9.2.4 Analog-to-Digital Converters (A/D)

Numerous schemes are used to accomplish analog-to-digital conversion. Only a few of the most commonly used will be examined: dual slope integration, successive approximation, flash, and tracking. Some others are voltage-to-frequency converters, staircase ramp or single slope, charge balancing or redistribution, switched capacitor, delta-sigma, and synchro or resolver.

Dual Slope A/D Converter

The dual slope A/D converter is shown in Fig. 9.5 in block diagram form. Its fundamental components are an integrator, electronically controlled switches, a comparator, a counter, a clock, a reference voltage, and some control logic.

Prior to the beginning of a conversion at $t = 0$, switch S_1 is set to the ground position and switch S_2 is closed which shorts the capacitor in the integrator. At $t = 0$ a conversion begins and switch S_2 is opened and switch S_1 is set to the center position which makes the input to the integrator V_{in}, the analog voltage to be digitized. Switch S_1 is held in the center position for an amount of time equal to T_i which is a constant predetermined time interval. Prior to time $t = 0$ the counter is set to 0 and when the switches are thrown at $t = 0$ the counter begins to count the clock pulses. The counter is designed to reset to zero at the end of the time interval T_i.

The value of the output of the integrator voltage at time $t = T_i$ is $V_{\text{in}} T_i/RC$ which is linearly proportional to the input voltage, V_i. The output voltage of the integrator is shown in Fig. 9.6. The output voltage as a function of time is plotted for three different values of V_{in}. The output voltage is a straight line with a slope proportional to V_{in}.

At time $t = T_i$, after the counter has been set to zero, the switch S_1 is set to connect the voltage $-V_{\text{ref}}$ to the input of the integrator which has the voltage $V_{\text{in}} T_i/RC$ stored on it. The integrator voltage then decreases as a linear function of time with slope $-V_{\text{ref}}/RC$. A comparator is used to determine when the output voltage of the integrator crosses zero. When it is zero the input to the counter from the clock is disabled. The state of the counter is then the digitized value, b_k, of V_{in} since its count is then determined by T_r, the amount of time required for the integrator voltage to go from its value at $t = T_i$ to zero. The time

9.2. THEORY

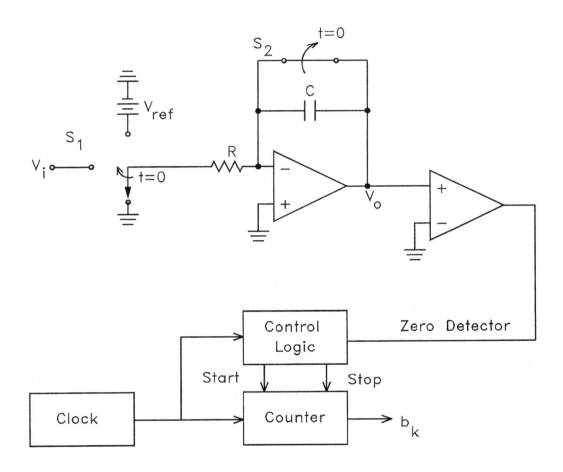

Figure 9.5: Dual slope A/D converter.

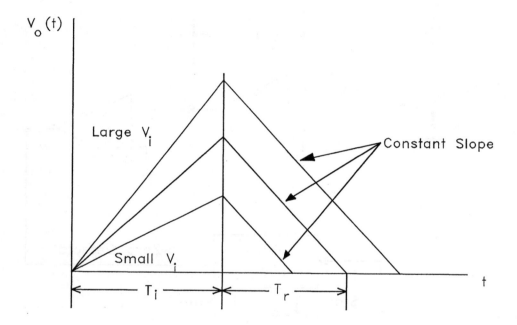

Figure 9.6: Integrator voltage for dual slope A/D converter.

9.2. THEORY

interval T_r is then given by

$$T_r = \frac{V_{\text{in}}}{V_{\text{ref}}} T_i \tag{9.4}$$

which accomplished the analog-to-digital conversion since the times T_i and T_r are digital quantities (the number of pulses counted by the counter times the period of the clock) while the voltages are analog quantities (the uncertainty in this measurement of T_r is the period of the clock which means that the resolution of this converter is a function of the clock period). If N_i is the count at which the counter resets to 0 and N_r is the count of the counter when counter is disabled when the comparator detects zero crossing for the integrator output voltage, then the above equation becomes

$$N_r = \frac{V_{\text{in}}}{V_{\text{ref}}} N_i \tag{9.5}$$

where the brackets indicate the first integer less than the quantity inside the brackets.

The most common use of this type of A/D is in digital voltmeters because of the noise rejection characteristics of the integrator. One of the more troublesome noises that can appear at the input of a digital voltmeter is 60 Hz AC power line ripple and its harmonics. By selecting the time interval T_i to be the period of a 60 Hz sinusoidal, 16.67 ms, any 60 Hz noise and its harmonics will be integrated out at the end of the time interval T_i. If the counter were designed to count 3,000 pulses and then reset to zero at $t = T_i$, then the clock frequency would be 179.96 kHz.

If the reference voltage were picked to be 30 mV, the count N_r would be the digitized value of the input voltage. For example if the input voltage were 10 mV, then the count N_r would be 1,000 which is the input voltage except for the decimal place. If a decade (four BCD counters) rather than binary counter were used, then this would be a 3 and 1/2 digit digital voltmeter since the last three decimal digits could range from 0 to 9 but the most significant decimal digit could only range from 0 to 3.

Ranges larger than 30 mV could be accommodated by placing an input attenuator at the input to the A/D converter. Since the largest value T_r can have is T_i this means that if a zero crossing is not obtained after $2T_i$, then the input voltage V_{in} is larger than V_{ref} and an over-range indication could be given by the control logic.

The conversion time for this A/D converter ranges from T_i to $2\,T_i$. If T_i is picked as 16.67 ms, this is too long for some applications. This type of A/D converter is used in digital voltmeters where speed is not a high priority but accuracy and noise rejection are, i.e. measurement of DC voltages. The accuracy of this A/D converter is determined by the precision of the clock frequency and the reference voltage.

Successive Approximation A/D Converter

A block diagram of a successive approximation A/D converter is shown in Fig. 9.7. The components of this converter are a successive approximation register (SAR), a clock, a comparator, a D/A converter, and some control logic. For an N bit A/D converter, an N bit D/A is required and an N stage register. This is a feedback system in which the output of the D/A converter is compared with the voltage V_{in} to be digitized until the difference between the two is sufficiently small.

The conversion begins a start input to the control logic. When this happens the first clock pulse sets the state of the SAR to the code word $(100\cdots00)$ for which the most significant digit is one and all the rest are zero. This makes the input to the D/A the code word which produces an output voltage $V_o = V_{\text{ref}}/2$. The output of the comparator then determines whether V_{in} is greater than or less than one-half of V_{ref}. If V_{in} is greater than $V_{\text{ref}}/2$ then $B_1(MSB)$ is set to 1, and if the converse is true then B_1 is set to 0.

On the next clock pulse, the input to the D/A converter is set to $(B_1 1\,0\,0\cdots00)$, and the voltage V_o is compared to V_{in}. If V_{in} is greater than V_o then B_2 is set to 1 and if the converse is true B_2 is set to 0. This process continues until the value of all N bits has been determined. The status output line then indicates that the A/D has been completed. If at the end of the determination of the LSB, B_N, the voltage V_{in} is still larger than V_o an overflow indication is given.

This type of A/D converter is analogous to weighing an unknown weight with a balance pan type of scale. The unknown weight would be place in one pan and then the heaviest standard weight would be

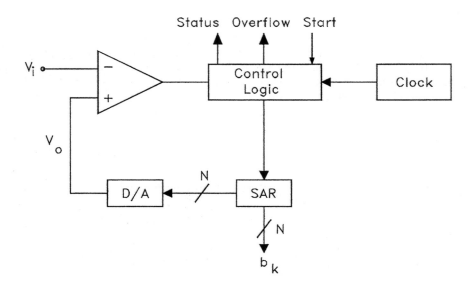

Figure 9.7: Successive approximation A/D converter.

placed in the other pan. If the unknown weight were larger than the heaviest standard weight, then the heaviest standard weight would be left in the pan ($MSB = 1$) and if the converse were true then the heaviest standard weight would be removed ($MSB = 0$). The next heaviest weight would be added to the pan and the process repeated until the lightest weight was reached. The weight of the lightest weight would represent the resolution of this analog-to-digital conversion.

The conversion time of this A/D is NT where T is the period of the clock. If a high speed clock is used, this A/D can have a relatively short conversion time. The actual value of the conversion time in an actual successive approximation A/D would be slight larger than NT due to the fact that the comparator and D/A converter have non-negligible settling times. Also in a practical converter, a type D latch flip-flop would be used to deglitch the output of the D/A and, since the voltage V_{in} must remain constant during the conversion cycle or an error will occur, a sample-and-hold circuit would precede the input to the A/D. This is a widely used scheme for A/D conversion and is usually placed on a single monolithic integrated circuit.

Flash A/D Converter

A block diagram of a flash or parallel A/D converter is shown in Fig. 9.8. The component parts required for an N bit A/D flash converter is $2^N - 1$ comparators, 2^N equal resistors, and some combinational logic.

This comparator simply uses the 2^N resistors to form a ladder voltage divider which divides the reference voltage into 2^N equal intervals and the $2^N - 1$ comparators to determine in which of these 2^N voltage intervals the input voltage V_{in} lies. The combinational logic then translates the information provided by the output of the comparators into the code word b_k.

This converter does not inherently require a clock so the conversion time is essentially set by the settling time of the comparators and propagation time of the combinational logic. This means that this is the fastest of the A/D converters and is used when minimum conversion time is required. In practice, a sample-and-hold circuit would be used at the input and D type latch flip-flops at the output. A two cycle clock would be used to sample the input on the first cycle and open the latches on the second half cycle after everything

9.2. THEORY

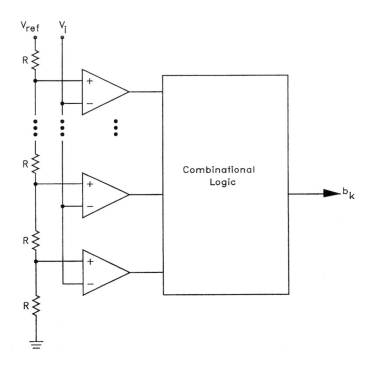

Figure 9.8: Flash A/D converter.

180 CHAPTER 9. ANALOG TO DIGITAL AND DIGITAL TO ANALOG CONVERSION SYSTEMS

has settled.

Tracking A/D Converter

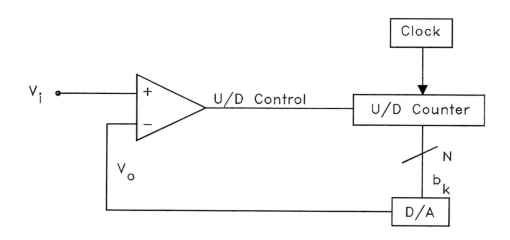

Figure 9.9: Tracking A/D converter.

A block diagram of a tracking A/D converter is shown in Fig. 9.9. The components of this converter are a comparator, an up-down counter, a clock, and a D/A converter. This is another feedback A/D converter which uses a comparator to examine the input voltage, V_{in}, and the output voltage V_o of a D/A converter. If an N bit D/A converter is used, then, in general, this results in an $(N-1)$ bit A/D converter. (The LSB of the D/A alternates between 1 and 0 on alternate clock cycles and the output of the A/D converter is the first $N-1$ most significant bits of the D/A converter.) If the state of the counter is sampled when the counter is in its count down state, then this can be used as an N bit A/D converter.

The counter in this converter is always counting the pulses produced by the clock. When the input voltage is larger than the D/A output voltage the comparator cause the counter to count up and when the converse is true the counter counts down. The analog to digital conversion is performed by the comparator; the comparator output indicates whether the voltage V_{in} is larger or smaller than V_o. When the magnitude of the difference between V_{in} and V_o is less than $V_{ref}/2^{N-1}$ where V_{ref} is the reference voltage of the D/A converter, then the conversion is complete.

The LSB of the D/A oscillates between 0 and 1 on alternate clock cycles. If V_{in} is larger than V_o but this difference is less than the resolution of the A/D converter ($V_{ref}/2^{N-1}$), then the counter counts up on the next clock cycle which makes V_{in} less than V_o on the next clock cycle. This process of counting up and down by one LSB of the D/A converter repeats ad infinitum.

The conversion time of this A/D converter depends on the difference between V_{in} and V_o. The maximum value of the conversion time is $2^N T$ when it is in its acquisition mode and T when it is in its tracking mode where T is the period of the clock. By picking the clock frequency to be sufficiently high, the conversion time can be made reasonably small.

This is the A/D converter that will be the subject of this experiment. It will be constructed in discrete form so that the performance of each component can be examined. This would be a poor choice for a practical A/D converter in that it is susceptible to noise pickup and some of the sophisticated circuitry that should be used, such as a sample and hold circuit, will not be employed.

9.2. THEORY

9.2.5 Components of A/D Converters

The various components that will be used to construct the tracking A/D converter will now be examined. These are the clock, the comparator, the counter, and the D/A converter.

Clock

The system clock that will be used in this experiment is produced by the 555 timer. This 8 pin minidip IC is one of the industry standard timers. It can be used as a one-shot multivibrator or as an oscillator or an astable multivibrator; in this experiment it will only be used as an oscillator. When the power supply voltage is selected to be 5 Volts it produces a signal which can be used as the clock signal for a digital system using TTL logic.

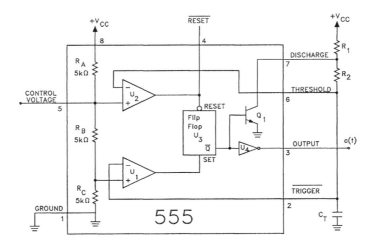

Figure 9.10: 555 timer/oscillator.

A block diagram of the 555 is shown in Fig. 9.10. This is also knows as a MC14555 in the parlance of Motorola where the prefix "MC" stands for Motorola certified. The rectangle represents the components contained inside the integrated circuit; the resistors R_1 and R_2 as well as the timing capacitor C_T are external components used to set the frequency of oscillation. Pins 2 and 6 are connected together, pin 1 is grounded, and pins 4 and 5 are not connected to anything.

Three internal $5\,\mathrm{k\Omega}$ resistors are used to establish a voltage divider which sets the voltage at the non-inverting input of the comparator U_2 at $2V_{cc}/3$ and the non-inverting input of comparator U_1 at $V_{cc}/3$ where V_{cc} is applied at pin 8 and pin 1 is grounded. (The control voltage input at pin 5 can be used to shift these voltages. No connection will be made for this experiment.) When power is first turned on, comparator U_1 sets the SR flip-flop U_3 which makes the $\bar{Q}$ output low which makes the output of the timer at pin 3 high (because of the inverter U_4) and turns the transistor Q_1 off. The capacitor C_T begins charging toward V_{cc} through the series combination of R_1 and R_2. When the capacitor voltage reaches the voltage $2V_{cc}/3$, the comparator U_2 resets the RS flip-flop which makes the $\bar{Q}$ output high which causes the output at pin 3 to go low and turns the transistor Q_1 on which places pin 7 at the ground potential and causes the capacitor C_T to begin discharging through the resistor R_2. When the capacitor voltage drops to $V_{cc}/3$, the comparator U_1 then sets the flip-flop which turns the transistor Q_1 off again and the capacitor again begins charging to $2V_{cc}/3$. Thus the capacitor voltage charges and discharges between the levels $V_{cc}/3$ and $2V_{cc}/3$ which causes the output at pin 3 to switch between low and high. This is then a relaxation oscillator since the oscillations are established by the charging and discharging of a capacitor.

The frequency of oscillation can be obtained by determining the amount of time, t_1, required for the capacitor voltage to charge from $V_{cc}/3$ to $2V_{cc}/3$ and the amount of time, t_2, required for the capacitor to discharge from $2V_{cc}/3$ to $V_{cc}/3$. The sum of these two time intervals is the period and the frequency is the reciprocal of the period. These times are given by

$$t_1 = 0.693(R_1 + R_2)C_T \text{ and } t_2 = 0.693\, R_2 C_T \tag{9.6}$$

which makes the period $T = 0.693(R_1 + 2R_2)C_T$ and the frequency f

$$f = \frac{1.44}{(R_1 + 2\,R_2)C_T} \tag{9.7}$$

which states that the frequency of oscillation is set by the resistors R_1 and R_2 and the timing capacitor C_T. The duty cycle is the percentage of a period that the output is high and is given by

$$\text{Duty Cycle} = \frac{t_1}{t_1 + t_2} = \frac{R_1 + R_2}{R_1 + 2R_2} \tag{9.8}$$

which is not a function of C_T. The duty cycle could be made 50% if R_2 were picked to be large compared with R_1. The duty cycle is normally not an important consideration for the clock signal of a digital system.

Comparator

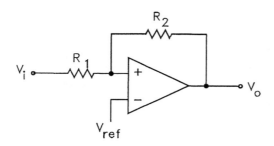

Figure 9.11: Comparator.

A comparator is a device that determines which of two input signals is the larger. A diagram of such a device is shown in Fig. 9.11. If the resistor $R_1 = 0$ and $R_2 = \infty$, then V_o would be V_U when V_{in} is greater than V_{ref} and V_o would be V_L when V_{ref} is greater than V_{in}. This can be thought of as the fundamental element in the analog-to-digital conversion since its input is an analog quantity (the difference between two voltages) and its output is a digital quantity (one of two discrete voltage levels).

The resistors R_1 and R_2 are required to give the comparator hysteresis so that the output of the comparator switches at a different level depending on whether V_{in} is above or below V_{ref}. They provide positive feedback from the output to the input of the comparator. The transfer function of the comparator with hysteresis is shown in Fig. 9.12. The transition voltages V_1 and V_2 are given by

$$V_1 = \frac{V_{\text{ref}}(R_1 + R_2) - V_U R_1}{R_2} \tag{9.9}$$

$$V_2 = \frac{V_{\text{ref}}(R_1 + R_2) - V_L R_1}{R_2} \tag{9.10}$$

which lie above and below V_{ref}. If the output voltage is V_U and input voltage is greater than V_{ref} and decreasing, the output doesn't switch to V_L until the input voltage drops to V_1 which is less than V_{ref}.

9.2. THEORY

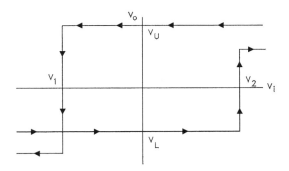

Figure 9.12: Transfer characteristics of a comparator.

Conversely, if the output is V_L and the input is less than V_{ref} and increasing, the output doesn't switch to V_U until V_{in} equals V_2 which is larger than V_{ref}. The hysteresis is the difference between V_2 and V_1 and is given by

$$\Delta V_h = \frac{R_1}{R_2}(V_U - V_L) \tag{9.11}$$

which represents the amount of tolerance that the comparator has to a small input noise voltage superimposed on the desired input. If hysteresis were not used, then a small noise voltage on the input could cause rapid switching of the output known as "chattering". A capacitor can be placed in parallel with the resistor R_2 or from the output to ground to speed up the switching.

A 741 op-amp could be used as the comparator but it wouldn't be the most efficient. This is a device designed to be operated in its linear mode and it would be exceptionally slow in switching between two discrete voltage levels. A device that is specifically intended for this application will be used, the LM311 comparator. The pinouts for this device will be given in the procedure section. The LM311 is an open collector devices which required that a load resistor be connected from the output to the power supply for V_U. Since this is being used to interface two analog voltages to a TTL logic devices, the voltages V_L and V_U will be chosen to be $V_L = 0$ and $V_U = +5\,\text{V}$. A connection diagram will be given in the procedure section.

Counter

The counter that will be used for this experiment will be the 74LS191 4 bit binary up-down TTL counter. It will count the clock pulses produced by the 555 timer delayed by a flip flop. It has an input that determines whether the counter is counting up or down; when this input is high the counter counts down and when it is low the counter counts up. The pin connections for this device are given in the procedure section.

For this counter, the control input that determines whether it is counting up or down must remain stable while the clock is low. Also, the enable input must remain constant while the clock is low. Should this not be the case, an error may result in the count.

D/A Converter

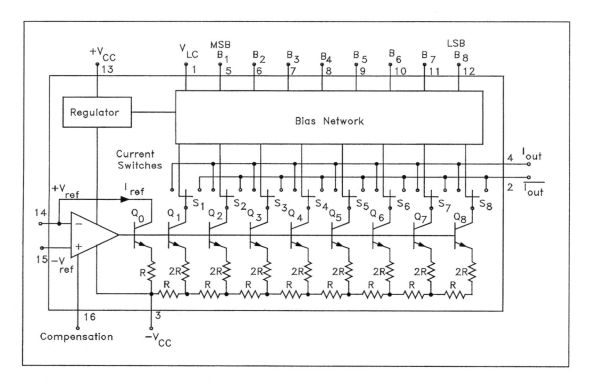

Figure 9.13: DAC-08 D/A converter.

The D/A converter that will be used in this experiment will be the DAC-08. It is an 8 bit TTL device that is intended to be directly interfaced with an 8 bit microprocessor. The MSB is pin 5 and the LSB is pin 12 as shown in the diagram in Fig. 9.13.

The DAC-08 is an $R-2R$ ladder D/A. A voltage is applied at pin 14 which produces the reference current flowing into pin 14. This current I_{ref} is the current $2I$ discussed previously; it supplies the input current to the $R-2R$ ladder. The output current I_{out} flowing into pin 4 is given by

$$I_{\text{out}} = I_{\text{ref}} \sum_{i=1}^{8} \frac{B_i}{2^i} \tag{9.12}$$

where B_1 is the MSB and B_8 is the LSB. The optimum value of I_{ref} is 2 mA. The accuracy of the current I_{ref} is the fundamental factor determining the accuracy of the A/D converter to be constructed using this device.

9.2.6 Bipolar DAC

A bipolar *DAC* may be obtained using a DAC-08 as shown in Fig. 9.14. The reference voltage is V_{ref}. The reference current is

$$I_{ref} = \frac{V_{ref}}{R} \tag{9.13}$$

since pin 14 is at the ground potential. (Pin 15 is grounded and pins 14 and 15 at the input terminals for an op amp that is internal to the DAC-08.)

9.3. PRELIMINARY DERIVATIONS

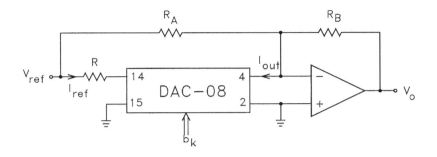

Figure 9.14: Bipolar DAC using DAC-08

The output voltage is given by

$$V_o = -V_{ref}\frac{R_B}{R_A} + I_{out}R_B \qquad (9.14)$$

If the input digital code is all zeroes (eight zeroes), the output voltage is

$$V_z = -V_{ref}\frac{R_B}{R_A} \qquad (9.15)$$

which is known as the zero code voltage. If the input digital code is all ones (eight ones), the output voltage is

$$V_1 = -V_{ref}\frac{R_B}{R_A} + I_{ref}(1 - 2^{-N})R_B \qquad (9.16)$$

which is known as the one code voltage. Normally, a symmetric swing is desired so that $V_1 = -V_z$. The range voltage is given by

$$V_r = V_1 - V_z + \Delta V = I_{ref}R_B \qquad (9.17)$$

and the resolution is given by

$$\Delta V = \frac{V_r}{2^N} \qquad (9.18)$$

which corresponds to an increment of the input digital code by the least significant bit.

9.3 Preliminary Derivations

None.

9.4 Preliminary Calculations

1. Determine the values of the resistors R_1 and R_2 and the timing capacitor to produce the clock frequency specified by the laboratory instructor for the 555 timer. Since this is to be the clock signal for a TTL logic circuit, use a power supply voltage of 5 V.

2. Determine the values of the resistors R_A and R_B required for the D/A circuit required to produce the zero code voltage, V_z, and the range voltage, V_{ref}, specified by the laboratory instructor.

9.5 Preliminary SPICE Simulations

Simulate the 555 Timer circuit designed in the Preliminary Calculations section. Use a power supply voltage of 5 V. The SPICE input deck should call the 555 timer as a subcircuit call as

x p1 p2 p3 p4 p5 p6 p7 p8 555d
.lib eval.lib

where pi's are the node number for the 8 pins of the 555 timer.

9.6 Experimental Procedures

9.6.1 Power Supply

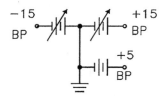

Triple DC Power Supply Connections

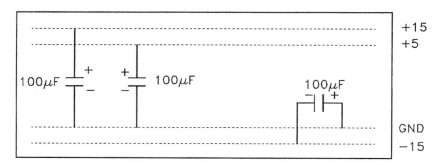

Breadboard with Power Supply Decoupling Capacitors

Figure 9.15: Power supply connections.

This circuit will require a triple power supply. The power supply connections are shown in Fig. 9.15. The COMMON on the power supply is to be connected to the ground bus on the breadboard. The ground binding post on the DC power supply is not to be connected to anything. The fixed 5 V supply is for the TTL logic and the +15 V and −15 V are required for the comparator, the op-amp and the DAC-08.

9.6.2 Breadboard

The breadboard topology is shown in Fig. 9.15. There must be a bus for each of the three DC voltages and ground. A 100 μF decoupling capacitor must be used on each of the three power supply buses; since these capacitors are electrolytic the polarity indicated must be adhered to. Do not place a 100 Ω in series with the +5 V power supply because it might drop the voltage on the TTL $+V_{CC}$ below the minimum level.

9.6. EXPERIMENTAL PROCEDURES

Circuits will be constructed in sequence and the resultant circuit will be moderately large. Care should be taken to optimize the circuit construction: short leads and the IC's should be as close together as possible.

9.6.3 Clock

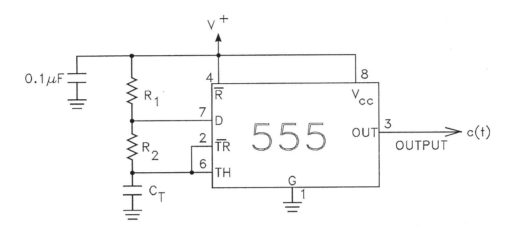

Figure 9.16: 555 clock oscillator.

Assemble the 555 clock oscillator shown in Fig. 9.16 using the 555 timer. Select the resistors R_1 and R_2 along with the timing capacitor determined in the preliminary calculations section to produce the system clock frequency specified by the laboratory instructor. Neither resistor R_1 nor R_2 should be less than $1\,k\Omega$. Since this will produce the basic system clock signal for a TTL logic system, V^+ should be chosen as $+5\,V$ since this will produce a clock signal with a lower level of $0\,V$ and an upper level of $+5\,V$.

Examine with the oscilloscope the waveform being produced at pin 3 of the 555 and sketch or print it. Also sketch or print the waveform being produced at pin 2.

Leave this circuit in place since it will be used as the system clock for the remainder of the experiment.

9.6.4 Four Bit Binary Counter

Assemble the four bit binary counter shown in Fig. 9.17 using the 74LS191 TTL counter. Pin 16 must be connected to the $+5\,V$ bus and pin 8 is ground. The enable input at pin 4 is grounded and the load input at pin 11 is connected to the $+5\,V$ supply. (The bubbles by pins 4 and 11 means that these inputs are active low. Nothing is to be loaded so pin 11 is tied to the $+5\,V$ supply. Each clock pulse is to be counted so pin 4 is grounded.) For this part the RCO (ripple count out) on pin 13 will not be connected to anything.

The input on pin 14 is the clock input which is obtained from pin 3 on the 555.

Pin 5 is the input that determines whether the counter counts up or down. If the input is low, the counter counts up and if the input is high the counter counts down. This is indicated with the symbol $\overline{U}/D$ at pin 5. Ground pin 5 for this step.

188   CHAPTER 9. ANALOG TO DIGITAL AND DIGITAL TO ANALOG CONVERSION SYSTEMS

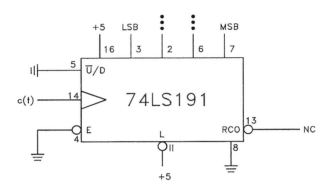

Figure 9.17: Four bit binary counter.

The pins 3 through 7 are the outputs of the counter. Pin 7 is the MSB and pin 3 is the LSB. As indicated on the diagram, pins 2 and 6 are in the sequence from the LSB to the MSB. These outputs will be connected to the input of the DAC-08 in the next step.

9.6.5 Four Bit D/A Converter

Assemble the four bit D/A converter shown in Fig. 9.18. (The DAC-08 is an 8 bit D/A. A four bit D/A is obtained by grounding the first four most significant bits on pins 5 through 8.)

The D/A converter examined in the theory section had a unipolar output, i.e. the output voltage was positive or zero for all of the input code words. The zero code word (all $B_i = 0$) produced the voltage $V_z = 0$ and the one code word (all $B_i = 1$) produced the output voltage $V_1 = V_r(1 - 2^{-N})$ where $N = 4$. The op-amp in Fig. 9.18 will be used to obtain a D/A converter with a bipolar output, i.e. V_z will be negative and V_1 will be positive.

Use the values for the resistors R_A and R_B that were determined in the preliminary calculations section to produce a D/A converter with the zero voltage, V_z, and one code voltage, V_1, specified by the laboratory instructor. The zero voltage is produced when all the B_i's are zero which means that

$$V_z = -\frac{R_B}{R_A} \times 10 \quad \quad (9.19)$$

since the voltage at the junction of the 1 $k\Omega$ and 3.3 $k\Omega$ resistors is approximately 10 V which is the reference voltage. This set the ratio of the resistors R_A and R_B. When the pins 9 through 12 on the DAC-08 are high or logical one's, the output voltage is

$$V_1 = V_z + I_{out} R_B \quad \quad (9.20)$$

$$I_{out} R_B = V_r(1 - 2^{-N}) \quad \quad (9.21)$$

where

$$I_{out} = I_{ref}\left(\frac{1}{32} + \frac{1}{64} + \frac{1}{128} + \frac{1}{256}\right) \quad \quad (9.22)$$

and $I_{ref} = 10/5.1$ mA, V_r is the range voltage, and $N = 4$. Once V_1 and V_z are specified, this sets R_B and since the ratio of R_A and R_B is known this sets R_A.

The 4 output bits of the counter are to be connected as inputs to the DAC08 as shown, i.e. pin 3 of the counter is connected to pin 12 of the DAC08, pin 2 of the counter is connected to pin 11 of the DAC08, pin

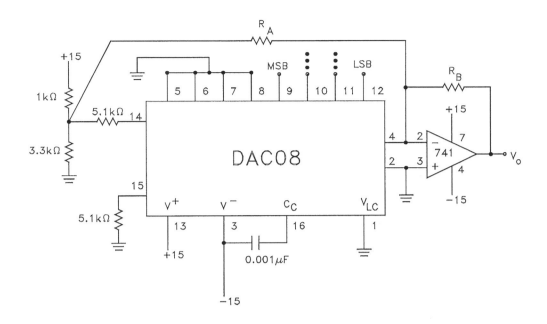

Figure 9.18: Four bit DAC.

6 of the counter is connected to pin 10 of the DAC08, and pin 7 of the counter is connected to pin 9 of the DAC08.

The control input on pin 5 of the counter should be grounded. Examine the output of the op-amp with the oscilloscope and sketch or print it. A stair-step pattern extending from V_z to one step short of V_1 should be present. Record the height and width of the steps and the highest and lowest voltages produced at the output of the op-amp. The height of the steps is the resolution of the D/A.

9.6.6 Four Bit A/D System

Assemble the four bit A/D converter system shown in Fig. 9.19. The digital outputs of the ADC are the four bits at the output of the counter which are the four bits at the input to the 4 bit D/A converter.

The resistors R_1 and R_2 should be selected to produce a hysteresis less than 0.01 of the resolution of the D/A converter. The control input at pin 5 of the counter should be connected to the Q output of the D flip flop U3 which is pin 5 of this IC.

Note the pin connections for the LM311 comparator. Although the same circuit symbol is used for the comparator and the 741 type op-amps, the pin connections are considerably different. The 1 kΩ resistor from the output of the comparator to the +5 V supply is a pullup resistor for the comparator; it sets the upper voltage V_U to +5 V. Pin 1 is grounded on this device which sets the lower voltage V_L to 0 V. Thus, the comparator's outputs are at TTL level. (Be certain that the jumper that grounded pin 5 on the counter for the previous step has been removed. The input for pin 5 for the counter is pin 5 on the 74LS74 D flip flop U3.) The 0.001 μF capacitor speeds up the switching of the comparator between 5 and 0 V.

Three D flip-flops are used to prevent counting errors with the 74LS191 counter. Both the $\overline{U/D}$ and $\overline{ENABLE}$ inputs must remain constant while the clock input to the counter is low or errors will occur. (The nomenclature for the output of the 555 has been changed in Fig. 9.5. The output of the 555, which is the system clock, is $e(t)$; the output of flip-flop U1 is $c(t)$; and the output of flip-flop U2 is $cd(t)$.) Flip-flops U1

CHAPTER 9. ANALOG TO DIGITAL AND DIGITAL TO ANALOG CONVERSION SYSTEMS

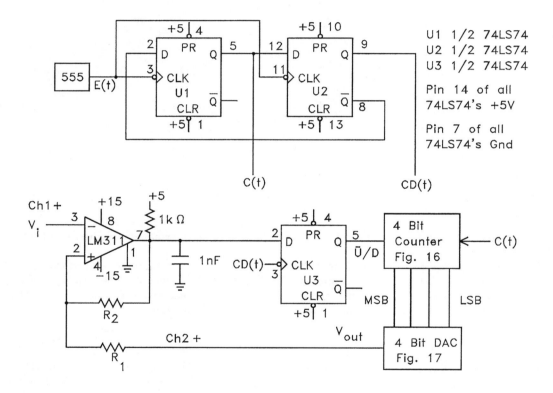

Figure 9.19: Four bit A/D.

9.6. EXPERIMENTAL PROCEDURES

and U2 are connected as a divide by four counter with the 555 output, $e(t)$, as the clock input for these two flip-flops. The outputs of flip-flops U1 and U2 are square waves with a frequency one fourth of the 555 that are 90° out of phase. The output of flip-flop U1, $c(t)$, is used as the clock input for the 74LS191 counter.

Flip-flop U3 is used as a digital sample-and-hold. An output is past from the D input to the Q output of U3 only when the clock input to U3, $cd(t)$ makes a transition from the low to the high state. (The 74LS74 is a positive edge triggered flip-flop.) Since $cd(t)$ and $c(t)$ are 90° out of phase, the $\overline{U}/D$ input to the 74LS191 can occur only when the clock input to the counter, $c(t)$, is high.

The 74LS74 is a dual D positive edge triggered flip-flop. There are two per integrated circuit. It is assumed in Fig. 9.19 that flip-flops U1 and U2 are on the same integrated circuit. Flip-flop U3 is on a separate integrated circuit. Pin 14 of each must be connected to the +5 V bus and pin 7 to the ground bus.

Connect the two channels of the oscilloscope as shown in Fig. 9.19. Ch1 is connected to the input of the ADC system and Ch2 is connected to the output of the D/A converter. Use the vertical position controls to set the center horizontal line on the CRT graticule as the reference for zero volts for both Ch1 & Ch2. Set the input coupling selector to DC for both channels (the default setting) and the Volts/Div control to 2 volts for both channels. Set the trigger source for the oscilloscope to Ch2.

Apply a DC voltage to the input of the ADC by connecting the wiper of a 10 kΩ pot to pin 3 of the comparator and connecting one of the two ends to +15 V bus and the remaining end to the −15 V bus. Connect the digital voltmeter to monitor the input voltage to the ADC; connect the LO input of the DVM to the ground bus and the HI input to pin 3 of the comparator. Apply a negative voltage that is sufficiently negative so that the waveform on channel 2 is negative slope or descending stair-step. Increase the input voltage to the ADC until the waveform switches from a stair-step to a square wave with a peak-to-peak value equal to the resolution voltage of the D/A converter and record this voltage. Continue to increase the voltage V_{in} and record the voltages at which the output voltage "steps up" by an amount equal to the resolution voltage of the D/A above the input voltage. When the input is sufficiently large, the output will switch to a positive slope or ascending stair-step.

A total of $16 = 2^4$ voltage intervals should have been obtained in the previous step. Beginning with the most negative voltage at which the output switched from a stair-step to a square wave, assign an increasing 4 bit natural binary code. The output code will be monitored in the next step with LED state indicators.

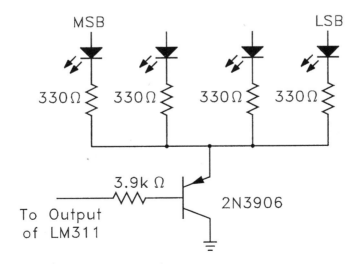

Figure 9.20: LED state indicators.

Connect four LED state indicators to the output of the D/A converter as shown in Fig. 9.20. The 330 Ω resistors are dropping resistors for the LED diodes. The transistor samples the output of the counter or converter when the output of the counter is in its count down state. Make certain the diodes are inserted with the correct polarity. Determine which side of each diode is the anode by using the DMM as an ohmmeter. The forward biased diode will yield the lower resistance indication. (A simple way of determining which side of the diode is the anode or p side is to apply $+5\,\text{V}$ from the power supply to the series combination of the diode and the $330\,\Omega$ resistor to determine if it will light.)

Vary the pot setting to make the input voltage V_{in} equal to center voltage for each of the sixteen code words for this ADC system. Record the output binary code indicated by the LED state monitors.

Disconnect the pot and power supply from the input of the ADC. Set the trigger source on the oscilloscope to Ch1. Connect the output of the function generator to the input of the ADC and set it to produce a triangular wave with a DC level of zero, a frequency of 50 Hz, and a peak-to-peak value equal to 0.95 of the reference voltage of the ADC system. Sketch the waveforms that appear on Chs 1 and 2. Vary the frequency above and below 50 Hz and sketch or print representative waveforms. Increase the frequency of the function generator until the output can no longer track the input and record this frequency. Change the input to a square wave and repeat. Increase the clock frequency by a factor of ten (simply decrease the timing capacitor in the 555 by a factor of ten) and repeat.

Remove the LED's and $330\,\Omega$ resistors. Change the timing capacitor in the 555 to its original value.

9.6.7 Eight Bit Binary Counter

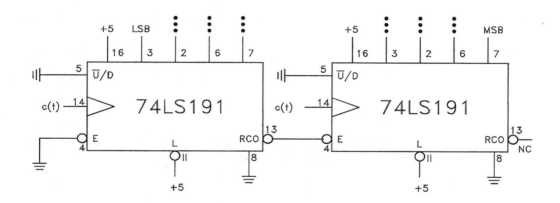

Figure 9.21: Eight bit binary counter system.

Construct the eight bit binary counter shown in Fig. 9.21 by cascading two 74LS191 4 bit binary up-down counters. Pins 16 on both IC's must be connected to the $+5\,\text{V}$ power supply and pins 8 must be grounded.

The clock is connected to pin 14 on each of the two IC's. The ripple count out of the first IC is connected to the enable input of the second counter and the ripple count out of the second counter is not connected to anything. When the first counter reaches the count of all 1's the ripple count out goes low and enables the second counter to count one clock pulse. Therefore, this is an 8 bit binary up-down counter.

Pin 11, the load input, on each IC is connected to the $+5\,\text{V}$ supply. For this step pin 5 on each IC is grounded which makes each counter count up.

The outputs are to be connected to the DAC08 in the next step. The LSB of the first counter on pin 3 is the LSB of the system and the MSB on the second counter is the MSB of the system. The intervening bits are shown in Fig. 9.21.

9.6.8 Eight Bit D/A Converter

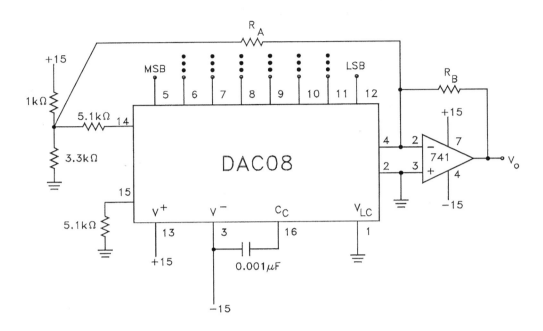

Figure 9.22: Eight bit DAC.

Assemble the 8 bit D/A converter shown in Fig. 9.22. The resistors R_A and R_B are not the same as for the 4 bit D/A previously constructed and must be recomputed.

Calculate the values of R_A and R_B to produce the zero code voltage and reference voltage specified by the laboratory instructor. When the input code word is all zeros the output of the D/A is

$$V_o = V_z = -\frac{R_B}{R_A} \times 10 \tag{9.23}$$

which sets the ratio of R_A and R_B as before. When the input code word is all 1s the output V_1 of the D/A converter is

$$V_1 = V_z + V_r(1 - 2^{-N}) \tag{9.24}$$

$$I_{out} R_B = V_r(1 - 2^{-N}) \tag{9.25}$$

where

$$I_{out} = I_{ref}\left(1 - 2^{-8}\right) = \frac{10}{5.1}\left(1 - 2^{-8}\right) \text{ mA} \tag{9.26}$$

and N is now 8. Since V_{ref} has been specified, this sets R_B and hence R_A.

Connect the appropriate output on the eight bit counter to the appropriate input on the eight bit D/A converter. For the first 74LS191 the connection are: pin 3 of the counter to pin 12 of the DAC08, pin 2 of the counter to pin 11 of the DAC08, pin 6 of the counter to pin 10 of the DAC08, and pin 7 of the counter to pin 9 of the DAC08. For the second 74LS191 the connections are: pin 3 of the counter to pin 8 of the DAC08, pin 2 of the counter to 7 of the DAC08, pin 6 of the counter to pin 6 of the DAC08, and, finally, pin 7 of the counter to pin 5 of the DAC08.

Observe the waveform at the output of the op-amp. It should be a stair-step beginning at the zero code voltage and extending to the voltage $V_{\text{ref}}(1 - 2^{-8})$ with 256 steps. Record the step heights and widths.

9.6.9 Eight Bit ADC System

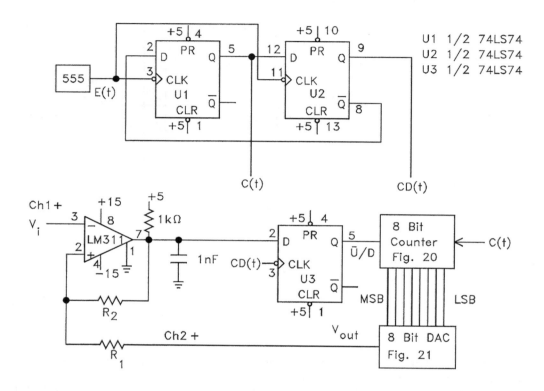

Figure 9.23: Eight bit ADC system.

Assemble the 8 bit ADC system shown in Fig. 9.23. Pin 5 on each of the 74LS191's is to be connected to pin 5 on the D flip flop U3.

Resistors R_1 and R_2 are to be chosen to produce a hysteresis of less than 0.01 of the resolution of the A/D converter.

Connected a triangular wave with a DC level of zero, a frequency of 50 Hz, and a peak-to-peak value of one half of the reference voltage to input of the ADC. Vary the frequency above and below 50 Hz and sketch representative waveforms. Increase the frequency until the ADC can no longer track the input waveform and record this frequency. Change the waveform to square and repeat. Increase the clock frequency by a factor of 10 and repeat.

9.7 Laboratory Report

Turn in all printouts made or sketches taken. Answer all questions posed in the procedure section as well as any supplementary questions formulated by the laboratory instructor.

Discuss anything unusual obtained such as glitches, code jumping, etc. and speculate as to the cause.

9.8 References

1. Sheingold, *Analog-Digital Conversion Notes*, Analog Devices, 1977.
2. Miner and Comer, *Physical Data Acquisition for Digital Processing*, Prentice-Hall, 1992.
3. Horowitz and Hill, *The Art of Electronics*, 2nd edition, Cambridge, 1989.
4. Barney, *Intelligent Instrumentation*, 2nd edition, Prentice-Hall, 1988.
5. Hoffman, *Archimedes' Revenge*, Fawcett, 1988.

Chapter 10

The Bipolar Junction Transistor

10.1 Object

The object of this experiment is to study the basic characteristics of the bipolar junction transistor (BJT). Values for the saturation current, the current gain, the Early voltage, and the base-to-emitter threshold, active, and saturation voltages are measured for typical NPN and PNP transistors.

10.2 Theory

Some of the background theory on modeling the BJT is covered in this section. This material is covered for reference purposes both for this experiment and for all following experiments on the BJT.

10.2.1 Notation

The notations used here for voltages and currents correspond to the following conventions: Dc bias values are indicated by an upper case letter with upper case subscripts, e.g. V_{CE}, I_C. Instantaneous values of small-signal variables are indicated by a lower-case letter with lower-case subscripts, e.g. v_{ce}, i_c. Total values are indicated by a lower-case letter with upper-case subscripts, e.g. v_{CE}, i_C. Circuit symbols for independent sources are circular and symbols for controlled sources have a diamond shape. Voltage sources have a $\pm$ sign within the symbol and current sources have an arrow.

10.2.2 Device Equations

Figure 10.1 shows the circuit symbols for the npn and pnp BJTs. In the active mode, the collector-base junction is reverse biased and the base-emitter junction is forward biased. For the npn device, the active-mode collector and base currents are given by

$$i_C = I_S \exp\left(\frac{v_{BE}}{V_T}\right) \qquad i_B = \frac{i_C}{\beta} \tag{10.1}$$

where V_T is the thermal voltage, I_S is the saturation current, and β is the base-to-collector current gain. These are given by

$$V_T = \frac{kT}{q} = 0.025 \text{ V for } T = 290 \text{ K} = 25.85 \text{ mV for } T = 300 \text{ K} \tag{10.2}$$

$$I_S = I_{S0}\left(1 + \frac{v_{CE}}{V_A}\right) \tag{10.3}$$

$$\beta = \beta_0 \left(1 + \frac{v_{CE}}{V_A}\right) \tag{10.4}$$

where V_A is the Early voltage and I_{S0} and β_0, respectively, are the zero bias values of I_S and β. Because $I_S/\beta = I_{S0}/\beta_0$, it follows that i_B is not a function of v_{CE}. The equations apply to the pnp device if the subscripts BE and CE are reversed.

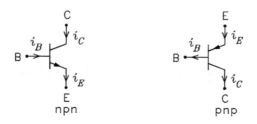

Figure 10.1: BJT circuit symbols.

The emitter-to-collector current gain α is defined as the ratio i_C/i_E. To solve for this, we can write

$$i_E = i_B + i_C = \left(\frac{1}{\beta} + 1\right) i_C = \frac{1+\beta}{\beta} i_C \tag{10.5}$$

It follows that

$$\alpha = \frac{i_C}{i_E} = \frac{\beta}{1+\beta} \tag{10.6}$$

Thus the currents are related by the equations

$$i_C = \beta i_B = \alpha i_E \tag{10.7}$$

10.2.3 Transfer and Output Characteristics

The transfer characteristics are a plot of the collector current i_C as a function of the base-to-emitter voltage v_{BE} with the collector-to-emitter voltage v_{CE} held constant. From Eqs. 10.1 and 10.3, we can write

$$i_C = I_{S0} \left(1 + \frac{v_{CE}}{V_A}\right) \exp\left(\frac{v_{BE}}{V_T}\right) \tag{10.8}$$

It follows that i_C varies exponentially with v_{BE}. A plot of this variation is given in Fig. 10.2. It can be seen from the plot that the collector current is essentially zero until the base-to-emitter voltage reaches a threshold value. Above this value, the collector current increases rapidly. The threshold value is typically in the range of 0.5 to 0.6 V. For high current transistors, it is usually smaller. The plot shows a single curve. If v_{CE} is increased, the current for a given v_{BE} is larger. However, the displacement between the curves is so small that it can be difficult to distinguish between them. The small-signal transconductance g_m defined below is the slope of the transfer characteristics curve evaluated at the quiescent or dc operating point for a device.

The output characteristics are a plot of the collector current i_C as a function of the collector-to-emitter voltage v_{CE} with the base current i_B held constant. From Eqs. 10.1 and 10.4, we can write

$$i_C = \beta_0 \left(1 + \frac{v_{CE}}{V_A}\right) i_B \tag{10.9}$$

It follows that i_C varies linearly with v_{CE}. A plot of this variation is given in Fig. 10.3. For small v_{CE} such that $0 \leq v_{CE} < v_{BE}$, Eq. (10.9) does not hold. In the region in Fig. 10.3 where this holds, the bjt is saturated. The small-signal collector-to-emitter resistance r_0 defined below is the reciprocal of the slope of the transfer characteristics curve evaluated at the quiescent or dc operating point for a device.

10.2. THEORY

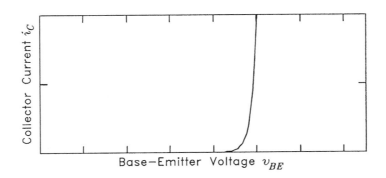

Figure 10.2: BJT transfer characteristics.

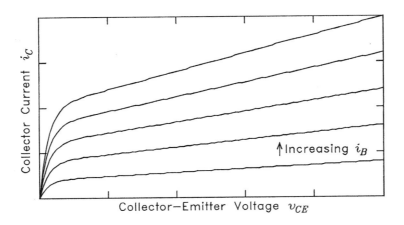

Figure 10.3: BJT output characteristics.

Calculation of V_A, β_0, and I_{S0} From the Characteristics

From the values for i_C, i_B, v_{CE}, and the slope $m = \Delta i_C / \Delta v_{CE}$ at a point on the output characteristics, it follows that V_A and β_0 can be calculated from the equations

$$V_A = \frac{i_C}{m} \tag{10.10}$$

$$\beta_0 = \frac{i_C/i_B}{1 + v_{CE}/V_A} \tag{10.11}$$

From a point on the transfer characteristics, the value of I_{S0} is given by

$$I_{S0} = \frac{i_C}{1 + v_{CE}/V_A} e^{-v_{BE}/V_T} \tag{10.12}$$

10.2.4 Hybrid-π Model

Let each current and voltage be written as the sum of a dc component and a small-signal ac component as follows:

$$i_C = I_C + i_c \qquad i_B = I_B = i_b \tag{10.13}$$

$$v_{BE} = V_{BE} + v_{be} \qquad v_{CE} = V_{CE} + v_{ce} \tag{10.14}$$

If the ac components are sufficiently small, we can write

$$i_c = \frac{\partial I_C}{\partial V_{BE}} v_{be} + \frac{\partial I_C}{\partial V_{CE}} v_{ce} \qquad i_b = \frac{\partial I_B}{\partial V_{BE}} v_{be} \tag{10.15}$$

where the derivatives are evaluated at the dc bias values. Let us define the transconductance g_m, the collector-to-emitter resistance r_0, and the base-to-emitter resistance r_π as follows:

$$g_m = \frac{\partial I_C}{\partial V_{BE}} = \frac{I_S}{V_T} \exp\left(\frac{V_{BE}}{V_T}\right) = \frac{I_C}{V_T} \tag{10.16}$$

$$r_0 = \left(\frac{\partial I_C}{\partial V_{CE}}\right)^{-1} = \left[\frac{I_{S0}}{V_A} \exp\left(\frac{V_{BE}}{V_T}\right)\right]^{-1} = \frac{V_A + V_{CE}}{I_C} \tag{10.17}$$

$$r_\pi = \left(\frac{\partial I_B}{\partial V_{BE}}\right)^{-1} = \left[\frac{I_{S0}}{\beta_0 V_T} \exp\left(\frac{V_{BE}}{V_T}\right)\right]^{-1} = \frac{V_T}{I_B} \tag{10.18}$$

The collector and base currents can thus be written

$$i_c = i_c' + \frac{v_{ce}}{r_0} \qquad i_b = \frac{v_\pi}{r_\pi} \tag{10.19}$$

where

$$i_c' = g_m v_\pi \qquad v_\pi = v_{be} \tag{10.20}$$

The small-signal circuit which models these equations is given in Fig. 10.4(a). This is called the hybrid-π model. The resistor r_x, which does not appear in the above equations, is called the base spreading resistance. It represents the resistance of the connection to the base region inside the device. Because the base region is very narrow, the connection exhibits a resistance which often cannot be neglected.

The small-signal base-to-collector ac current gain β is defined as the ratio i_c'/i_b. It is given by

$$\beta = \frac{i_c'}{i_b} = \frac{g_m v_\pi}{i_b} = g_m r_\pi = \frac{I_C}{V_T} \frac{V_T}{I_B} = \frac{I_C}{I_B} \tag{10.21}$$

Note that i_c differs from i_c' by the current through r_0. Therefore, $i_c/i_b \neq \beta$ unless $r_0 = \infty$.

10.2. THEORY

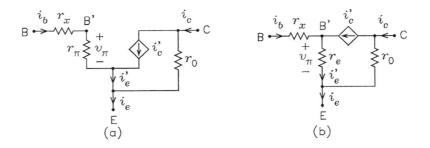

Figure 10.4: (a) Hybrid-π model. (b) T model.

10.2.5 T Model

The T model replaces the resistor r_π in series with the base with a resistor r_e in series with the emitter. This resistor is called the emitter intrinsic resistance. The current i'_e can be written

$$i'_e = i_b + i'_c = \left(\frac{1}{\beta} + 1\right) i'_c = \frac{1+\beta}{\beta} i'_c = \frac{i'_c}{\alpha} \tag{10.22}$$

where α is the small-signal emitter-to-collector ac current gain given by

$$\alpha = \frac{\beta}{1+\beta} \tag{10.23}$$

Thus the current i'_c can be written

$$i'_c = \alpha i'_e \tag{10.24}$$

The voltage v_π can be related to i'_e as follows:

$$v_\pi = i_b r_\pi = \frac{i'_c}{\beta} r_\pi = \frac{\alpha i'_e}{\beta} r_\pi = i'_e \frac{\alpha r_\pi}{\beta} = i'_e \frac{r_\pi}{1+\beta} = i'_e r_e \tag{10.25}$$

It follows that the intrinsic emitter resistance r_e is given by

$$r_e = \frac{v_\pi}{i'_e} = \frac{r_\pi}{1+\beta} = \frac{V_T}{(1+\beta) I_B} = \frac{V_T}{I_E} \tag{10.26}$$

The T model of the BJT is shown in Fig. 10.4(b). The currents in both models are related by the equations

$$i'_c = g_m v_\pi = \beta i_b = \alpha i'_e \tag{10.27}$$

10.2.6 Simplified T Model

Figure 10.5 shows the T model with a Thévenin source in series with the base. We wish to solve for an equivalent circuit in which the source $i'_c = \alpha i'_e$ connects from the collector node to ground rather than from the collector node to the B' node.

The first step is to replace the source $\alpha i'_e$ with two identical series sources with the common node grounded. The circuit is shown in Fig. 10.6(a). The object is to absorb the left $\alpha i'_e$ source into the base-emitter circuit. For the circuit, we can write

$$v_e = v_{tb} - \frac{i'_e}{1+\beta}(R_{tb} + r_x) - i'_e r_e = v_{tb} - i'_e \left(\frac{R_{tb} + r_x}{1+\beta} + r_e\right) \tag{10.28}$$

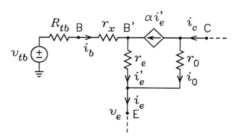

Figure 10.5: T model with Thévenin source connected to the base.

Let us define the resistance r'_e by

$$r'_e = \frac{R_{tb} + r_x}{1+\beta} + r_e = \frac{R_{tb} + r_x + r_\pi}{1+\beta} \qquad (10.29)$$

With this definition, v_e is given by

$$v_e = v_{tb} - i'_e r'_e \qquad (10.30)$$

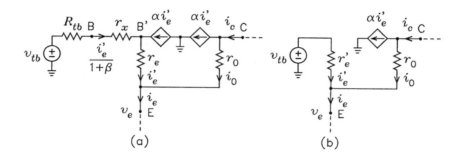

Figure 10.6: (a) Circuit with the i'_c source replaced by identical series sources. (b) Simplified T model.

The circuit which models Eq. (10.30) is shown in Fig. 10.6(b). We will call this the simplified T model. It predicts the same emitter and collector currents as the circuit in Fig. 10.5. Note that the resistors R_{tb} and r_x do not appear in this circuit because they are contained in r'_e.

10.2.7 Norton Collector Circuit

The Norton equivalent circuit seen looking into the collector can be used to solve for the response of the common-emitter and common-base stages. Fig. 10.7(a) shows the bjt with Thévenin sources connected to its base and emitter. With the collector grounded, the collector current is called the short-circuit output current or $i_{c(sc)}$. The current source in the Norton collector circuit has this value. To solve for this current, we use the simplified T model in Fig. 10.7(b). The current $i_{c(sc)}$ can be solved for by superposition of v_{tb} and v_{te}.

With $v_{te} = 0$, it follows from Fig. 10.7(b) that

$$i_{c(sc)} = \alpha i'_e + i_0 = \alpha i'_e - i'_e \frac{R_{te}}{r_0 + R_{te}} = \frac{v_{tb}}{r'_e + R_{te} \| r_0} \left(\alpha - \frac{R_{te}}{r_0 + R_{te}} \right) \qquad (10.31)$$

10.2. THEORY

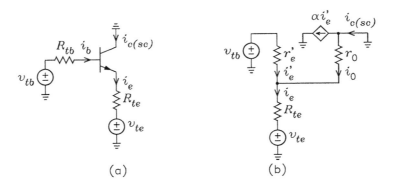

Figure 10.7: (a) BJT with Thevenin sources connected to the base and the emitter. (b) Simplified T model.

where current division has been used to express i_0 as a function of i'_e. With $v_{tb} = 0$, we have

$$i_{c(sc)} = \alpha i'_e + i_0 = \alpha i_e \frac{r_0}{r_0 + r'_e} + i_e \frac{r'_e}{r_0 + r'_e} = -\frac{v_{te}}{R_{te} + r'_e \| r_0} \frac{\alpha r_0 + r'_e}{r_0 + r'_e} \tag{10.32}$$

where current division has been used to express i'_e and i_0 as functions of i_e. These equations can be combined to obtain

$$i_{c(sc)} = \frac{v_{tb}}{r'_e + R_{te} \| r_0} \left(\alpha - \frac{R_{te}}{r_0 + R_{te}} \right) - \frac{v_{te}}{R_{te} + r'_e \| r_0} \frac{\alpha r_0 + r'_e}{r_0 + r'_e} \tag{10.33}$$

This equation is of the form

$$i_{c(sc)} = G_{mb} v_{tb} - G_{me} v_{te} \tag{10.34}$$

where

$$G_{mb} = \frac{1}{r'_e + R_{te} \| r_0} \left(\alpha - \frac{R_{te}}{r_0 + R_{te}} \right) = \frac{\alpha}{r'_e + R_{te} \| r_0} \frac{r_0 - R_{te}/\beta}{r_0 + R_{te}} \tag{10.35}$$

$$G_{me} = \frac{1}{R_{te} + r'_e \| r_0} \frac{\alpha r_0 + r'_e}{r_0 + r'_e} = \frac{\alpha}{R_{te} + r'_e \| r_0} \frac{r_0 + r'_e/\alpha}{r_0 + r'_e} \tag{10.36}$$

Figure 10.8(a) shows the simplified T model with $v_{tb} = v_{te} = 0$ and a test source connected to the collector. The resistance seen looking into the collector is given by $r_{ic} = v_t/i_c$. The resistor in the collector Norton equivalent circuit has this value. To solve for r_{ic}, we can write

$$i_c = \alpha i'_e + i_0 = -\alpha i_0 \frac{R_{te}}{r'_e + R_{te}} + i_0 = \frac{v_t}{r_0 + r'_e \| R_{te}} \left(1 - \frac{\alpha R_{te}}{r'_e + R_{te}} \right) \tag{10.37}$$

where current division has been used to express i'_e as a function of i_0. It follows that r_{ic} is given by

$$r_{ic} = \frac{v_t}{i_c} = \frac{r_0 + r'_e \| R_{te}}{1 - \alpha R_{te}/(r'_e + R_{te})} \tag{10.38}$$

The Norton equivalent circuit seen looking into the collector is shown in Fig. 10.8(b).

For the case $r_0 \gg R_{te}$ and $r_0 \gg r'_e$, we can write

$$i_{c(sc)} = G_m (v_{tb} - v_{te}) \tag{10.39}$$

where

$$G_m = \frac{\alpha}{r'_e + R_{te}} \tag{10.40}$$

The value of $i_{c(sc)}$ calculated with this approximation is simply the value of $\alpha i'_e$, where i'_e is calculated with r_0 considered to be an open circuit. The term "r_0 approximation" is used in the following when r_0 is neglected in calculating $i_{c(sc)}$ but not neglected in calculating r_{ic}.

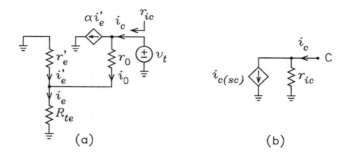

Figure 10.8: (a) Circuit for calculating r_{ic}. (b) Norton collector circuit.

10.2.8 Thévenin Emitter Circuit

The Thévenin equivalent circuit seen looking into the emitter is useful in calculating the response of common-collector stages. Fig. 10.9(a) shows the bjt symbol with a Thévenin source connected to the base. The resistor R_{tc} represents the external load resistance in series with the collector. With the emitter open circuited, the emitter voltage is called the open-circuit emitter voltage $v_{e(oc)}$. The voltage source in the Thévenin emitter circuit has this value. To solve for this voltage, we use the simplified T model in Fig. 10.9(b).

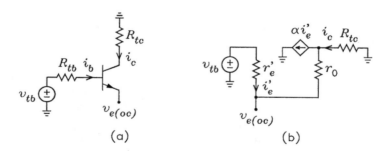

Figure 10.9: (a) BJT with Thévenin source connected to the base. (b) T model circuit for calculating $v_{e(oc)}$.

The current i'_e can be solved for by superposition of v_{tb} and $\alpha i'_e$. It is given by

$$i'_e = \frac{v_{tb}}{r'_e + r_0 + R_{tc}} + \alpha i'_e \frac{R_{tc}}{r'_e + r_0 + R_{tc}} \tag{10.41}$$

where current division has been used for the $\alpha i'_e$ term. This can be solved for i'_e to obtain

$$i'_e = \frac{v_{tb}}{r'_e + r_0 + (1-\alpha)R_{tc}} = \frac{v_{tb}}{r'_e + r_0 + R_{tc}/(1+\beta)} \tag{10.42}$$

The open-circuit emitter voltage is given by

$$v_{e(oc)} = v_{tb} - i'_e r'_e = v_{tb}\frac{r_0 + R_{tc}/(1+\beta)}{r'_e + r_0 + R_{tc}/(1+\beta)} = \frac{v_{tb}}{1 + r'_e/[r_0 + R_{tc}/(1+\beta)]} \tag{10.43}$$

The resistance seen looking into the emitter can be solved for as the ratio of the open-circuit emitter voltage to the short-circuit emitter current. The circuit for calculating the current is shown in Fig. 10.10(a).

10.2. THEORY

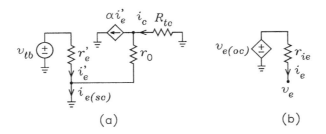

Figure 10.10: (a) Circuit for calculating $i_{e(sc)}$. (b) Thévenin emitter circuit.

By superposition of i'_e and $\alpha i'_e$, we can write

$$i_{e(sc)} = i'_e - \alpha i'_e \frac{R_{tc}}{r_0 + R_{tc}} = i'_e \frac{r_0 + (1-\alpha) R_{tc}}{r_0 + R_{tc}} = \frac{v_{tb}}{r'_e} \frac{r_0 + R_{tc}/(1+\beta)}{r_0 + R_{tc}} \qquad (10.44)$$

where current division has been used for the $\alpha i'_e$ term. The resistance seen looking into the emitter is given by

$$r_{ie} = \frac{v_{e(oc)}}{i_{e(sc)}} = \frac{r'_e \| (r_0 + R_{tc})}{1 - \alpha R_{tc}/(r'_e + r_0 + R_{tc})} = \frac{r'_e}{1 + (r'_e - \alpha R_{tc})/(r_0 + R_{tc})} \qquad (10.45)$$

The Thévenin equivalent circuit seen looking into the emitter is shown in Fig. 10.10(b).

10.2.9 Thévenin Base Circuit

Although the base is not an output terminal, the Thévenin equivalent circuit seen looking into the base is useful in calculating the base current. Fig. 10.11(a) shows the BJT symbol with a Thévenin source connected to its emitter. Fig. 10.11(b) shows the T model for calculating the open-circuit base voltage. Because $i_b = 0$, it follows that $i'_e = 0$. Thus, by voltage division, $v_{b(oc)}$ is given by

$$v_{b(oc)} = v_{te} \frac{r_0 + R_{tc}}{R_{te} + r_0 + R_{tc}} = \frac{v_{te}}{1 + R_{te}/(r_0 + R_{tc})} \qquad (10.46)$$

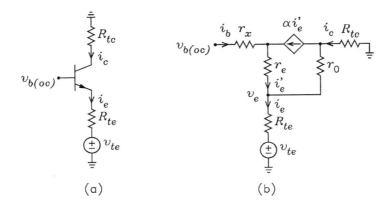

Figure 10.11: (a) BJT with Thevenin source connected to the emitter. (b) T model for calculating $v_{b(oc)}$.

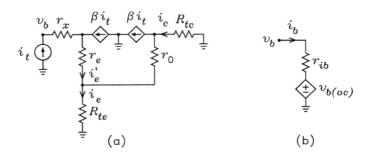

Figure 10.12: (a) Circuit for calculating v_b. (b) Thévenin base circuit.

The resistance seen looking into the base can be calculated by setting $v_{te} = 0$ and connecting a test current source to the base. The resistance is given by $r_{ib} = v_b/i_t$, where i_t is the test current. Fig. 10.12(a) shows the T circuit for calculating v_b, where the current source βi_t has been divided into identical series sources to simplify use of superposition. By superposition of i_t and the two βi_t sources, we can write

$$v_b = i_t r_x + (i_t + \beta i_t)\left[r_e + R_{te} \| (r_0 + R_{tc})\right] - \beta i_t \frac{R_{tc} R_{te}}{R_{tc} + r_0 + R_{te}} \tag{10.47}$$

This can be solved for $r_{ib} = v_b/i_t$ to obtain

$$\begin{aligned} r_{ib} &= r_x + (1+\beta)\left[r_e + R_{te} \| (r_0 + R_{tc})\right] - \frac{\beta R_{tc} R_{te}}{R_{tc} + r_0 + R_{te}} \\ &= r_x + (1+\beta) r_e + R_{te} \frac{(1+\beta) r_0 + R_{tc}}{r_0 + R_{te} + R_{tc}} \\ &= r_x + r_\pi + R_{te} \frac{(1+\beta) r_0 + R_{tc}}{r_0 + R_{te} + R_{tc}} \end{aligned} \tag{10.48}$$

The Thévenin base circuit is shown in Fig. 10.12(b).

10.2.10 Summary of Models

10.3 Small-Signal High-Frequency Models

Figure 10.14 shows the hybrid-π and T models for the BJT with the base-emitter capacitance c_π and the base-collector capacitance c_μ added. The capacitor c_{cs} is the collector-substrate capacitance which in present in monolithic integrated-circuit devices but is omitted in discrete devices. These capacitors model charge storage in the device which affects its high-frequency performance. The capacitors are given by

$$c_\pi = c_{je} + \frac{\tau_F I_C}{V_T} \tag{10.49}$$

$$c_\mu = \frac{c_{jc}}{[1 + V_{CB}/\phi_C]^{m_c}} \tag{10.50}$$

$$c_{cs} = \frac{c_{jcs}}{[1 + V_{CS}/\phi_C]^{m_c}} \tag{10.51}$$

where I_C is the dc collector current, V_{CB} is the dc collector-base voltage, V_{CS} is the dc collector-substrate voltage, c_{je} is the zero-bias junction capacitance of the base-emitter junction, τ_F is the forward transit time of the base-emitter junction, c_{jc} is the zero-bias junction capacitance of the base-collector junction,

10.3. SMALL-SIGNAL HIGH-FREQUENCY MODELS

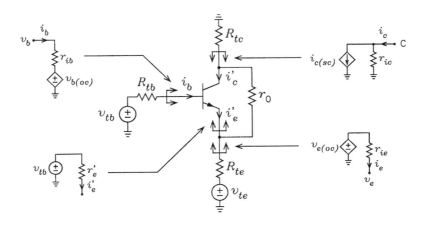

Figure 10.13: Summary of the small-signal equivalent circuits.

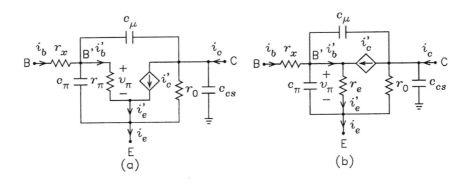

Figure 10.14: High-frequency small-signal models of the BJT. (a) Hybrid-π model. (b) T model.

c_{jcs} is the zero-bias collector-substrate capacitance, ϕ_C is the built-in potential, and m_c is the junction exponential factor. For integrated circuit lateral pnp transistors, c_{cs} is replaced with a capacitor c_{bs} from base to substrate, i.e. from the B node to ground.

In these models, the currents are related by

$$i'_c = g_m v_\pi = \beta i'_b = \alpha i'_e \qquad (10.52)$$

These relations are the same as those in Eq. (10.27) with i_b replaced with i'_b.

10.4 Preliminary Derivations

(a) Use Eqs. 10.1 through 10.4 to derive Eqs. 10.5 through 10.12. (b) Calculate the partial derivatives in Eqs. 10.16 through 10.18 to verify the equations for g_m, r_π, and r_o. (c) Use the relations between i'_b and i'_e and between r_π and r_e to verify that the voltage drops across r_π and r_e in the circuits of Fig. 10.4 are equal.

10.5 Preliminary SPICE Simulations

(a) Use SPICE to calculate the output characteristics of a PNP BJT having the parameters $I_{SO} = 1.83 \times 10^{-15}$ A, $\beta_O = 200$, and $V_A = 50$ V. Sweep the collector-to-emitter voltage from 0 V to 9.8 V in steps of 0.1 V and the base current from 10 μA to 90 μA in steps of 10 μA. Use the graphics post processor of SPICE to display the characteristic curves. (b) Use SPICE to calculate the transfer characteristics for the transistor of the preceding part for the collector-to-base voltage $V_{CB} = 5$ V. Sweep the base-to-emitter voltage from 0 V to 0.75 V in steps of 0.01 V. Use the graphics post processor of SPICE to display the transfer characteristics for the range $0 \leq I_C \leq 10$ mA.

10.6 Experimental Procedures

10.6.1 Junction Resistance Measurement

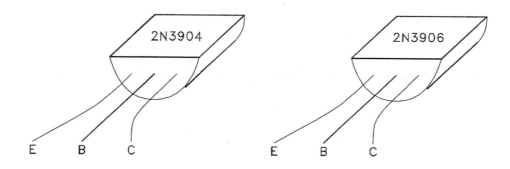

Figure 10.15: Lead arrangement for the 2N3904 and 2N3906 transistors.

The transistors specified for the experiment are the 2N3904 (NPN) and the 2N3906 (PNP). The physical arrangement of the leads for both transistors is shown in Fig. 10.15. For each transistor, use the digital multimeter to measure the resistance between two leads at a time for all permutations of the leads. (The digital multimeter should be set to the "diode test" setting.) From the measurements, verify that the transistors are fabricated as back-to-back diodes. (When using the multimeter to measure the resistance of

10.6. EXPERIMENTAL PROCEDURES

a reverse biased junction, the resistance of the human body can affect the measurements. Therefore, the transistor leads and the metal terminals of the multimeter leads should not be touched with the hands when making the resistance measurements.)

10.6.2 Preparation

Prepare the electronic breadboard to provide busses for the positive and negative power supply rails and the circuit ground. Each power supply rail should be decoupled with a 100 Ω, 1/4 W resistor and a 100 μF, 25 V (or greater) capacitor. The resistors are connected in series with the external power supply leads and the capacitors are connected from power supply rail to ground on the circuit side of the resistors. The capacitors must be installed with the proper polarity to prevent reverse polarity breakdown.

10.6.3 Threshold Measurement

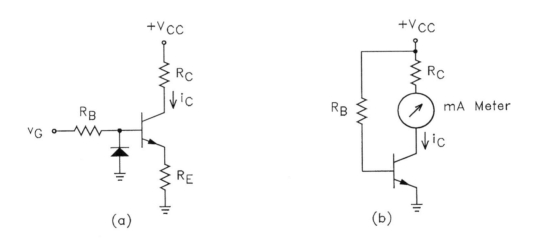

Figure 10.16: (a) Circuit for measuring the junction voltages of the BJT. (b) Circuit for measuring the Early voltage and the current gain.

The object of this step is to use the oscilloscope to measure the threshold or cutin voltage of the 2N3904 transistor. **(a)** Assemble the circuit of Fig. 10.16a on the breadboard. The diode is in the circuit to protect the transistor from accidental reverse breakdown of the base-to-emitter junction. It is specified to be a 1N4148 type. (The band on the diode case is the cathode or N side.) The circuit element values are $R_B = R_C = R_E = 1$ $k\Omega$. The power supply voltage is $V_{CC} = +15$ V. **(b)** Set the function generator for a minimum amplitude 1 kHz sine wave output with a DC offset of zero. Connect the generator to the circuit input. Connect the channel 1 input of the dual channel oscilloscope to the base terminal of the transistor and the channel 2 input to the emitter terminal. The oscilloscope should be DC coupled with a vertical gain of 0.5 V per division. **(c)** Slowly increase the output level of the function generator until the tip of the sine wave is just barely visible on the oscilloscope on the emitter waveform. Record the two voltage waveforms. Use the oscilloscope to measure the peak value of the base voltage. Record this as the cutin or threshold voltage of the transistor.

10.6.4 Junction Voltage Measurement

The object of this step is to use the oscilloscope to measure the active base-to-emitter junction voltage for several values of the emitter current. The circuit is the same as that used in step 10.6.3, i.e. the circuit in Fig. 10.16a. **(a)** Increase the amplitude of the function generator output until the peak emitter voltage is $1\ V$. This corresponds to a peak emitter current of 1 mA. Sketch the oscilloscope voltage waveforms. Read the difference voltage between the base and the emitter voltage at the peak of the sine wave. Record this voltage as the active base-to-emitter voltage for an emitter current of 1 mA. **(b)** Repeat the preceding step for peak emitter currents of 2 mA, 3 mA, 4 mA, and 5 mA. Record the measured base-to-emitter voltages.

10.6.5 Verge of Saturation Measurement

The transistor is saturated when the base-to-collector junction becomes forward biased. The object of this step is to measure the base-to-emitter junction voltage at the verge of saturation. The circuit is the same as that used in step 10.6.4. **(a)** Connect the dual channel oscilloscope so that the channel 1 input is connected to the collector terminal of the transistor and the channel 2 input is connected to the base. Set the gain for both channels to $2\ V$ per division. With the input coupling selector for both channels of the oscilloscope set to the ground position, adjust the vertical position of the traces so that they overlap on the fourth major division below the center line on the oscilloscope screen. Unground both channels and set them to DC coupling. **(b)** While observing the base and collector voltage waveforms, slowly increase the function generator output until the negative peak of the collector waveform just touches the positive peak of the base waveform. This corresponds to $V_{CB} = 0$ at the peaks. Sketch the voltage waveforms. **(c)** Without changing the function generator settings, disconnect the channel A input of the oscilloscope from the collector terminal of the transistor and connect it to the emitter terminal. Record the base and emitter voltage waveforms. Read the difference voltage between the base and the emitter voltage waveforms at the peak of the sine wave. Record this voltage as the base-to-emitter voltage for the verge of saturation.

10.6.6 Hard Saturation

The object of this step is to measure the voltages across the transistor when it is in hard saturation. The circuit is the same as that used in step 10.6.5. **(a)** Connect the dual channel oscilloscope so that the channel 1 input is connected to the base terminal of the transistor and the channel 2 input is connected to the collector terminal. Set the gain for both channels to $2\ V$ per division. With the input coupling selector for both channels of the oscilloscope set to the ground position, adjust the vertical position of the traces so that they overlap on the fourth major division below the center line on the oscilloscope screen. Unground the two channels and set them to DC coupling. **(b)** While observing the base and collector voltage waveforms, slowly increase the function generator output until the negative peak of the collector waveform drops below the positive peak of the base waveform and becomes inverted, i.e. the inverted collector waveform begins to show a positive going peak caused by the base-to-collector diode being forward biased. Sketch the oscilloscope voltage waveforms. Read the difference voltage between the base and the collector waveforms at the peak of the sine wave. Record this voltage as the saturation base-to-collector voltage. **(c)** Without changing the settings of the function generator, disconnect the channel 1 input of the oscilloscope from the base terminal of the transistor and connect it to the emitter terminal. Sketch the collector and emitter voltage waveforms. Read the difference voltage between the collector and the emitter voltage waveforms at the peak of the emitter voltage. Record this voltage as the saturation collector-to-emitter voltage. **(d)** Without changing the settings of the function generator, disconnect channel 2 of the oscilloscope from the collector terminal of the transistor and connect it to the base terminal. Sketch the oscilloscope voltage waveforms. Read the difference voltage between the base and the emitter voltage waveforms at the peak of the sine wave. Record this voltage as the saturation base-to-emitter voltage.

10.6.7 Early Voltage Measurement

The object of this step of the experiment is to measure the Early voltage V_A for the transistor. **(a)** On the breadboard, assemble the circuit of Fig. 10.16b, where the element values are $R_B = 200\ k\Omega$ and $R_C = 1\ k\Omega$. For V_{CC}, the value $+15\ V$ is specified. **(b)** Use the multimeter to measure the DC collector current in the transistor. If the current is greater than 10 mA, increase R_B in 5% steps until a current just less than 10 mA is obtained. Record this current. **(c)** Connect a short circuit jumper wire across R_C and record the new collector current. The current will change by only a small amount. However, the change should be easy to measure. Care must be taken to take the current reading at the time that the short circuit jumper is connected to eliminate errors caused by temperature drift of the transistor. It should be observed that the current drifts to a higher value after the jumper is connected because of the temperature increase due to the increase in power dissipation in the transistor. **(d)** Remove the short circuit jumper across R_C and connect it across the breadboard terminals to which the digital multimeter connects. Disconnect the multimeter and use it to measure the BJT base-to-emitter voltage v_{BE}. **(e)** Calculate $m = \Delta i_C/\Delta v_{CE}$ from the measured data, where $\Delta v_{CE} = i_C R_C$ and i_C is the collector current with the jumper not connected across R_C. Use Eq. 10.10 to calculate V_A for the transistor, where $v_{CE} = V_{CC} - i_C R_C$.

10.6.8 Saturation Current Measurement

Use the measured data taken in the preceding part and Eq. 10.12 to calculate the zero-bias saturation current I_{S0} of the transistor.

10.6.9 Current Gain Measurement

Use the measured data taken in part 10.6.7 to calculate the BJT base current from the relation

$$i_B = \frac{V_{CC} - v_{BE}}{R_B} \tag{10.53}$$

Use the calculated value of i_B, the measured data taken in part 10.6.7, and Eq. 10.11 to calculate the value of β_0 for the transistor.

10.6.10 PNP Transistor Measurement

Repeat parts 10.6.3 through 10.6.9 with the 2N3906 PNP transistor in place of the 2N3904 NPN transistor. For V_{CC}, use the value $-15\ V$ in the circuits of Fig. 10.16. In addition, the direction of the diode in Fig. 10.16a must be reversed. It should be found that the junction voltage values are similar for the two transistor types. However, the polarity of the voltages is reversed for the PNP type. The Early voltage for the PNP device is usually less than that for the NPN device. In addition, the current gain is generally higher for the PNP device.

10.6.11 Curve Tracer

Use the laboratory transistor curve tracer to obtain plots of the output characteristics, i.e. I_C versus V_{CE} for stepped values of the base current I_B, for both of the transistors used in this experiment.

10.7 Laboratory Report

The laboratory report should include:

- All preliminary derivations, calculations, and SPICE simulations
- The measured data for the two transistors should be presented in a tabular and graphical formats.

- Best estimates of the saturation current I_{SO}, the current gain β_O, and the Early voltage V_A for each transistor should be calculated and displayed in a tabular format.

- SPICE simulations to calculate and plot the output and transfer characteristics of the 2N3904 and 2N3906 BJT's measured in the laboratory.

A comparison of the measured parameters of the 2N3904 and 2N3906 BJT's with the SPICE default values.

10.8 References

1. E. J. Angelo, *Electronics: BJT's, FET's, and Microcircuits*, McGraw-Hill, 1969.
2. W. Banzhaf, *Computer-Aided Circuit Analysis Using SPICE*, Prentice-Hall, 1989.
3. M. N. Horenstein, *Microelectronics Circuits and Devices*, Prentice-Hall, 1990.
4. P. Horwitz & W. Hill, *The Art of Electronics*, 2nd edition, Cambridge University Press, 1989.
5. P. Horowitz & I. Robinson, *Laboratory Manual for The Art of Electronics*, Cambridge University Press, 1981.
6. J. H. Krenz, *An Introduction to Electrical and Electronic Devices*, Prentice-Hall, 1987.
7. R. Mauro, *Engineering Electronics*, Prentice-Hall, 1989.
8. F. H. Mitchell & F. H. Mitchell, *Introduction to Electronic Design*, Prentice-Hall, 1988.
9. Motorola, Inc., *Small-Signal Semiconductors*, DL 126, Motorola, 1987.
10. C. J. Savant, M. S. Roden, & G. L. Carpenter, *Electronic Circuit Design*, Benjamin Cummings, 1987.
11. A. S. Sedra & K. C. Smith, *Microelectronics Circuits*, 4th edition, Oxford, 1998.
12. D. L. Schilling & C. Belove, *Electronic Circuits: Discrete and Integrated*, McGraw-Hill, 1968.
13. P. W. Tuinenga, *SPICE*, Prentice-Hall, 1988.

Chapter 11

The Common-Emitter Amplifier

11.1 Object

The object of this experiment is to study a classical circuit of analog electronics – the common emitter BJT amplifier. The dc biasing, midband voltage gain, midband input and output resistances, and frequency response are examined.

11.2 Theory

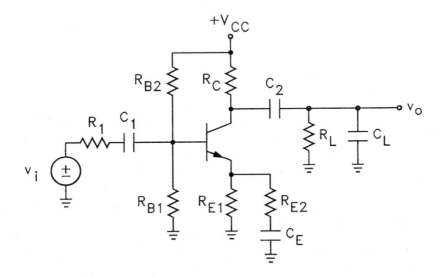

Figure 11.1: Circuit diagram of the NPN common-emitter BJT amplifier.

11.2.1 Circuit Description

Fig. 11.1 gives the circuit diagram of the classical common-emitter (CE) BJT amplifier. The circuit is called a CE amplifier because the emitter is connected through a RC network to the circuit ground which is common to both the input and the output. In other words, the emitter is neither the input nor the output terminal. The active device is a NPN BJT which is assumed to be biased in its active mode. In this mode the base-to-emitter junction is forward biased and the collector-to-base junction is reverse biased. The CE amplifier is commonly used as a voltage gain stage in amplifier circuits.

In the circuit of Fig. 11.1, the generator is represented by a Thévenin equivalent source having an open circuit output voltage v_i and an output resistance R_1. The load is modeled as a parallel resistor R_L and capacitor C_L. The BJT base bias is provided by a voltage divider consisting of resistors R_{B1} and R_{B2} connected between the power supply rail and the base. Resistor R_C is the collector bias resistor. Resistor R_{E1} is the emitter bias resistor. Capacitor C_1 is a dc blocking capacitor which prevents dc current from flowing in the generator. It is also called an input coupling capacitor because it couples the generator to the BJT base. Capacitor C_2 prevents dc current from flowing in the load. It is also called an output coupling capacitor because it couples the BJT collector to the load resistor R_L. Capacitor C_E is called an emitter bypass capacitor because it bypasses emitter signal current to ground through R_{E2}. At dc, C_E is an open circuit so that the dc emitter current must flow through R_{E1}. This improves the stability of the BJT bias current.

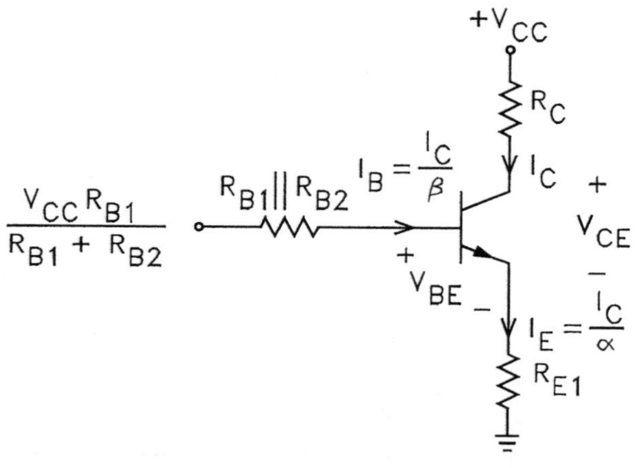

Figure 11.2: Base equivalent circuit for the CE amplifier.

11.2.2 dc Bias Equation

Assume that the transistor is biased in the active mode. In the active mode, the base-to-emitter junction is forward biased and the collector-to-base junction is reverse biased, i.e. $V_{BE} > 0$ and $V_{CB} > 0$. Let I_C, I_B, and I_E, respectively, be the dc bias currents in the BJT collector, base, and emitter. These are related by the equations $I_C = \alpha I_E = \beta I_B$, where α is the emitter-to-collector current gain and β is the base-to-collector current gain. The current gains are related by the equation $\alpha = \beta/(1+\beta)$. Fig. 11.2 shows the bias equivalent circuit for the CE amplifier. This is obtained by replacing all capacitors with open circuits

11.2. THEORY

and by making a Thévenin equivalent circuit seen looking out of the base. For the circuit, the equation

$$\frac{R_{B1}}{R_{B1}+R_{B2}}V_{CC} = \frac{I_C}{\beta}R_{B1}\|R_{B2} + V_{BE} + \frac{I_C}{\alpha}R_{E1} \tag{11.1}$$

holds. This equation can be solved for I_C to obtain

$$I_C = \frac{\dfrac{V_{CC}R_{B1}}{R_{B1}+R_{B2}} - V_{BE}}{\dfrac{R_{B1}\|R_{B2}}{\beta} + \dfrac{R_{E1}}{\alpha}} \tag{11.2}$$

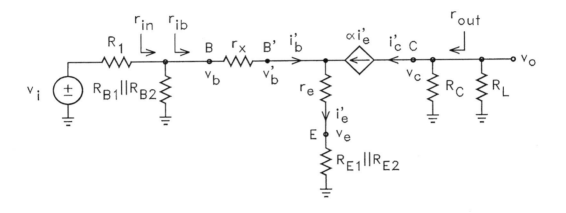

Figure 11.3: Midband small-signal voltage gain.

11.2.3 Midband Small-Signal Voltage Gain

The midband frequency range is defined as the frequency band for which the coupling and bypass capacitors can be considered to be signal short circuits and C_L and the internal capacitances inside the transistor (c_π and c_μ in the small-signal model) can be considered to be open circuits. Fig. 11.3 shows the small-signal circuit which is used to solve for the midband voltage gain. The BJT has been replaced by its T model. It is assumed that the collector-to-emitter resistor r_0 in the model is large enough so that it can be neglected. The intrinsic emitter resistance is given by $r_e = V_T/I_E$, where V_T is the thermal voltage and I_E is the quiescent emitter current.

Fig. 11.4 shows the circuit of Fig. 11.3 with the external base circuit replaced by a Thévenin equivalent circuit. In this circuit, v_{tb}, R_{tb}, and R_{te} are given by

$$v_{tb} = v_i \frac{R_{B1}\|R_{B2}}{R_1 + R_{B1}\|R_{B2}} \tag{11.3}$$

$$R_{tb} = R_1\|R_{B1}\|R_{B2} \tag{11.4}$$

$$R_{te} = R_{E1}\|R_{E2} \tag{11.5}$$

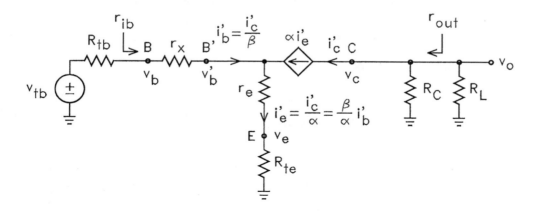

Figure 11.4: Circuit of Fig. 11-3 with Thévenin equivalent circuit seen looking out of base.

For the circuit of Fig. 11.4, the relationships

$$v_{tb} = \frac{i'_c}{\beta}(R_{tb} + r_x) + \frac{i'_c}{\alpha}(r_e + R_{te}) \tag{11.6}$$

hold. This equation can be solved for i'_c/v_{tb} to obtain

$$\frac{i'_c}{v_{tb}} = \frac{1}{\frac{R_{tb} + r_x}{\beta} + \frac{r_e + R_{te}}{\alpha}} = \frac{\alpha}{r'_e + R_{te}} \tag{11.7}$$

where r'_e is defined by the relation

$$r'_e = \frac{\alpha}{\beta}(R_{tb} + r_x) + r_e = \frac{R_{tb} + r_x}{1 + \beta} + r_e \tag{11.8}$$

The voltage gain v_o/v_i can be written as the product of three terms as follows:

$$\frac{v_o}{v_i} = \frac{v_{tb}}{v_i} \times \frac{i'_c}{v_{tb}} \times \frac{v_o}{i'_c} = \frac{R_{B1} \| R_{B2}}{R_1 + R_{B1} \| R_{B2}} \times \frac{\alpha}{r'_e + R_{te}} \times (-R_C \| R_L) \tag{11.9}$$

11.2.4 Approximate Voltage-Gain Expressions

The large-β approximation to Eq. 11.9 is commonly used for design purposes. This approximation is obtained by letting $\beta \to \infty$ (and $\alpha \to 1$) to obtain

$$\frac{v_o}{v_i} \simeq \frac{R_{B1} \| R_{B2}}{R_1 + R_{B1} \| R_{B2}} \times \frac{1}{r_e + R_{te}} \times (-R_C \| R_L) \tag{11.10}$$

If $R_1 \ll R_{B1} \| R_{B2}$, this can be further simplified to

$$\frac{v_o}{v_i} \simeq \frac{1}{r_e + R_{te}} \times (-R_C \| R_L) \tag{11.11}$$

This is the negative of the ratio of the small-signal midband resistance seen looking out of the collector to the small-signal midband resistance of the emitter circuit.

11.2. THEORY

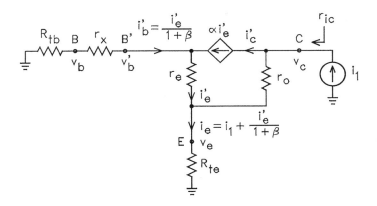

Figure 11.5: Small-signal circuit used to calculate r_{ic}.

11.2.5 Modeling the Effect of r_0

If the collector-to-emitter resistor r_0 is included in the T model, denote by r_{ic} the small-signal resistance seen looking into the collector with the generator zeroed. Fig. 11.5 shows the circuit which is used to calculate r_{ic}. The circuit shows a current source driving the collector node of the BJT. The relationships

$$v_c = (i_1 - \alpha i'_e)r_0 + \left[i_1 + \frac{i'_e}{1+\beta}\right] R_{te} \tag{11.12}$$

$$\frac{i'_e}{1+\beta}(R_{tb} + r_x) + i'_e r_e + \left[i_1 + \frac{i'_e}{1+\beta}\right] R_{te} = 0 \tag{11.13}$$

hold. These equations and the relation $\alpha = \beta/(1+\beta)$ can be solved for r_{ic} to obtain

$$r_{ic} = \frac{v_c}{i_1} = \frac{r_0 + r'_e \| R_{te}}{1 - \dfrac{\alpha R_{te}}{r'_e + R_{te}}} \tag{11.14}$$

where r'_e is defined by Eq. 11.8. In these equations, r_0 is given by $r_0 = (V_{CB} + V_A)/I_C$ and V_A is the BJT Early voltage. The effect of r_{ic} on the small-signal gain of the CE amplifier can be accounted for in Eqs. 11.9 through 11.11 by replacing $(-R_C \| R_L)$ with $(-r_{ic} \| R_C \| R_L)$.

11.2.6 Input Resistance

The midband small-signal input resistance to the CE amplifier is the resistance r_{in} in Fig. 11.3 seen by the generator. It is given by

$$r_{in} = R_{B1} \| R_{B2} \| r_{ib} \tag{11.15}$$

where r_{ib} is the small-signal resistance seen looking into the BJT base. To solve for r_{ib}, the circuit of Fig. 11.4 will be used. The relationship

$$v_b = i'_b r_x + (1+\beta)i'_b(r_e + R_{te}) \tag{11.16}$$

holds. This equation can be solved for the small-signal resistance seen looking into the base to obtain

$$r_{ib} = \frac{v_b}{i'_b} = r_x + (1+\beta)(r_e + R_{te}) \tag{11.17}$$

11.2.7 Output Resistance

The midband small-signal output resistance is the resistance seen looking into the output lead of the amplifier, i.e. the source resistance seen by the load. It is given by

$$r_{\text{out}} = r_{ic} \| R_C \cong R_C \tag{11.18}$$

where the approximation is valid if $r_{ic} \gg R_C$. Thus the output resistance is approximately equal to the collector bias resistor.

11.2.8 Design Criteria

In a normal design procedure, the load resistance R_L and the generator output resistance R_1 are given. The other resistors in the circuit are then calculated to meet the desired specifications. These specifications include the midband small-signal voltage gain, the input and output resistances, and any specifications on the peak clipping levels for the output signal. An amplifier that is designed for ac signals is normally designed so that the peak clipping of the output signal occurs symmetrically on the positive and negative peaks of the output voltage swing. In any design procedure, the amplifier should be designed to make the performance as insensitive to variations in the parameters of the transistor as possible.

11.2.9 Biasing for Equal Voltage Drops

A rule of thumb that is often used for biasing the circuit in Fig. 11.1 is to design for equal quiescent dc voltage drops across the collector resistor R_C, between the collector-to-emitter nodes of the transistor, and across the resistor R_{E1}. It follows that this voltage drop must be $V_{CC}/3$. Once the collector current is specified, the bias resistors can be calculated. The resistors R_{B1} and R_{B2} are chosen to set the collector current and to set the input resistance to the circuit.

Design Example 1

Let $R_L = 2$ kΩ, $R_1 = 50$ Ω, and $V_{CC} = +15$ V be given. It is desired to design a CE amplifier with the midband small-signal voltage gain $v_o/v_i = -5$. The circuit is to be biased for equal quiescent dc voltage drops across R_C, between the BJT collector-to-emitter nodes, and across R_{E1}. For the BJT parameters, it is given that $\beta = 99$, $r_x = 0$, $r_0 = \infty$, $V_{BE} = 0.65$ V, and $V_T = 25.9$ mV. Choose $I_E = 2.5$ mA as the bias current for the transistor. For R_C, R_{E1}, and r_e, $R_C = 5/(0.99 \times 2.5m) = 2.02$ kΩ, $R_{E1} = 5/2.5m = 2$ kΩ, and $r_e = 25.9m/2.5m = 10.4$ Ω is obtained.

The voltage across R_{B1} is given by $I_E R_{E1} + V_{BE} = 5.65$ V. Choose the current through R_{B1} to be $19 I_B = 19 \times 2.5m/100 = 0.475$ mA. It follows that $R_{B1} = 5.65/0.475m = 11.9$ kΩ. The current through R_{B2} is $20 I_B = 20 \times 2.5m/100 = 0.5$ mA. It follows that $R_{B2} = (15 - 5.65)/0.5m = 18.7$ kΩ. For a voltage gain of -5, Eq. 11.9 gives

$$-5 = \frac{7.27k}{50 + 7.27k} \times \frac{0.99}{\dfrac{50 \| 7.27k}{100} + 10.4 + R_{te}} \tag{11.19}$$

Solution for R_{te} yields $R_{te} = 187$ Ω. Because $R_{te} = R_{E1} \| R_{E2}$ and $R_{E1} = 2$ kΩ, it follows that $R_{E2} = 2k \times 187/(2k - 187) = 206$ Ω.

11.2.10 Biasing for Symmetrical Clipping

A second method for biasing the circuit in Fig. 11.1 is to select the circuit components for symmetric clipping. Clipping occurs when the transistor enters either cutoff or saturation so that the output signal cannot respond to further increases in the amplitude of the input signal. For the NPN BJT, cutoff occurs with a negative going input signal which reverse biases the transistor base-to-emitter junction. Saturation occurs with a positive going input signal which forward biases the base-to-collector junction. If a symmetric signal

11.2. THEORY

is applied to the input, e.g. a sine wave, and the amplitude is slowly increased, symmetric clipping occurs if the BJT simultaneously reaches the verge of cutoff and the verge of saturation for the same amplitude input signal. With symmetric ac input signals, biasing for symmetric clipping is the preferred biasing method.

The clipping behavior of the CE amplifier can be analyzed by calculating the peak output voltages at which the transistor is at the verge of cutoff and at the verge of saturation. Because the bypass and coupling capacitors have dc voltage drops across them, the model that is used for each is an uncharged capacitor in series with a battery equal to the dc voltage drop across the charged capacitor. At midband frequencies, each uncharged capacitor is considered to be a short circuit. Thus the capacitors can be represented by a dc battery having a voltage equal to the voltage across the charged capacitor.

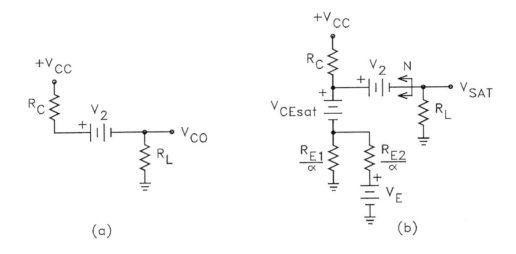

Figure 11.6: (a) Circuit for the transistor at the verge of cutoff. (b) Circuit for the transistor at the verge of saturation.

For the positive output clipping level, the BJT is at the verge of cutoff and the equivalent circuit shown in Fig. 11.6a applies. Because the total instantaneous collector current is zero at the verge of cutoff, the output voltage across R_L is independent of the transistor and the circuits connected to its base and emitter leads. The dc voltage drop V_2 across the output coupling capacitor C_2 is given by

$$V_2 = V_{CC} - I_C R_C \tag{11.20}$$

where I_C is the quiescent collector bias current. It follows from the circuit that the peak output voltage when the transistor is at the verge of cutoff is given by

$$V_{CO} = (V_{CC} - V_2)\frac{R_L}{R_L + R_C} = I_C(R_C \| R_L) \tag{11.21}$$

This represents the maximum possible positive voltage swing at the output of the CE amplifier.

For the negative output clipping level, the transistor is at the verge of saturation and the equivalent circuit shown in Fig. 11.6b applies. The emitter resistors are scaled by the factor $1/\alpha$ in order for the voltage drops across them to be correct. This must be done because the BJT base and collector currents flow into the external emitter circuit in the original circuit while only the collector current flows into that circuit in the figure. The battery V_{CEsat} represents the collector-to-emitter saturation voltage. At the verge of saturation, $V_{CB} = 0$ so that $V_{CEsat} = V_{BE}$ for the following calculations. The dc voltage drop V_E across the emitter bypass capacitor C_E is given by

$$V_E = I_E R_{E1} \tag{11.22}$$

where I_E is the quiescent emitter bias current. It follows from the circuit that the output voltage V_{SAT} when the transistor is at the verge of saturation is given by

$$V_{SAT} = -\left[V_{CC} - V_{CEsat} - I_C\left(R_C + \frac{R_{E1}}{\alpha}\right)\right] \frac{R_C \| R_L}{R_C \| R_L + \frac{R_{E1} \| R_{E2}}{\alpha}} \tag{11.23}$$

This represents the maximum possible negative voltage swing at the output of the CE amplifier. (The derivation of Eqn. 11.23 may be expedited by finding a Norton equivalent with respect to the load resistor, R_L, as shown in Fig. 11.6.).

Symmetric clipping occurs when $V_{C0} = -V_{SAT}$. This condition places the transistor at the verge of cutoff and the verge of saturation for the same amplitude input signal. It follows that the value of I_C which gives symmetric clipping is given by

$$I_C = \frac{V_{CC} - V_{CEsat}}{R_C \| R_L + R_C + \frac{R_{E1} + R_{E1} \| R_{E2}}{\alpha}} \tag{11.24}$$

Design Example 2

As an example of biasing for symmetric clipping, let $R_L = 2$ kΩ, $R_1 = 50$ Ω, and $V_{CC} = +15$ V be given. The midband small-signal voltage gain is specified to be -5. For the BJT parameters, it is given that $\beta = 99$, $r_x = 0$, $r_0 = \infty$, $V_{BE} = 0.65$ V, and $V_T = 25.9$ mV. Choose $R_C = 4.3$ kΩ and $R_{E1} = 1$ kΩ. Assume that the approximate gain relation in Eq. 11.11 can be used to solve for $r_e + R_{te}$, where at midband $R_{te} = R_{E1} \| R_{E2}$. From Eq. 11.11, the expression

$$-5 \simeq \frac{1}{r_e + R_{te}} \times (-4.3k\|2k) \tag{11.25}$$

is obtained. The solution for $r_e + R_{te}$ yields $r_e + R_{te} = 273$ Ω. Because r_e is small compared to $R_{te} = R_{E1}\|R_{E2}$, it is reasonable to assume that $R_{te} \simeq 273$ Ω. It follows the R_{E2} is given by $R_{E2} = 1k \times 273/(1k - 273) = 376$ Ω. Eq. 11.24 can be used to solve for the collector bias current I_C to obtain $I_C = 2.06$ mA. In this case, $r_e = 25.9m/(0.99 \times 2.06m) = 12.7$ Ω. Because this is much smaller than 273 Ω, the approximation used to solve for R_{E2} and I_C is valid.

To solve for the base bias resistors, the current in R_{B1} will be specified to be $19I_B$ and the current in R_{B2} to be $20I_B$. The voltage across R_{B1} is $0.65 + 2m \times 1k = 2.65$ V. The voltage across R_{B2} is $15 - 2.65 = 12.4$ V. It follows that R_{B1} and R_{B2} are given by $R_{B1} = 2.65/(19 \times 2.06m/99) = 6.70$ kΩ and $R_{B2} = 12.4/(20 \times 2m/99) = 29.8$ kΩ. Eq. 11.9 can be used to solve for the actual gain of the circuit to obtain

$$\frac{v_o}{v_i} = \frac{6.7k\|29.8k}{50 + 6.7k\|29.8k} \times \frac{0.99}{\frac{50\|6.7k\|29.8k}{100} + 273} \times (-4.3k\|2k) = -4.9 \tag{11.26}$$

This is only 2% low compared to the target value of -5. The gain can be "tweaked" by decreasing the value of R_{E2}.

11.2.11 Lower Cutoff Frequency

The three capacitors C_1, C_2, and C_E determine the amplifier lower cutoff frequency f_ℓ. This frequency is defined as the frequency where the voltage gain is reduced by a factor or 0.707 from its midband value, i.e. the gain is down by 3 dB. The cutoff frequency can be calculated from the time constants for the capacitors.

11.2. THEORY

It can be shown that C_1 sets a pole time constant τ_1, C_2 sets a pole time constant τ_2, and C_E sets both a pole time constant τ_{E1} and a zero time constant τ_{E2}. These time constants are given by

$$\tau_1 = [R_1 + R_{B1}\|R_{B2}\|r_{ib}]\, C_1 \tag{11.27}$$

$$\tau_2 = (r_{ic}\|R_C + R_L)C_2 \tag{11.28}$$

$$\tau_{E1} = [R_{E2} + r'_e\|R_{E1}]\, C_E \tag{11.29}$$

$$\tau_{E2} = (R_{E1} + R_{E2})C_E \tag{11.30}$$

The lower cutoff frequency is approximately given by

$$f_\ell = \frac{1}{2\pi}\sqrt{\frac{1}{\tau_1^2} + \frac{1}{\tau_2^2} + \frac{1}{\tau_{E1}^2} - \frac{2}{\tau_{E2}^2}} \tag{11.31}$$

11.2.12 Upper Cutoff Frequency

The amplifier upper cutoff frequency f_u is a function of the load capacitance C_L, the base-to-emitter diffusion capacitance c_π, and the base-to-collector junction capacitance c_μ. The latter two capacitors are internal to the transistor. The model equations for these are

$$c_\pi = \frac{\tau_F I_C}{V_T} \tag{11.32}$$

$$c_\mu = \frac{c_{jco}}{\left[1 + \dfrac{V_{CB}}{\Phi_C}\right]^{m_c}} \tag{11.33}$$

where τ_F is the forward transit time of the base-to-emitter junction, c_{jco} is the zero bias junction capacitance of the base-to-collector junction, Φ_C is the built-in potential (the SPICE default value is 0.75 V), and m_c is the junction exponential factor (the SPICE default value is 0.33).

The upper cutoff frequency can be calculated from the time constants for the capacitors. It can be shown that C_L sets a pole time constant τ_L, c_π sets a pole time constant τ_π, and c_μ sets a pole time constant $\tau_{\mu1}$ and a zero time constant $\tau_{\mu2}$. These time constants are calculated with C_1, C_2, and C_E shorted. They are given by

$$\tau_L = (r_{ic}\|R_C\|R_L)C_L \tag{11.34}$$

$$\tau_\pi = \frac{(R_{tb} + r_x + R_{te})r_e}{r'_e + R_{te}} c_\pi \tag{11.35}$$

$$\tau_{\mu1} = \left([(R_{tb} + r_x)\|r'_{ib}]\left[1 + \alpha\frac{r_{ic}\|R_C\|R_L}{r_e + R_{te}}\right] + r_{ic}\|R_C\|R_L\right)c_\mu \tag{11.36}$$

$$\tau_{\mu2} = \frac{r_e + R_{te}}{\alpha} c_\mu \tag{11.37}$$

where r'_{ib} is given by

$$r'_{ib} = r_{ib} - r_x = (1+\beta)(r_e + R_{E1}) \tag{11.38}$$

The upper cutoff frequency f_u is approximately given by

$$f_u = \frac{1}{2\pi\sqrt{\tau_L^2 + \tau_\pi^2 + \tau_{\mu1}^2 - 2\tau_{\mu2}^2}} \tag{11.39}$$

11.3 Preliminary Derivations

Derive Eqs. 11.1, 11.2, 11.14, 11.21, 11.23, and 11.24. It is suggested that Eq. 11.23 be written by superposition of V_{CC}, V_E, V_{CEsat}, and V_2 to find the Norton equivalent circuit with respect to the load resistor.

11.4 Preliminary Calculations

1. For the circuit of Fig. 11.1, it is given that $R_C = R_L = 3$ kΩ and $V_{CC} = 15$ V. Determine the values of R_{B1}, R_{B2}, R_{E1}, and R_{E2} so that the midband small-signal voltage gain is approximately -10, the quiescent collector current is 2 mA, the collector-to-emitter voltage is 6 V, and the quiescent dc voltage drop across R_{E1} is 3 V. In the calculations, use the BJT parameters $\beta = 99$, $r_x = 0$, $r_0 = \infty$, $V_T = 0.0259$ V, and $V_{BE} = 0.65$ V. The quiescent current in R_{B1} is specified to be $19 I_B$. The nearest 5% standard values should be specified for all resistors.

2. Repeat the design of the CE amplifier described in the preceding part with the exception that the amplifier is to be biased for symmetric clipping. The nearest 5% standard values should be specified for all resistors.

11.5 Preliminary SPICE Simulations

The transistor specified for this experiment is the 2N3904 NPN general purpose small-signal transistor. The ".MODEL" parameters to be used for the SPICE simulations are as follows: IS=2E-14 (I_{S0}), VA=170 (V_A), BF=200 (β_0), CJC=3.6E-12 (c_{jco}), TF=3E-10 (τ_F), and RB=10 (r_x). Although these parameters vary from transistor to transistor, these values are representative for the 2N3904. There are other SPICE ".MODEL" parameters for BJT, but these have little effect on the circuits that will be examined in this and subsequent experiments.

1. Write the code for the SPICE input deck to perform the ".AC" and ".TRAN" analyses on the circuit of Fig. 1 for the resistor values calculated in *Example 1* in the Theory section and the capacitor values $C_1 = 1$ μF, $C_2 = 10$ μF, $C_E = 100$ μF, and $C_L = 100$ pF. For the ".AC" analysis, use the frequency range from 10 Hz to 10 MHz with 20 points per decade. Display the Bode magnitude plot for v_o/v_i using a log scale for the vertical axis (use the y-axis menu in the PROBE graphics processor to convert to a log scale). Compare the calculated midband gain and lower cutoff frequency to the values predicted by the theory. For the ".TRAN" analysis, use a sine wave input signal for v_i having frequency of 1 kHz and a dc level of 0 V. The amplitude of the input signal should be varied so that the clipping behavior can be investigated. Several plots should be obtained to illustrate the following cases: an undistorted output, a distorted output prior to clipping (unequal positive and negative peaks), and a clipped output. In conjunction with the ".TRAN" analysis, perform a ".FOUR" (Fourier) analysis and determine the input signal level for which the total harmonic distortion (THD) is 10%.

2. Repeat step 1 for the component values given in *Example 2* of the Theory section. Use the same capacitor values used in step 1 of this section.

3. Repeat step 1 for the component values determined in step 1 of the Preliminary Calculations. Use the same capacitor values used in step 1 of this section.

4. Repeat step 1 for the component values determined in step 2 of the Preliminary Calculations. Use the same capacitor values used in step 1 of this section.

11.6 Experimental Procedures

The BJT specified for this experiment is the general purpose small-signal 2N3904 NPN transistor. The physical arrangement of the leads is shown in Fig. 11.7. If the package is held so that the flat face is toward the viewer, the leads are arranged from left to right as emitter, base, and collector.

11.6. EXPERIMENTAL PROCEDURES

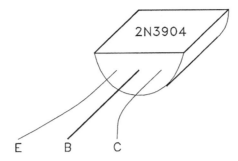

Figure 11.7: Lead arrangement for 2N3904 NPN BJT transistor.

The values specified for the capacitors in this experiment are: $C_1 = 1\ \mu\text{F}$, $C_2 = 10\ \mu\text{F}$, and $C_E = 100\ \mu\text{F}$. These capacitors are electrolytic types that must be inserted into the breadboard with the proper polarity to prevent reverse polarity breakdown. Capacitor C_L in the circuit diagram models the capacitance of the oscilloscope leads and the oscilloscope input amplifier. It is not a capacitor that is inserted into the breadboard.

11.6.1 Preparation

Prepare the electronic breadboard to provide busses for the positive power supply rail and the circuit ground. The power supply rail should be decoupled with a 100 Ω resistor and a 100 μF, 25 V (or greater) capacitor. The resistor is connected in series with the external power supply lead and the capacitor is connected from power supply rail to ground on the circuit side of the resistor. The capacitor must be installed with the proper polarity to prevent reverse polarity breakdown.

11.6.2 Bias Measurement

Assemble the circuit shown in Fig. 11.1 using the resistor values calculated in Example 1 of the Theory section, i.e. $R_{B1} = 12$ kΩ, $R_{B2} = 18$ kΩ, $R_C = 2$ kΩ, $R_L = 2$ kΩ, $R_{E1} = 2$ kΩ, and $R_{E2} = 200\ \Omega$. Use the digital voltmeter to measure the dc voltage across the collector resistor. Use this reading along with the measured resistance value to calculate the quiescent collector current. If this value is not reasonably close to 2.5 mA, examine the circuit for errors.

11.6.3 Gain Measurement

(a) With the signal generator connected to the circuit input, set the generator to produce a sine-wave output with a frequency of 1 kHz, a dc level of zero, and a peak-to-peak ac amplitude of 0.5 V. Use the dual channel oscilloscope to observe the input and output signals. (If the oscilloscope has a bandwidth limit feature, it should be enabled.) Both channels should be dc coupled. To prevent possible oscillation problems when the oscilloscope test leads are connected to the circuit, it is recommended that a 100 Ω resistor be used in series with the test lead for each channel. Set the oscilloscope for XY operation and find the frequency at which the Lissajous figure collapses to a straight line. This is the center frequency of the midband region. Measure the midband small-signal voltage gain to determine if it is reasonably close to -5.
(b) Vary the frequency of the signal generator and determine the two frequencies at which the magnitude of the gain is reduced by a factor of 0.707 from the midband value, i.e. the gain is decreased by 3 dB.
(c) Measure and record the small-signal voltage gain versus frequency over the frequency range from f_1 to

f_2, where f_1 is one-half the lower frequency at which the gain is down by 0.707 and f_2 is twice the upper frequency at which the gain is down by 0.707. The measurement frequencies should be chosen so that they are approximately equally spaced on a log frequency scale. The bandwidth limit feature of the oscilloscope should be disabled for the high frequency measurements. Due to the performance limitations of the function generator and/or the oscilloscope, it may not be possible to obtain the upper frequency at which the gain has been reduced by 0.707. If this is the case, measure the gain at the maximum frequency for which the function generator/oscilloscope combination permits a valid gain measurement to be made. Compare the measured lower cutoff frequency to that predicted by the theory.

11.6.4 Clipping

Set the frequency of the function generator to 1 kHz. Increase the input signal level until the output signal begins to show distortion. Continue to increase the amplitude of the input signal until hard clipping of either the positive or negative peak occurs. Sketch the waveform and record the amplitude of the input signal. Increase the amplitude of the input until both peaks show clipping. Sketch the waveform and record the amplitude of the input signal.

11.6.5 Input Resistance Measurement

Connect a resistor $R_A = 10$ kΩ in series with the input between the generator and the input coupling capacitor C_1. Connect a temporary short circuit across this resistor. Set the function generator for a 1 kHz sine wave having an amplitude that does not produced a clipped output signal. With a 1 kHz input signal, measure the amplitude of the output signal and record this voltage as v_1. Remove the short circuit across R_A and measure the amplitude of the output signal. Record this voltage as v_2. The experimentally measured value of the input resistance of the amplifier is then given by

$$r_{\text{in}} = \frac{R_A}{\frac{v_1}{v_2} - 1} \tag{11.40}$$

After r_{in} is determined, remove the resistor R_A.

11.6.6 Output Resistance Measurement

Disconnect the function generator from the circuit input and ground the input node. Remove the load resistor R_L from the circuit output and connect a resistor $R_A = 10$ kΩ in series with the output lead. Connect the function generator to the open end of R_A and adjust its output for a 1 kHz sine-wave with a peak voltage of $1\ V$ and a dc level of zero. Use the oscilloscope to measure the amplitude of the voltage on each side of R_A. Denote the voltage on the generator side by v_1 and the voltage on the amplifier side as v_2. Calculate the output resistance of the circuit from the expression

$$r_{\text{out}} = \frac{R_A}{\frac{v_1}{v_2} - 1} \tag{11.41}$$

11.6.7 Symmetric Clipping

Change the value of resistors R_{B1}, R_{B2}, R_C, R_L, R_{E1}, and R_{E2} to those used in *Example 2* of the Theory section. These values are $R_{B1} = 6.7$ kΩ, $R_{B2} = 30$ kΩ, $R_C = 4.3$ kΩ, $R_L = 2$ kΩ, $R_{E1} = 1$ kΩ, and $R_{E2} = 360$ Ω. Repeat steps 13.6.2 through 14.6.2.

11.7 Laboratory Report

The laboratory report should include:

- all preliminary derivations, calculations, and SPICE simulations
- plots for each of the frequency response measurements made in step 15.7.5 of the Procedure section. The plots should be made by plotting the magnitude of the gain versus frequency on log-log graph paper
- Sketches which illustrate the experimentally determined clipping behavior of the amplifier
- a comparison between the theoretically predicted results, the simulation results, and the experimentally measured results

11.8 References

1. E. J. Angelo, *Electronics: BJTs, FETs, and Microcircuits*, McGraw-Hill, 1969.
2. W. Banzhaf, *Computer-Aided Circuit Analysis Using SPICE*, Prentice-Hall, 1989.
3. M. N. Horenstein, *Microelectronics Circuits and Devices*, Prentice-Hall, 1990.
4. P. Horwitz and W. Hill, *The Art of Electronics*, 2nd edition, Cambridge University Press, 1989.
5. P. Horowitz and I. Robinson, *Laboratory Manual for The Art of Electronics*, Cambridge University Press, 1981.
6. J. H. Krenz, *An Introduction to Electrical and Electronic Devices*, Prentice-Hall, 1987.
7. R. Mauro, *Engineering Electronics*, Prentice-Hall, 1989.
8. F. H. Mitchell and F. H. Mitchell, *Introduction to Electronic Design*, Prentice-Hall, 1988.
9. Motorola, *Small-Signal Semiconductors*, DL 126, Motorola, 1987.
10. C. J. Savant, M. S. Roden, and G. L. Carpenter, *Electronic Circuit Design*, Benjamin Cummings, 1987.
11. A. S. Sedra and K. C. Smith, *Microelectronics Circuits*, 4th edition, Oxford, 1998.
12. D. L. Schilling and C. Belove, *Electronic Circuits: Discrete and Integrated*, McGraw-Hill, 1968.
13. P. W. Tuinenga, *SPICE*, Prentice-Hall, 1988.

Chapter 12

The Common-Base and Cascode Amplifiers

12.1 Object

The object of this experiment is to examine the classical BJT common-base and cascode amplifiers. The biasing, small-signal voltage gain, input resistance, output resistance, frequency response, and large-signal clipping behavior are studied.

12.2 Theory

12.2.1 The Common-Base Amplifier

Fig. 12.1 shows the circuit diagram of the classical common-base (CB) BJT amplifier. The active device in the circuit is a NPN BJT which is assumed to be biased in the active mode. In this mode, the base-to-emitter junction is forward biased and the collector-to-base junction is reverse biased. The input signal is applied to the emitter of the transistor and the output is taken from the collector. In contrast for the CE amplifier, the signal is applied to the base and the output is taken from the collector. Because the small-signal input resistance to the emitter is much smaller than the small-signal input resistance to the base, it follows that the CB amplifier has a much lower input resistance than the CE amplifier. For this reason, the CB amplifier is not a good choice for a voltage amplifier if the signal source has a high output impedance. The CB amplifier is often used as a transresistance amplifier, i.e. an amplifier which converts an input current to an output voltage. In this case, the low input resistance to the CB amplifier is desirable.

The signal generator in Fig. 12.1 has an open-circuit output voltage v_g and an output resistance R_g. The generator is coupled to the emitter circuit through coupling capacitor C_1. The collector output signal is coupled to the load through the output coupling capacitor C_2. The output signal is the voltage across the load resistor R_L. The value of each coupling capacitor is chosen so that the capacitive reactance is negligible in the frequency band of interest. In other words, the capacitors present an infinite impedance at DC and negligible impedance at signal frequencies. Capacitor C_B is used to provide a signal ground reference for the base of the transistor. Its value is also chosen so that its reactance is negligible at signal frequencies. The load capacitor C_L represents the capacitance of the oscilloscope and the interconnecting cable between the oscilloscope and the circuit. This capacitor can be considered to be an open circuit at signal frequencies.

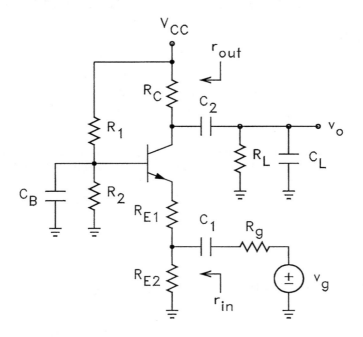

Figure 12.1: Circuit diagram of the common-base amplifier.

12.2. THEORY

Collector Bias Current

The quiescent bias current in the BJT in Fig. 12.1 is calculated with all capacitors replaced by open circuits. The collector current is given by the following bias equation:

$$I_C = \frac{\frac{V_{CC} R_2}{R_1 + R_2} - V_{BE}}{\frac{R_1 \| R_2}{\beta} + \frac{R_{E1} + R_{E2}}{\alpha}} \tag{12.1}$$

where V_{BE} is the quiescent base-to-emitter voltage, α is the emitter-to-collector current gain, and β is the base-to-collector current gain.

Midband Voltage Gain

The midband range is defined as the frequency band where the small-signal voltage gain of the circuit is a maximum. In this band, the coupling and bypass capacitors can be assumed to be signal short circuits. In addition, the load capacitor and the capacitances inside the transistor which cause the high-frequency gain to decrease can be assumed to be open circuits. The midband small-signal voltage gain is given by the following expression:

$$\begin{aligned}\frac{v_o}{v_g} &= \frac{v_{te}}{v_g} \times \frac{i_c}{v_{te}} \times \frac{v_o}{i_c} \\ &= \frac{R_{E2}}{R_g + R_{E2}} \times \frac{-\alpha}{r'_e + R_{te}} \times (-R_C \| R_L \| r_{ic})\end{aligned} \tag{12.2}$$

where r'_e and R_{te} are defined by the relations

$$r'_e = \frac{r_x}{1+\beta} + r_e \tag{12.3}$$

$$R_{te} = R_{E1} + R_{E2} \| R_g \tag{12.4}$$

In these equations, r_x is the BJT small-signal base spreading resistance, $r_e = V_T/I_E$ is the intrinsic emitter resistance, V_T is the thermal voltage, and r_{ic} is the small-signal resistance seen looking into the collector. The latter is given by

$$r_{ic} = \frac{r_0 + r'_e \| R_{te}}{1 - \frac{\alpha R_{te}}{r'_e + R_{te}}} \tag{12.5}$$

where $r_0 = (V_{CB} + V_A)/I_C$ is the small-signal collector-to-emitter resistance and V_A is the BJT Early voltage.

Midband Input and Output Resistances

The midband small-signal input resistance and output resistance of the common-base amplifier are given by

$$r_{in} = R_{E2} \| (R_{E1} + r'_e) \tag{12.6}$$

$$r_{out} = R_C \| r_{ic} \tag{12.7}$$

Approximate Design Equations

If r_0 is taken to be an open circuit, β is considered to be large, and α is considered to be unity, the expressions for the gain, input resistance, and output resistance reduce to

$$\frac{v_o}{v_g} \cong \frac{R_{E2}}{R_g + R_{E2}} \times \frac{1}{r_e + R_{te}} \times (R_C \| R_L \| r_{ic}) \tag{12.8}$$

$$r_{in} \cong R_{E2} \| (R_{E1} + r_e) \tag{12.9}$$

$$r_{out} \cong R_C \tag{12.10}$$

For purpose of design, these approximate formulas give acceptable results in most applications.

Lower Cutoff Frequency

The lower cutoff frequency of the CB amplifier is a function of the input coupling capacitor C_1 and the output coupling capacitor C_2. The time constant for C_1 depends on the state of C_B. For a worst case analysis, we assume that C_B is a short circuit. The time constants for C_1 and C_2, respectively, are given by

$$\tau_1 = [R_g + (R_{E1} + r'_e) \| R_{E2}] C_1 \tag{12.11}$$

$$\tau_2 = (R_C + R_L) C_2 \tag{12.12}$$

where r'_e for C_B a short circuit is given by Eq. 12.3. The lower cutoff frequency of the amplifier is approximately given by

$$f_\ell \simeq \frac{1}{2\pi} \sqrt{\frac{1}{\tau_1^2} + \frac{1}{\tau_2^2}} \tag{12.13}$$

Upper Cutoff Frequency

The upper cutoff frequency is a function of the load capacitance C_L, the base-to-emitter diffusion capacitance c_π, and the base-to-collector junction capacitance c_μ. The latter two capacitors are internal to the transistor. The model equations for these are

$$c_\pi = \frac{\tau_F I_C}{V_T} \tag{12.14}$$

$$c_\mu = \frac{c_{jco}}{\left[1 + \dfrac{V_{CB}}{\Phi_C}\right]^{m_c}} \tag{12.15}$$

where τ_F is the forward transit time of the base-to-emitter junction, c_{jco} is the zero bias junction capacitance of the base-to-collector junction, Φ_C is the built-in potential (the SPICE default value is 0.75 V), and m_c is the junction exponential factor (the SPICE default value is 0.33). The time constants for C_L, c_π, and c_μ are calculated with C_1, C_2, and C_B shorted. In addition, the time constants due to c_π and c_μ are often calculated under the assumption that $r_x \simeq 0$. The time constants are given by

$$\tau_L = (R_L \| R_C \| r_{ic}) C_L \tag{12.16}$$

$$\tau_\pi \simeq [r_e \| (R_{E1} + R_{E2} \| R_g)] c_\pi \tag{12.17}$$

$$\tau_\mu \simeq (R_L \| R_C \| r_{ic}) c_\mu \tag{12.18}$$

The upper cutoff frequency is approximately given by

$$f_u \simeq \frac{1}{2\pi \sqrt{\tau_L^2 + \tau_\pi^2 + \tau_\mu^2}} \tag{12.19}$$

12.2. THEORY

Design Considerations

In CB amplifier design, values for the desired midband voltage gain v_o/v_g, the desired lower cutoff frequency f_ℓ, the generator output resistance R_g, the load resistance R_L, and the load capacitance C_L are normally specified. If the quiescent collector current in the transistor is not specified, it must be chosen. The current will affect both the small-signal performance and the large-signal performance of the circuit. Once the specifications are established, values for the other components in the circuit can be calculated.

12.2.2 The Cascode Amplifier

Fig. 12.2 shows the circuit diagram of the classical cascode amplifier. This is a two-stage amplifier consisting of a common-emitter input stage followed by a common-base output stage. The term *cascode* originated with vacuum tube circuits and denotes a contraction of "cascade common-cathode common-grid". In vacuum tube circuits, a cascode amplifier is a common-cathode stage followed by a common-grid stage.

The cascode amplifier is commonly used in wideband amplifier design where the shunt input capacitance to a high voltage gain stage is to be minimized. This can lead to an extension of the upper half-power or -3 dB cutoff frequency of the amplifier. By the Miller theorem, the part of the base input capacitance to Q_1 due to the collector-to-base capacitance $c_{\mu 1}$ of Q_1 is given by $(1 - A_v)c_{\mu 1}$, where A_v is the midband small-signal voltage gain across $c_{\mu 1}$. (Note that A_v is negative.) Because A_v is proportional to the collector load resistance, it follows that the input capacitance due to $c_{\mu 1}$ is minimized if the collector load resistance for Q_1 is minimized. This is accomplished by using the emitter input resistance r'_{e2} of Q_2 as the collector load resistance for Q_1. For the cascode amplifier, it can be shown that A_v is inverting and that $|A_v| \leq 1$ so that $(1 - A_v)c_{\mu 1}$ can never be larger than $2c_{\mu 1}$. In contrast, $(1 - A_v)c_\mu$ for a high gain CE amplifier is much larger so that a high gain CE amplifier has a much higher input capacitance than a cascode amplifier.

Midband Voltage Gain

The midband small-signal voltage gain of the cascode amplifier is given by

$$\frac{v_o}{v_g} = \frac{v_{tb}}{v_g} \times \frac{i_{c1}}{v_{tb}} \times \frac{i_{c2}}{i_{c1}} \times \frac{v_o}{i_{c2}} \qquad (12.20)$$

$$= \frac{R_1 \| R_2}{R_g + R_1 \| R_2} \times \frac{\alpha_1}{r'_{e1} + R_{E1}} \times \alpha_2 \times (-R_C \| R_L)$$

r'_{e1} is defined by the relation

$$r'_{e1} = \frac{R_g \| R_1 \| R_2 + r_{x1}}{1 + \beta_1} + r_{e1} \qquad (12.21)$$

and $r_{e1} = V_T/I_{E1}$.

Input and Output Resistances

The input and output resistances of the cascode amplifier are given by

$$r_{in} = R_1 \| R_2 \| r_{ib1} \qquad (12.22)$$

$$r_{out} = R_C \qquad (12.23)$$

where r_{ib1} is the small-signal resistance seen looking into the base of Q_1. This is given by

$$r_{ib1} = r_{x1} + (1 + \beta_1)(r_{e1} + R_{E1}) \qquad (12.24)$$

Note that the midband small-signal collector input resistance r_{ic2} has been omitted from the voltage gain and output resistance expressions. This is because the collector of Q_1 looks like a current source driving the emitter of Q_2 so that $r_{ic2} \simeq r_{02}/(1 - \alpha_2) = (1 + \beta_2)r_{02}$. This is so large for practical purposes that r_{ic2} can be replaced by an open circuit.

CHAPTER 12. THE COMMON-BASE AND CASCODE AMPLIFIERS

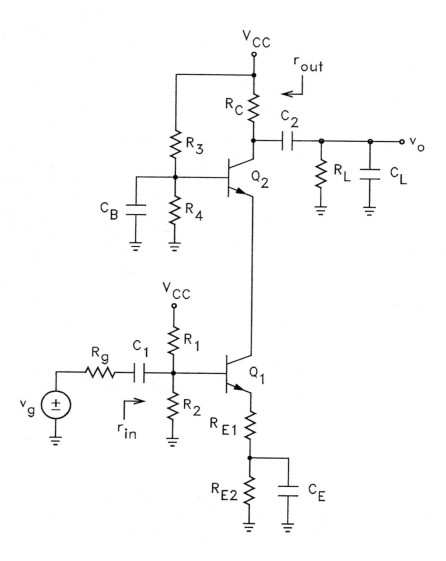

Figure 12.2: Circuit diagram of the cascode amplifier.

12.2. THEORY

Approximate Design Equations

If β is considered to be large and α is considered to be unity, the expressions for the gain and input resistance reduce to

$$\frac{v_o}{v_g} \cong \frac{R_1\|R_2}{R_g + R_1\|R_2} \times \frac{1}{r_{e1} + R_{E1}} \times (-R_C\|R_L) \tag{12.25}$$

$$r_{in} \cong R_1\|R_2 \tag{12.26}$$

For purpose of design, these approximate formulas give acceptable results in most applications.

Lower Cutoff Frequency

The lower half-power or -3 dB cutoff frequency of the cascode amplifier is a function of the input coupling capacitor C_1, the output coupling capacitor C_2, and the emitter bypass capacitor C_E. Capacitors C_1 and C_2 both set high-pass poles in the voltage gain transfer function while C_E sets both a pole and a zero. The worst case time constant for C_1 is calculated with C_E a short circuit. The worst case pole time constant for C_E is calculated with C_1 a short circuit. The time constants are given by

$$\tau_1 = (R_g + R_1\|R_2\|r_{ib1})\,C_1 \tag{12.27}$$

$$\tau_2 = (R_C + R_L)\,C_2 \tag{12.28}$$

$$\tau_{E1} = [(r'_{e1} + R_{E1})\,\|R_{E2}]\,C_E \tag{12.29}$$

$$\tau_{E2} = R_{E2}C_E \tag{12.30}$$

where τ_{E1} is the pole time constant for C_E, τ_{E2} is the zero time constant, and r_{ie1} is given by Eq. 12.21. The lower cutoff frequency of the amplifier is approximately given by

$$f_\ell \simeq \frac{1}{2\pi}\sqrt{\frac{1}{\tau_1^2} + \frac{1}{\tau_2^2} + \frac{1}{\tau_{E1}^2} - \frac{2}{\tau_{E2}^2}} \tag{12.31}$$

Upper Cutoff Frequency

The upper cutoff frequency is a function of the load capacitance C_L and the capacitors c_π and c_μ internal to each transistor given by Eqs. 12.14 and 12.15. The time constants for C_L, $c_{\pi1}$, $c_{\pi2}$, $c_{\mu1}$, and $c_{\mu2}$ are calculated with C_1, C_2, and C_E shorted. In addition, the time constants due to $c_{\pi2}$ and $c_{\mu2}$ for the CB stage are often calculated under the assumption that $r_{x2} \simeq 0$. The time constants are given by

$$\tau_L = (R_L\|R_C)\,C_L \tag{12.32}$$

$$\tau_{\pi1} = \frac{r_{e1}}{r'_{e1} + R_{E1}}(R_1\|R_2\|R_g + r_{x1} + R_{E1})\,c_{\pi1} \tag{12.33}$$

$$\tau_{\mu1} = \left\{[(R_1\|R_2\|R_g + r_{x1})\,\|r'_{ib1}]\left[1 + \frac{\alpha_1 r_{e2}}{r_{e1} + R_{E1}}\right] + r_{e2}\right\}c_{\mu1} \tag{12.34}$$

$$\tau_{\pi2} = r_{e2}c_{\pi2} \tag{12.35}$$

$$\tau_{\mu2} = (R_L\|R_C)c_{\mu2} \tag{12.36}$$

where r'_{ib1} is given by

$$r'_{ib1} = r_{ib1} - r_{x1} = (1 + \beta_1)(r_{e1} + R_{E1}) \quad (12.37)$$

The upper cutoff frequency is approximately given by

$$f_u \simeq \frac{1}{2\pi\sqrt{\tau_L^2 + \tau_{\pi 1}^2 + \tau_{\mu 1}^2 + \tau_{\pi 2}^2 + \tau_{\mu 2}^2}} \quad (12.38)$$

Design Considerations

In any cascode amplifier design, values for the desired midband voltage gain v_o/v_g, the desired lower cutoff frequency f_ℓ, the generator output resistance R_g, the load resistance R_L, and the load capacitance C_L are normally specified. If the quiescent collector currents in the transistors are not specified, they must be chosen. The currents will effect both the small-signal performance and the large-signal performance of the circuit. Once the specifications are established, values for the other components in the circuit can be calculated.

12.3 Preliminary Derivations

1. (a) Derive Eq. 12.2 for the voltage gain of the CB amplifier. (b) Derive Eq. 12.6 for the input resistance to the CB amplifier, including the expression for r_{ie} in Eq. 12.3.

2. (a) Derive Eq. 12.20 for the voltage gain of the cascode amplifier. (b) Derive Eq. 12.22 for the input resistance to the cascode amplifier, including the expression for r_{ib1} in Eq. 12.24.

12.4 Preliminary Calculations

1. For the circuit shown in Fig. 12.1, it is given that $R_C = 2$ kΩ, $R_L = 5.6$ kΩ, $R_{E1} = 180$ Ω, $R_g = 50$ Ω, $R_{E2} = 1.8$ kΩ, $R_1 = 18$ kΩ, $R_2 = 11$ kΩ, and $V_{CC} = 15$ V. (a) Calculate the quiescent collector current and the quiescent collector voltage. If C_1, C_2, and C_B are signal short circuits at midband, calculate the small-signal midband voltage gain and input resistance. For the calculations, assume that $V_{BE} = 0.7$ V, $\beta = 200$, and $r_x = 10$ Ω. (b) Calculate the quiescent DC voltage that appears across each of the three capacitors in Fig. 12.1.

2. For $R_C = 2$ kΩ, $R_L = 5.6$ kΩ, and $V_{CC} = +15$ V, calculate the value of the other resistors in Fig. 12.1 which will give a small-signal voltage gain of 8. Bias the transistor so that the quiescent collector voltage is 8 V and the quiescent emitter voltage is 2 V. Assume that $V_{BE} = 0.7$ V, $\beta = 200$, and $r_x = 10$ Ω.

12.5 Preliminary SPICE Simulations

The transistor type that is specified for this experiment is the 2N3904 NPN general purpose transistor. The .MODEL parameters for the SPICE analyses are IS = 2E − 14 (I_{SO}), VA = 170 (V_A), BF = 200 (β_O), CJC = 3.6E − 12 (c_{jco}), TF = 3E − 10 (τ_F), and RB = 10 (r_x or r_B). Although these parameters vary from transistor to transistor, they are representative for the 2N3904.

1. For the circuit of Fig. 12.1, the component values given in part 1 of the preliminary calculations are to be used. In addition, it is given that $C_1 = 100$ μF, $C_B = 1$ μF, $C_2 = 10$ μF, $C_L = 100$ pF, and $V_{CC} = 15$ V. The SPICE analyses to be performed on the circuit are the ".AC" and ".TRAN" analyses. For the ".AC" analysis, use the frequency range from 10 Hz to 10 MHz with 20 points per decade. For the ".TRAN" analysis, use a sine-wave input signal with a frequency of 1 kHz, a DC level of zero, and at least 20 points per cycle. (a) Write the code for the SPICE input deck to calculate and graphically display the small-signal AC voltage gain as a function of frequency. Determine the midband gain and the lower and upper cutoff frequencies from the gain plot. (b) Write the SPICE code to calculate and graphically display the transient sine-wave output voltage as a function of time. The amplitude of the input signal should be varied so that the clipping behavior can be investigated. At least three plots should be obtained to illustrate

12.6. EXPERIMENTAL PROCEDURES

an undistorted output, a distorted output that is not hard clipped, and a hard clipped output. (c) Perform a Fourier analysis (.FOUR) in conjunction with a transient analysis and determine the input signal level that produces a total harmonic distortion of 10% at the output.

2. (a) For the element values of the preceding part, use a SPICE operating point (".OP") analysis to determine the DC voltage across each of the three capacitors in the circuit of Fig. 12.1. Compare the voltages calculated by SPICE to those determined in the Preliminary Calculations section. (b) Replace the three capacitors in the SPICE code with DC voltage sources having a value equal to the quiescent voltage drop across each capacitor calculated by SPICE and given in the .OUT file. Perform a ".DC" SPICE analysis to display the plot of v_O vs. v_G for v_G in the range from -2 to $+2$ V with a step size that gives 20 points per volt.

3. For the circuit shown in Fig. 12.2 it is given that $R_{E1} = 100$ Ω, $R_{E2} = 1$ kΩ, $R_C = 6.2$ kΩ, $R_L = 10$ kΩ, $R_1 = 130$ kΩ, $R_2 = 18$ kΩ, $R_3 = 110$ kΩ, $R_4 = 36$ kΩ, $R_g = 50$ Ω, $C_1 = C_2 = 10$ μF, $C_B = 1$ μF, $C_E = 100$ μF, and $C_L = 100$ pF. (a) Write the code for the SPICE input deck to calculate and graphically display the small-signal AC voltage gain as a function of frequency over the frequency range from 10 Hz to 10 MHz. Determine the midband gain and the lower and upper cutoff frequencies of the circuit from the gain plot. (b) Write the SPICE code to calculate and graphically display the transient sine-wave output voltage as a function of time. The amplitude of the input signal should be varied so that the clipping behavior can be investigated. At least three plots should be obtained to illustrate an undistorted output, a distorted output that is not hard clipped, and a hard clipped output. (c) Perform a Fourier analysis (".FOUR") in conjunction with a transient analysis and determine the input signal level that produces a total harmonic distortion of 10% at the output.

12.6 Experimental Procedures

The transistor specified for this experiment is the 2N3904 NPN. The physical arrangement of the leads is shown in Fig. 7 of Experiment 11.

12.6.1 Preparation

Prepare the electronic breadboard to provide busses for the positive power supply rail and the circuit ground. The power supply rail should be decoupled with a 100 Ω resistor and a 100 μF, 25 V (or greater) capacitor. The resistor is connected in series with the external power supply lead and the capacitor is connected from the power supply rail to ground on the circuit side of the resistor. The capacitor must be installed with proper polarity to prevent reverse polarity breakdown.

12.6.2 Bias Measurement

Assemble the circuit shown in Fig. 12.1. The element values are: $R_C = 2$ kΩ, $R_L = 5.6$ kΩ, $R_{E1} = 180$ Ω, $R_g = 50$ Ω, $R_{E2} = 1.8$ kΩ, $R_1 = 18$ kΩ, $R_2 = 11$ kΩ, $C_1 = 100$ μF, $C_2 = 10$ μF, and $C_B = 1$ μF. The supply voltage is specified to be $V_{CC} = 15$ V. The electrolytic capacitors must be installed with the proper polarity. Use the digital voltmeter to measure the voltage across the collector resistor. From this reading and the resistance, calculate the collector current. If this is not reasonably close to 2.5 mA, examine the circuit for errors.

12.6.3 Gain Measurement

Use the function generator for the signal source. Set the generator to produce a sine-wave output with a DC level of 0 V and a frequency of 1 kHz. Set the generator output level to obtain an undistorted sine-wave output from the circuit having a peak level of about 2 V. Use the dual channel oscilloscope to measure the levels of the input and output signals. (If the oscilloscope has a bandwidth limit feature, it should be turned on.) Both channels of the oscilloscope should be DC coupled. Measure the midband voltage gain and the lower and upper half-power or -3 dB cutoff frequencies. Measure and record the small-signal voltage gain as a function of frequency over the range from 10 Hz to 1 MHz. The measurement frequencies should be chosen so that they are approximately equally spaced on a logarithmic scale. (Turn the bandwidth limit feature off for high frequency measurements.)

12.6.4 Clipping

Set the frequency of the generator to 1 kHz. Increase the input signal level until the circuit output signal begins to distort. Continue to increase the amplitude of the input signal until clipping of either the positive or negative peak occurs. Record the output signal and the amplitude of the input signal. Increase the amplitude of the input until both peaks show clipping. Record the output signal and the amplitude of the input signal.

12.6.5 Design Circuit

Assemble the circuit of Fig. 12.1 with the element values calculated in step 2 of the Preliminary Calculations. Use the nearest 5% value for each resistor. Repeat steps 13.6.2 through 12.6.4 of the Procedure section for this circuit.

12.6.6 Cascode Amplifier

(a) Construct the cascode amplifier shown in Fig. 12.2. The circuit element values are specified as: $R_{E1} = 100\ \Omega$, $R_{E2} = 1$ kΩ, $R_C = 6.2$ kΩ, $R_L = 10$ kΩ, $R_1 = 130$ kΩ, $R_2 = 18$ kΩ, $R_3 = 110$ kΩ, $R_4 = 36$ kΩ, $C_1 = C_2 = 10\ \mu$F, $C_B = 1\ \mu$F, and $C_E = 100\ \mu$F. (b) Use the digital multimeter to measure the voltage across R_C and R_{E2}. Use the measured voltages to calculate the quiescent currents I_{C2} and I_{E1}. The currents should be approximately 1 mA. If they are not, examine the circuit for errors. (c) Perform the same measurements on the cascode amplifier that are specified for the CB amplifier in steps 15.7.5 and 12.6.4. If the function generator output is too high to obtain an undistorted output signal from the circuit, assemble an L-pad attenuator (a two resistor voltage divider) between the generator output and the circuit input consisting of a series 10 kΩ and a 100 Ω shunt resistor. Consider the input voltage to the circuit to be the output voltage from the attenuator.

12.7 Laboratory Report

The laboratory report should include

- all preliminary derivations, calculations, and SPICE simulations

- Plots of the frequency response measurements made in step 15.7.5 of the procedure. The plots should be made by plotting the magnitude of the gain versus frequency on log-log graph paper or the dB gain versus frequency on linear-log graph paper

- Sketches which illustrate the experimentally observed clipping behavior of the amplifier

- a comparison between the theoretically predicted results, the simulation results, and the experimentally measured results

12.8 References

1. E. J. Angelo, *Electronics: BJTs, FETs, and Microcircuits*, McGraw-Hill, 1969.
2. W. Banzhaf, *Computer-Aided Circuit Analysis Using SPICE*, Prentice-Hall, 1989.
3. H. N. Horenstein, *Microelectronics Circuits and Devices*, Prentice-Hall, 1990.
4. P. Horwitz and W. Hill, *The Art of Electronics*, 2nd edition, Cambridge University Press, 1989.

12.8. REFERENCES

5. P. Horowitz and I. Robinson, *Laboratory Manual for The Art of Electronics,* Cambridge University Press, 1981.
6. J. H. Krenz, *An Introduction to Electrical and Electronic Devices,* Prentice-Hall, 1987.
7. R. Mauro, *Engineering Electronics,* Prentice-Hall, 1989.
8. F. H. Mitchell and F. H. Mitchell, *Introduction to Electronic Design,* Prentice-Hall, 1988.
9. Motorola, *Small-Signal Semiconductors,* DL 126, Motorola, 1987.
10. C. J. Savant, M. S. Roden, and G. L. Carpenter, G. L., *Electronic Circuit Design,* Benjamin Cummings, 1987.
11. A. S. Sedra, and K. C. Smith, *Microelectronics Circuits,* 4th edition, Holt, Oxford, 1998.
12. D. L. Schilling, and C. Belove, *Electronic Circuits: Discrete and Integrated,* McGraw-Hill, 1968.
13. P. W. Tuinenga, *SPICE,* Prentice-Hall, 1988.

Chapter 13

The Common-Collector Amplifier

13.1 Object

The object of this experiment is to examine the third of the three classical single-stage BJT amplifier configurations—the common-collector or emitter-follower amplifier. The small-signal voltage gain, frequency response, and large-signal performance characteristics will be studied.

13.2 Theory

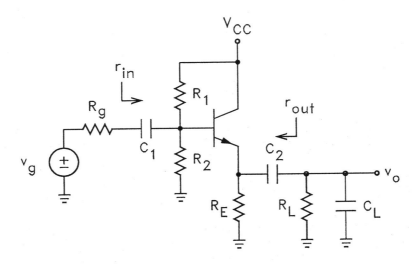

Figure 13.1: Circuit diagram of the common-collector amplifier.

13.2.1 The Common-Collector Amplifier

Fig. 13.1 gives the circuit diagram of the classical BJT common-collector (CC) or emitter-follower amplifier. The active device is a NPN transistor which is assumed to be biased in its active mode. In this mode the base-to-emitter junction is forward biased and the collector-to-base junction is reverse biased. The CC amplifier has a voltage gain that is approximately unity. In addition, it has a high input impedance and a low output impedance. This makes the CC amplifier an ideal buffer stage, i.e. a stage which can be used to interface a low impedance load to a high impedance source. The term emitter follower (from the vacuum tube circuit called the cathode follower) is used because the voltage at the emitter of the transistor follows the voltage at the base.

The signal generator in Fig. 13.1 is the voltage source v_g which has an output resistance R_g. The generator is coupled to the base of the transistor through coupling capacitor C_1. The load is the resistor R_L which is coupled to the emitter of the transistor through the capacitor C_2. The load capacitor C_L in the circuit is included to represent the input capacitance of the oscilloscope and the interconnecting cable between the oscilloscope and the circuit. The base bias resistors R_1 and R_2 and the emitter bias resistor R_E set the quiescent DC collector current in the transistor.

DC Bias Equation

The quiescent collector current in the CC amplifier of Fig. 1 can be obtained from the equation

$$\frac{V_{CC}R_2}{R_1+R_2} = \frac{I_C}{\beta}R_1\|R_2 + V_{BE} + \frac{I_C}{\alpha}R_E \tag{13.1}$$

where V_{BE} is the quiescent base-to-emitter voltage, α is the emitter-to-collector current gain, and β is the base-to-collector current gain.

Small-Signal Gain

The midband frequency range is the band of frequencies where the small-signal voltage gain of the CC amplifier is a maximum. In this band, the coupling capacitors C_1 and C_2 can be considered to be signal short circuits. In addition, the load capacitor C_L and the capacitances inside the transistor which cause the high-frequency gain to decrease (c_π and c_μ in the hybrid-π model) can be assumed to be open circuits. The midband small signal voltage gain is given by

$$\frac{v_o}{v_i} = \frac{R_1\|R_2}{R_g + R_1\|R_2} \times \frac{R_E\|R_L}{r'_e + R_E\|R_L} \tag{13.2}$$

where r'_e is the small-signal resistance seen looking into the emitter. It is given by

$$r'_e = \frac{R_1\|R_2\|R_g + r_x}{1+\beta} + r_e \tag{13.3}$$

where r_x is the BJT small-signal base spreading resistance and r_e is the intrinsic emitter resistance given by $r_e = V_T/I_E$.

Input and Output Resistances

The midband small-signal input resistance r_{in} and output resistance r_{out} of the CC amplifier are given by

$$r_{in} = R_1\|R_2\|r_{ib} \tag{13.4}$$

$$r_{out} = R_E\|r'_e \tag{13.5}$$

where r_{ib} is the small-signal resistance seen looking into the base. It is given by

$$r_{ib} = r_x + (1+\beta)(r_e + R_E\|R_L) \tag{13.6}$$

13.2. THEORY

Approximate Design Equations

If R_g is small compared to $R_1\|R_2$ and r'_e is small compared to $R_E\|R_L$, it follows from Eq. 13.2 that the small-signal voltage gain of the CC amplifier is approximately unity. For design purposes, the large–β approximations are often used. In this case, the expressions for the small-signal voltage gain, input resistance, and output resistance reduce to

$$\frac{v_o}{v_i} \cong \frac{R_1\|R_2}{R_g + R_1\|R_2} \times \frac{R_E\|R_L}{r_e + R_E\|R_L} \tag{13.7}$$

$$r_{in} \cong R_1\|R_2 \tag{13.8}$$

$$r_{out} \cong R_E\|r_e \tag{13.9}$$

Condition for Symmetrical Clipping

The CC amplifier of Fig. 13.1 is often designed for symmetrical clipping of the output signal. When symmetrical clipping is desired, it can be shown that the quiescent collector bias current must satisfy the relation

$$I_C = \frac{R_L + R_E}{2R_L + R_E} \frac{V_{CC} - V_{CEsat}}{R_E} \tag{13.10}$$

where V_{CEsat} is the collector-to-emitter saturation voltage. This voltage is often approximated by zero in calculations.

Lower Cutoff Frequency

The lower cutoff frequency of the CC amplifier in Fig. 13.1 is a function of the coupling capacitors C_1 and C_2. The worst case time constant for each capacitor is calculated with the other capacitor a short circuit. The time constants for C_1 and C_2, respectively, are given by

$$\tau_1 = (R_g + R_1\|R_2\|r_{ib})C_1 \tag{13.11}$$

$$\tau_2 = (R_E\|r'_e + R_L)C_2 \tag{13.12}$$

where r_{ib} and r'_e, respectively, are given by Eqs. 13.6 and 13.3. The lower half-power or -3 dB cutoff frequency is given approximately by

$$f_\ell \simeq \frac{1}{2\pi}\sqrt{\frac{1}{\tau_1^2} + \frac{1}{\tau_2^2}} \tag{13.13}$$

Because the time constants are worst case, it follows that the actual lower cutoff frequency is greater than or equal to the value predicted by this equation.

13.2.2 Cascade CE-CC Amplifier

Eqs. 13.4 and 13.5 show that the emitter follower can have a high input resistance and a low output resistance. In addition, Eq. 13.2 shows that the voltage gain is non-inverting and can be approximated by unity. These characteristics qualify the CC amplifier as a buffer amplifier, i.e. a stage which can be used to interface a low resistance load to a high resistance source. Such an application is shown in Fig. 13.2. The figure shows a CE amplifier cascaded with a CC amplifier. If the CC stage were not used, i.e. Q_2 and R_E removed and the two points indicated with an "X" connected, the collector load resistance for Q_1 would be $R_C\|R_L$. With the CC stage included, the collector load resistance for Q_1 is $R_C\|r_{ib2}$, where r_{ib2} is the small-signal resistance seen looking into the base of Q_2 given by Eq. 13.6. Because r_{ib2} can be much larger than R_L, it follows that the voltage gain of the circuit is increased with the addition of the CC stage, even though the CC stage by itself has a voltage gain of approximately unity.

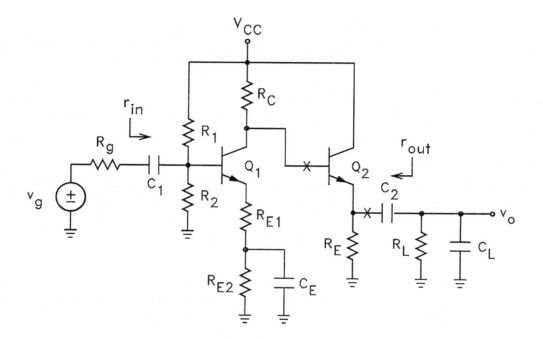

Figure 13.2: Circuit diagram of the CE/CC cascade amplifier.

13.2. THEORY

DC Bias Equations

For the cascade CE-CC amplifier of Fig. 13.2, the quiescent DC collector bias currents can be calculated from the following two bias equations:

$$\frac{V_{CC}R_1}{R_1 + R_2} = \frac{I_{C1}}{\beta_1}R_1\|R_2 + V_{BE1} + \frac{I_{C1}}{\alpha_1}(R_{E1} + R_{E2}) \tag{13.14}$$

$$(V_{CC} - I_{C1}R_C) = \frac{I_{C2}}{\beta_2}R_C + V_{BE2} + \frac{I_{C2}}{\alpha_2}R_E \tag{13.15}$$

Small-Signal Voltage Gain

It can be shown that the small-signal midband voltage gain of the CE-CC amplifier is given by

$$\frac{v_o}{v_g} = -\frac{R_1\|R_2}{R_g + R_1\|R_2} \times \frac{\alpha_1}{r'_{e1} + R_{E1}} \times R_C\|r_{ic1} \times \frac{R_E\|R_L}{r'_{e2} + R_E\|R_L} \tag{13.16}$$

where r'_{e1} is the midband small-signal resistance seen looking into the emitter of Q_1, r_{ic1} is the midband small-signal resistance seen looking into the collector of Q_1, and r'_{e2} is the midband small-signal resistance seen looking into the emitter of Q_2. These resistances are given by

$$r'_{e1} = \frac{R_g\|R_1\|R_2 + r_{x1}}{1 + \beta_1} + r_{e1} \tag{13.17}$$

$$r_{ic1} = \frac{r_{o1} + r'_{e1}\|R_{E1}}{1 - \dfrac{\alpha_1 R_{E1}}{r'_{e1} + R_{E1}}} \tag{13.18}$$

$$r'_{e2} = \frac{R_C\|r_{ic1} + r_{x2}}{1 + \beta_2} + r_{e2} \tag{13.19}$$

Approximate Design Equation

If the large–β approximations are assumed to hold and it is assumed that $r_{ic1} = \infty$, Eq. 13.16 reduces to

$$\frac{v_o}{v_g} \cong -\frac{R_1\|R_2}{R_g + R_1\|R_2} \times \frac{R_C}{r_{e1} + R_{E1}} \times \frac{R_E\|R_L}{r_{e2} + R_E\|R_L} \tag{13.20}$$

This approximate equation is useful for design purposes.

Input and Output Resistances

The small-signal input resistance r_{in} and output resistance r_{out} of the CE-CC amplifier given by

$$r_{in} = R_1\|R_2\|r_{ib1} \tag{13.21}$$

$$r_{out} = R_E\|r'_{e2} \tag{13.22}$$

where r_{ib1} is the small-signal resistance seen looking into the base of Q_1. This is given by

$$r_{ib1} = r_{x1} + (1 + \beta_1)(r_{e1} + R_{E1}) \tag{13.23}$$

Lower Cutoff Frequency

The lower cutoff frequency of the CE-CC amplifier in Fig. 13.2 is a function of the coupling capacitors C_1 and C_2. The worst case time constant for each capacitor is calculated with the other capacitor a short circuit. The time constants for C_1 and C_2, respectively, are given by

$$\tau_1 = (R_g + R_1 \| R_2 \| r_{ib1}) C_1 \tag{13.24}$$

$$\tau_2 = (R_E \| r'_{e2} + R_L) C_2 \tag{13.25}$$

where r'_{e2} is given by Eq. 13.19 and r_{ib1} is given by Eq. 13.23. The lower half-power or −3 dB cutoff frequency is given approximately by

$$f_\ell \simeq \frac{1}{2\pi} \sqrt{\frac{1}{\tau_1^2} + \frac{1}{\tau_2^2}} \tag{13.26}$$

Because the time constants are worst case, it follows that the actual lower cutoff frequency is greater than or equal to the value predicted by this equation.

13.3 Preliminary Derivations

Derive Eq. 13.10 for the quiescent collector bias current which gives symmetrical clipping. The equation is derived by determining the output voltages when the transistor is at the verge of saturation and at the verge of cutoff and equating one to the negative of the other. For the calculations, the base current in the transistor can be neglected and the coupling capacitor C_2 can be modeled as a constant DC voltage source having a value given by $I_E R_E \simeq I_C R_E$.

13.4 Preliminary Calculations

For the circuit given in Fig. 13.1 it is given that: $R_L = 620 \ \Omega$, $R_g = 50 \ \Omega$, $R_E = 1.1 \ k\Omega$, $V_{CC} = 15 \ V$, $V_{BE} = 0.7 \ V$, $r_x = 10$, and $\beta = 200$.

1. Calculate the values of R_1 and R_2 required to set the quiescent DC collector current to $I_C = 1$ mA. Choose the current through R_2 to be $20 I_B$, where I_B is the quiescent DC base current. Calculate the midband small-signal input and output resistances. If the base current can be neglected and C_2 can be modeled as a constant DC voltage source, calculate the value of the output voltage when the transistor is at the verge of cutoff and at the verge of saturation.

2. Repeat step 1 for $I_C = 5$ mA.

3. Repeat step 1 for I_C equal to the value required for symmetric clipping.

13.5 Preliminary SPICE Simulations

The transistor that is specified for this experiment is the 2N3904 general purpose small-signal transistor. The ".MODEL" parameters to be used for the SPICE simulations are: IS = 2E − 14 (I_{SO}), VA = 170 (V_A), BF = 200 (β_O), CJC = 3.6E − 12 (c_{jco}), TF = 3E − 10 (τ_F), and RB = 10 (r_B or r_x). Although these parameters vary from transistor to transistor, they are representative for the 2N3904.

1. Write the code for the SPICE input deck to calculate and graphically display the AC small-signal voltage gain as a function of frequency for the circuit of Fig. 13.1 for each of the set of parameter values in the Preliminary Calculations section. The values to be used for the coupling capacitors are $C_1 = 10 \ \mu F$ and $C_2 = 100 \ \mu F$. For the load capacitor, use the value $C_L = 100 \ pF$. The power supply voltage is $V_{CC} = 15 \ V$. The SPICE analyses to be performed are the .AC and the .OP analyses. Use a frequency range from 10 Hz to 100 MHz with at least 20 points per decade. Obtain a printout of the ".OUT" file for the circuit. Examine the ".OUT" file to verify that the bias currents and voltages are correct.

2. (a) For the parameter values used in the preceding step, write the SPICE code to graphically display the time-domain sine-wave output signal as a function of time. The SPICE analysis to be performed is the ".TRAN" analysis. Use a sine-wave input signal with a frequency of 1 kHz, a DC level of $0\ V$, and at least 20 points per cycle. The amplitude of the input signal should be varied so that the clipping behavior of the amplifier can be examined. At least three plots should be obtained to illustrate an undistorted output, a distorted output that is not hard clipped, and a hard clipped output. **(b)** Perform a Fourier analysis (".FOUR") in conjunction with the transient analysis and determine the input signal level for which the total harmonic distortion is 10%.

3. (a) Write the code for the SPICE input deck to calculate and graphically display the small-signal voltage gain as a function of frequency for the circuit shown in Fig. 13.2. Both transistors are 2N3904's. The component values are: $C_1 = 10\ \mu F$, $C_2 = 100\ \mu F$, $R_{B1} = 24\ k\Omega$, $R_{B2} = 15\ k\Omega$, $R_C = 2\ k\Omega$, $R_{E1} = 180\ \Omega$, $R_{E2} = 1.8\ k\Omega$, $R_E = 680\ \Omega$, $R_L = 620\ \Omega$, $R_g = 50\ \Omega$, and $C_L = 100\ pF$. The analyses to be performed are the ".AC" and ".OP" analyses. Obtain a printout of the ".OUT" file and note the DC bias currents and voltages in the circuit. **(b)** Repeat the simulation with Q_2 and R_E omitted and the two points indicated by an "X" connected. By what factor does the CC stage increase the voltage gain of the circuit?

13.6 Experimental Procedures

The transistor type specified for this experiment is the 2N3904. The lead arrangement for this device is shown in Fig. 7 of Experiment 11.

13.6.1 Preparation

Prepare the electronic breadboard to provide busses for the positive power supply and the circuit ground. The power supply rail should be decoupled with a $100\ \Omega$ resistor and a $100\ \mu F$ $25\ V$ (or greater) capacitor. The resistor is connected in series with the external power supply lead and the capacitor is connected from the power supply rail to ground on the circuit side of the resistor. The capacitor must be installed correctly to prevent reverse polarity breakdown.

13.6.2 Bias Measurement

Assemble the circuit shown in Fig. 13.1. The component values are: $C_1 = 10\ \mu F$, $C_2 = 100\ \mu F$, $R_1 = 3.6\ k\Omega$, $R_2 = 12\ k\Omega$, $R_E = 1.1\ k\Omega$, and $R_L = 620\ \Omega$. The power supply voltage is $V_{CC} = 15\ V$. Omit C_L because this is not a component that is physically put on the breadboard. Measure and record the value of each resistor using the digital multimeter prior to assembling the circuit. Use the digital voltmeter to measure the voltage across R_E and use this value to calculate the quiescent emitter current. If the current is not reasonably close to 10 mA, examine the circuit for errors.

13.6.3 Gain Measurement

(a) For the signal source, use the function generator set to produce a sine wave with a frequency of 1 kHz, a DC level of $0\ V$, and a peak-to-peak voltage of $1\ V$. Use the dual channel oscilloscope to observe and measure the input and output signals. Both channels of the oscilloscope should be DC coupled. Measure the midband voltage gain as the gain in the frequency band in which the measured gain is a maximum. (This gain should be slightly less than unity.) **(b)** Vary the frequency of the input signal and determine the lower frequency f_ℓ and the upper frequency f_u at which the gain is reduced by a factor of 0.707, i.e. the gain is reduced by 3 dB. Record the two frequencies as the cutoff frequencies of the amplifier. **(c)** Measure and record the small-signal gain versus frequency from a lower frequency of $f_\ell/2$ to an upper frequency of $2f_u$. (Due to the performance limitations of the function generator and/or oscilloscope, it may not be possible to obtain the upper frequency. If this is the case, data should be recorded to the highest obtainable frequency.)

13.6.4 Clipping

Set the frequency of the function generator to 1 kHz. Increase the input signal level until the output signal begins to distort. Continue to increase the amplitude of the input signal until clipping of either the positive or negative peak occurs. Record the output signal waveform and the amplitude of the input signal. Increase the amplitude of the input until both peaks show clipping. Record the output signal waveform and the amplitude of the input signal.

13.6.5 Input Resistance Measurement

Set the amplitude of the function generator to a value that does not produce a distorted output. Connect a resistor $R_A = 100\ k\Omega$ in series with the function generator at the amplifier input. Connect a short-circuit jumper across this resistor and measure the amplitude of the amplifier output voltage. Record this as v_1. Remove the short circuit jumper and measure amplitude of the output voltage. Record this value as v_2. The experimentally measured value of the input resistance of the amplifier is given by

$$r_{in} = \frac{R_A}{\frac{v_1}{v_2} - 1} \qquad (13.27)$$

Remove the resistor R_A.

13.6.6 Output Resistance Measurement

Disconnect the function generator from the circuit input and replace it with a 51 Ω resistor connected to ground. Remove the load resistor R_L from the circuit and connect resistor $R_B = 1\ k\Omega$ in series with the output lead. Connect the function generator to the open end of R_B and adjust its output for a 1 kHz sine wave with a peak voltage of 1 V and a DC level of 0 V. Use the oscilloscope to the measure the amplitude of the voltage on each side of R_B. Denote the voltage on the generator side by v_1 and on the amplifier side by v_2. If the waveform for v_2 is undistorted, the output level of the generator can be increased to improve the accuracy of the measurement. However, do not use a level which results in a distorted waveform for v_2. Calculate the output resistance of the circuit from the following expression:

$$r_{out} = \frac{R_B}{\frac{v_1}{v_2} - 1} \qquad (13.28)$$

13.6.7 Design Circuits

Repeat steps 13.6.2 through 14.6.2 for the circuits of steps 1 and 2 in the Preliminary Calculations section.

13.6.8 Impedance Buffer

Assemble the circuit shown in Fig. 13.2. The power supply voltage is $V_{CC} = 15\ V$. Each transistor is a 2N3904. The component values are: $C_1 = 10\ \mu F$, $C_2 = 100\ \mu F$, $C_E = 100\ \mu F$, $R_1 = 24\ k\Omega$, $R_2 = 15\ k\Omega$, $R_C = 2\ k\Omega$, $R_{E1} = 180\ \Omega$, $R_{E2} = 1.8\ k\Omega$, $R_E = 680\ \Omega$, and $R_L = 620\ \Omega$. Adjust the function generator to produce a sine-wave output signal with a frequency of 1 kHz, a DC level of 0 V, and a peak-to-peak voltage of 0.5 V. Measure the small-signal voltage gain. Measure the upper and lower half-power or −3 dB cutoff frequencies as the two frequencies at which the gain is reduced by a factor of 0.707.

13.6.9 Common-Emitter Amplifier

For the circuit of the preceding part, remove Q_2 and resistor R_E and place a short circuit jumper between the two positions marked with an "X" in Fig. 13.2 . (This circuit then becomes a common-emitter amplifier with load resistor R_L.) Measure the small-signal voltage gain.

13.7 Laboratory Report

The laboratory report should include

- all preliminary derivations, calculations, and SPICE simulations
- plots of the measured frequency response. The plots should be made by plotting the magnitude of the gain versus frequency on log-log graph paper or the dB gain versus frequency on semi-log graph paper.
- sketches which illustrate the clipping behavior of the amplifiers
- a comparison between the theoretically predicted results, the simulated results, and the experimentally measured results

13.8 References

1. E. J. Angelo, *Electronics: BJTs, FETs, and Microcircuits,* McGraw-Hill, 1969.
2. W. Banzhaf, *Computer-Aided Circuit Analysis Using SPICE,* Prentice-Hall, 1989.
3. M. N. Horenstein, *Microelectronics Circuits and Devices,* Prentice-Hall, 1990.
4. P. Horwitz and W. Hill, *The Art of Electronics,* 2nd edition, Cambridge University Press, 1989.
5. P. Horowitz and I. Robinson, *Laboratory Manual for The Art of Electronics,* Cambridge University Press, 1981.
6. J. H. Krenz, *An Introduction to Electrical and Electronic Devices,* Prentice-Hall, 1987.
7. R. Mauro *Engineering Electronics,* Prentice-Hall, 1989.
8. F. H. Mitchell and F. H. Mitchell, *Introduction to Electronic Design,* Prentice-Hall, 1988.
9. Motorola, *Small-Signal Semiconductors,* DL 126, Motorola, 1987.
10. C. J. Savant, M. S. Roden, and G. L. Carpenter, *Electronic Circuit Design,* Benjamin Cummings, 1987.
11. A. S. Sedra and K. C. Smith, Microelectronic Circuits, 4th edition, Oxford, 1998.
12. D. L. Schilling and C. Belove, *Electronic Circuits: Discrete and Integrated,* McGraw-Hill, 1968.
13. P. W. Tuinenga, *SPICE,* Prentice-Hall, 1988.

Chapter 14

The Junction Field Effect Transistor

14.1 Object

The object of the experiment is to measure the static terminal characteristics of a n-channel junction field effect transistor (JFET), to calculate the fundamental parameters of this transistor from the measured terminal characteristics, and to use these parameters to design a small-signal amplifier, a voltage-controlled attenuator, and an electronic switch.

14.2 Theory

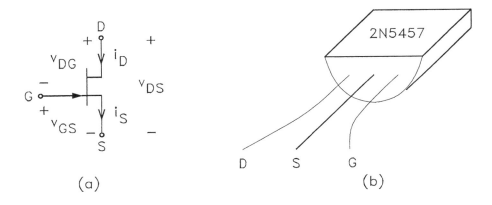

Figure 14.1: (a) Circuit symbol for the n-channel JFET. (b) Lead configuration for the 2N5457 n-channel JFET.

14.2.1 Terminal Characteristics

The circuit symbol for the n-channel JFET is shown in Fig. 14.1a. The device has three terminals designated the source, the drain, and the gate. The arrow points from the p-type semiconductor gate to the n-type channel. For proper operation, the gate-to-channel junction must be zero or reverse biased. This requires the gate-to-source voltage v_{GS} to be zero or negative. The gate voltage controls the flow of current from the drain to the source. When the gate-to-channel junction is reverse biased, the gate current is so small that it can be assumed to be zero for all practical purposes so that the drain and source currents are equal. The JFET is fabricated symmetrically with respect to the drain and source leads so that these two leads are interchangeable in a circuit.

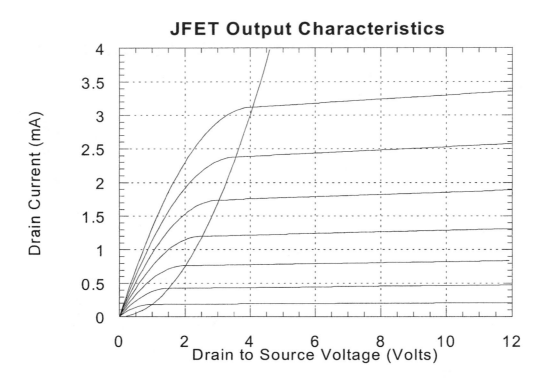

Figure 14.2: Plot of the drain current in mA versus the drain-to-source voltage for the hypothetical n-channel JFET.

A set of typical output characteristics and transfer characteristics for a n-channel JFET are shown in Figs. 14.2 and 14.3. Fig. 14.2 is a plot of the drain current i_D as a function of the drain-to-source voltage v_{DS} for the constant gate-to-source voltages $v_{GS} = 0, -0.5, -1, -1.5, -2, -2.5,$ and $-3\ V$, where the upper curve is for $v_{GS} = 0$. Fig. 14.3 is a plot of the drain current i_D as a function of the gate-to-source voltage v_{GS} for the constant drain-to-source voltages $v_{DS} = 4, 8,$ and $12\ V$, where the lower curve is for $v_{DS} = 4\ V$.

One of the characteristic parameters of any JFET is its threshold voltage V_{TO} which is negative. For the n-channel device, the drain current is zero if the gate-to-source voltage satisfies the relation $v_{GS} \leq V_{TO}$ and the device is said to be cut off. A typical value for the threshold voltage is $V_{TO} = -3\ V$. In this case, the drain current is zero if the gate-to-source voltage is more negative than $-3\ V$. Because the gate-to-source voltage cannot be positive, it follows that our hypothetical JFET must be operated with the gate-to-source voltage in the range of $-3\ V$ to $0\ V$.

There are two theoretical model equations for the drain current in the JFET. The equation which applies depends on voltages applied to device. A n-channel JFET is assumed with a gate-to-source voltage in the range $V_{TO} \leq v_{GS} \leq 0$. For $v_{DS} \leq v_{GS} - V_{TO}$, the device is said to be in its triode or linear region and the

14.2. THEORY

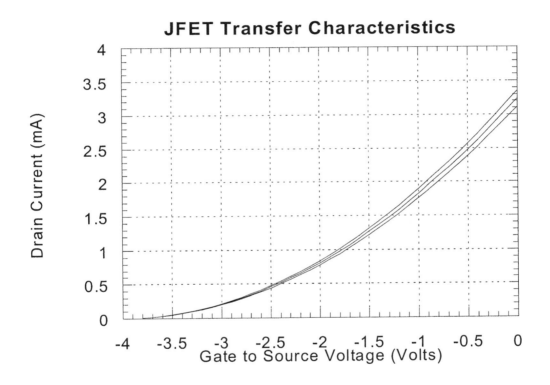

Figure 14.3: Plot of the drain current in mA versus the gate-to-source voltage for the hypothetical n-channel JFET.

drain current is given by

$$i_D = 2\beta \left[(v_{GS} - V_{TO})v_{DS} - \frac{v_{DS}^2}{2} \right] \tag{14.1}$$

where β (not to be confused with the current gain of the BJT) is the transconductance parameter and has the units of A/V^2. For $v_{DS} \geq v_{GS} - V_{TO}$, the device is said to be in its saturation or pinch-off region and the drain current is given by

$$i_D = \beta(v_{GS} - V_{TO})^2 \tag{14.2}$$

The transconductance parameter is a function of the drain-to-source voltage and is given by

$$\beta = \beta_O(1 + \lambda v_{DS}) \tag{14.3}$$

where β_O is the zero-bias value and λ is the channel-width modulation factor which has the units V^{-2}. The reciprocal of λ in the JFET model equations plays the same role as the Early voltage V_A in the BJT model equations. The parameters used for the device curves in Figs. 14.2 and 14.3 are $\beta_O = 0.1875$ mA, $V_{TO} = -4$ V, and $\lambda = 0.01$ V^{-2}.

The expressions for the drain current given by Eqs. 14.1 and 14.2 are known as the Schichman-Hodges formulation. Another commonly used formulation expresses the drain current in terms of the drain-to-source saturation current I_{DSS} and the pinch-off voltage V_P. For this formulation, the equivalent of Eq. 14.3 is

$$i_D = I_{DSS} \left[1 - \frac{v_{GS}}{V_P} \right]^2 \tag{14.4}$$

where $I_{DSS} = I_{DSSO}(1 + \lambda v_{DS})$. It follows that $V_P = V_{TO}$ and $I_{DSS} = \beta V_{TO}^2$. The Schichman-Hodges formulation is used in this laboratory manual because it is the formulation used in SPICE.

The boundary between the triode and the saturation regions in Fig. 14.2 is indicated by the parabolic curve. The equation for this curve is given by

$$i_D = \beta v_{DS}^2 \tag{14.5}$$

The saturation region is the region to the right of the curve. The triode region is the region to the left. When biased in the saturation region, the JFET can be used as an amplifying device. When it is biased in the triode region, it can be used as a voltage variable resistor. A third application of the JFET is to use it as a switch where it is used alternately between cutoff and the triode region. All three applications of the JFET are covered in this experiment.

The drain-to-source saturation current I_{DSS} is defined as the current which flows from the drain to the source with the gate-to-source voltage equal to zero. In Fig. 14.2, I_{DSS} is the value of I_D for the top curve corresponding to $v_{GS} = 0$. Eq. 14.2 can be used to relate I_{DSS} to β and V_{TO} to obtain

$$I_{DSS} = \beta V_{TO}^2 \tag{14.6}$$

Given the value of I_{DSS}, this equation provides one relationship which can be used to solve for β and V_{TO} for a particular JFET. To be able to solve for both β and V_{TO}, a second relationship is required. For this, it is convenient to determine the value of v_{GS} for which the drain current has the value I_{DSS}/n. (A convenient value for n is $n = 4$.) Let this drain voltage be denoted by $v_{GS} = V_{GS1}$. From Eq. 14.2 discloses that

$$\frac{I_{DSS}}{n} = \beta(V_{GS1} - V_{TO})^2 \tag{14.7}$$

Eqs. 14.6 and 14.7 can be solved simultaneously for V_{TO} and β to obtain

$$V_{TO} = \frac{V_{GS1}}{1 - \frac{1}{\sqrt{n}}} \tag{14.8}$$

14.2. THEORY

$$\beta = \left(1 - \frac{1}{\sqrt{n}}\right)^2 \frac{I_{DSS}}{V_{GS1}^2} \quad (14.9)$$

An alternate method for solving for V_{TO} is to use the transfer characteristics given in Fig. 14.3 to determine the value of v_{GS} for which i_D just reaches zero. However, because the slope of the curve is zero at this point, the value of V_{TO} is difficult to read with accuracy.

In the design of small-signal amplifiers, the biasing of a JFET is different from the biasing of a BJT. This is because the BJT is turned on by the bias whereas the JFET is turned off, or partially turned off, by the bias. That is, a gate-to-source bias voltage is applied to cause the drain current to be less than its value with zero bias. Once the BJT is turned on, its base-to-emitter voltage exhibits very little change with collector current so that the base-to-emitter voltage can be approximated by a constant (typically $0.7\ V$) in the BJT bias equation. In contrast, the JFET gate-to-source voltage exhibits considerable variation with the drain current. For this reason, it is important to know both β and V_{TO} for a particular JFET in order to properly bias it as a small-signal amplifier.

14.2.2 Output Characteristics Curve Tracer

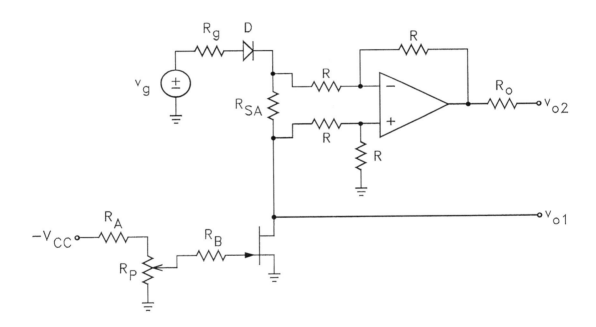

Figure 14.4: Circuit diagram of the output characteristics curve tracer.

The circuit shown in Fig. 14.4 can be used as a simple curve tracer for obtaining plots of the JFET drain current as a function of the drain-to-source voltage for constant values of gate-to-source voltage. The potentiometer R_P in the circuit sets the gate-to-source voltage. The function generator is used to sweep the drain-to-source voltage over the desired range. Because this voltage range is from zero to some positive voltage, the DC offset on the generator should be used to offset the output voltage to the desired range. The diode in the circuit prevents the drain-to-source voltage from being negative. The resistor R_{SA} is used to sample the drain current. The op amp and four resistors labeled R form a differential amplifier or diff amp which amplifies the difference voltage across R_{SA}. The output of the differential amplifier is given by $v_{O2} = -R_{SA}I_D$. Thus the drain current can be obtained by dividing $-v_{O2}$ by R_{SA}. The plot of the drain current as a function of the drain-to-source voltage for a constant gate-to-source voltage can be obtained with an oscilloscope set for $X - Y$ operation. The horizontal input is connected to the v_{O1} output of the circuit. The vertical input is connected to the v_{O2} output. Because of the minus sign in the equation for v_{O2}, the vertical input to the oscilloscope must be inverted to obtain the proper display.

14.2.3 Transfer Characteristics Curve Tracer

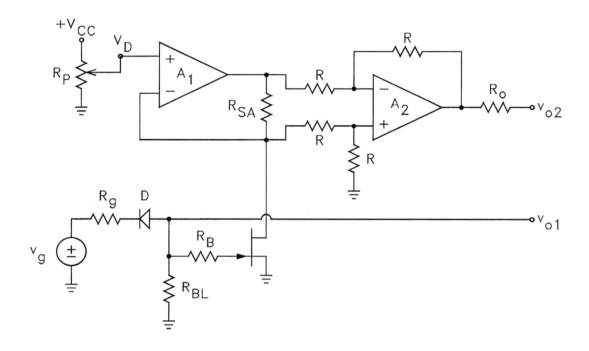

Figure 14.5: Circuit diagram of the transfer characteristics curve tracer.

The circuit shown in Fig. 14.5 can be used to obtain a plot of the drain current as a function of the gate-to-source voltage with the drain-to-source voltage held constant. The drain-to-source voltage is set with the potentiometer R_P. The gate-to-source voltage is swept through negative values by the function generator. The diode prevents the gate-to-source voltage from going positive. Op amp A_1 holds the drain voltage constant independently of the current through resistor R_{SA}. Op amp A_2 is connected as a differential amplifier which has an output voltage $v_{O2} = -I_D R_{SA}$. Thus the drain current can be obtained by dividing $-v_{O2}$ by R_{SA}. The plot of the drain current as a function of the gate-to-source voltage for a constant drain-to-source voltage can be obtained with an oscilloscope set for XY operation. The horizontal input is connected to the v_{O1} output of the circuit. The vertical input is connected to the v_{O2} output. Because of the minus sign in the equation for v_{O2}, the vertical input to the oscilloscope must be inverted to obtain the proper display.

14.2. THEORY

14.2.4 Common Source Amplifier

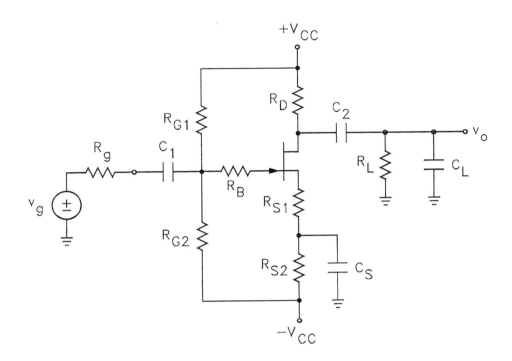

Figure 14.6: Circuit diagram of the classical common-source or CS JFET amplifier.

Fig. 14.6 shows the circuit diagram of the classical JFET common-source or CS amplifier. We assume that the JFET is biased in its saturation or pinch-off region. The circuit is called a CS amplifier because the source is connected to the common or ground terminal, *i.e.* the source is used for neither the input nor the output. The gate input to the JFET is driven from a generator with the open-circuit output voltage v_g and output resistance R_g. The generator is coupled to the gate through the coupling capacitor C_1. The gate series resistor R_B is included in the circuit to protect the JFET from damage in case the gate-to-channel junction is accidentally forward biased. Normally, this resistor would not be included in the circuit. The resistors R_{G1} and R_{G2} are used to set the DC bias voltage at the gate. The resistor R_D is used to set the DC voltage at the drain. The series source resistors $R_{S1} + R_{S2}$ are used to set the quiescent source current. Although the circuit can be designed to operate from a single power supply, a dual power supply is specified for the circuit to increase the maximum output signal swing. The output voltage is the voltage across the load resistor R_L which is coupled to the drain through coupling capacitor C_2. The capacitor C_L represents the capacitance of the oscilloscope input and the interconnecting cable between the circuit and the input to the oscilloscope. The capacitor C_S bypasses resistor R_{S2} in the signal circuit so that the signal current in the source flows through R_{S1} and C_S to ground.

The midband frequency range is the band of frequencies for which the coupling capacitors and bypass capacitor are short circuits and the capacitor C_L and the capacitors internal to the transistor (c_{gs} and c_{gd} in the hybrid−π model) are open circuits. In this frequency band, the small-signal gain is given by

$$\frac{v_o}{v_i} = -\frac{R_{G1}\|R_{G2}}{R_{G1}\|R_{G2} + R_g} \times \frac{R_L\|R_D\|r_{id}}{R_{S1} + r_s} \qquad (14.10)$$

where r_s is the source intrinsic resistance given by $r_s = 1/g_m$ and g_m is the transconductance given by

$$g_m = \left.\frac{\partial i_D}{\partial v_{GS}}\right|_{i_D=I_D} = 2\sqrt{\beta I_D} \qquad (14.11)$$

The resistance r_{id} is the small-signal resistance seen looking into the drain which is given by

$$r_{id} = \frac{r_o + r_s \| R_{S1}}{1 - \dfrac{R_{S1}}{r_s + R_{S1}}} \qquad (14.12)$$

where r_o is the small-signal drain-to-source resistance given by

$$r_o = \left[\left.\frac{\partial i_D}{\partial v_{DS}}\right|_{v_{DS}=V_{DS}}\right]^{-1} = \frac{1 + \lambda V_{DS}}{\lambda I_D} \qquad (14.13)$$

The input and output resistances are given by

$$r_{in} = R_{G1} \| R_{G2} \qquad (14.14)$$

$$r_{out} = R_D \| r_{id} \qquad (14.15)$$

In the design of the CS amplifier, the load resistance R_L and the desired midband small-signal voltage gain v_o/v_g are normally specified. In addition, the desired peak output clipping voltages might be given. The first step in the design is the specification of the quiescent DC drain current I_D and drain-to-source voltage V_{DS}. The value of I_D can be used to calculate the value of V_{GS} provided the parameters V_{TO} and β for the particular JFET are known. If the quiescent DC voltage at either the drain or source is specified, resistor R_D and the sum resistance $R_{S1} + R_{S2}$ can then be calculated. The resistors R_{G1} and R_{G2} are calculated to set the required quiescent DC voltage at the gate. Finally, Eq. 14.11 can be used to calculate R_{S1}.

Example 1 *Design a CS amplifier which has a small-signal midband voltage gain of -5. The transistor specified is the 2N5457 n-channel JFET with the parameters $V_{TO} = -3\ V$, $\beta_O = 3.333 \times 10^{-4}\ A/V^2$, and $\lambda \simeq 0$. Because $\lambda \simeq 0$, we have $\beta \simeq \beta_O$. The load resistance is $R_L = 100\ k\Omega$ and the load capacitance is $C_L = 100\ pF$. A dual power supply of $+15\ V$ and $-15\ V$ is specified.*

Solution. The first step in the design is to select the quiescent DC values for I_D, V_{DS}, the voltage across R_D, and the voltage across $R_{S1} + R_{S2}$. These will be chosen as follows: $I_D = I_{DSS}/2$, $V_{DS} = 15\ V$, $I_D R_D = 7.5\ V$, and $I_D(R_{S1} + R_{S2}) = 7.5\ V$. The current I_{DSS} is the drain-to-source current when the gate-to-source voltage is zero and is given by $I_{DSS} = \beta V_{TO}^2 = 3.333 \times 10^{-4} \times (-3)^2 = 3$ mA. The DC drain bias current is thus $I_D = 1.5$ mA.

The quiescent DC gate-to-source voltage V_{GS} is given by

$$V_{GS} = \sqrt{\frac{I_D}{\beta}} + V_{TO} = -0.879 \qquad (14.16)$$

Thus the gate voltage must have the value $V_G = V_{GS} + V_S = -0.879 - 7.5 = -8.38$. The voltage at the gate is given by either of the two relations

$$V_G = -V_{CC}\left[\frac{2R_{G1}}{R_{G1} + R_{G2}} - 1\right] = -V_{CC}\left[1 - \frac{2R_{G2}}{R_{G1} + R_{G2}}\right] \qquad (14.17)$$

These relations provide a basis for calculation of the gate bias resistors R_{G1} and R_{G2}. The sum resistance $R_{G1} + R_{G2}$ will be chosen to be $1\ M\Omega$. Thus we obtain $R_{G1} = 779\ k\Omega$ and $R_{G2} = 221\ k\Omega$. The resistor R_B prevents excessive current from flowing into the gate if the gate-to-source junction is accidentally forward

14.2. THEORY

biased. A reasonable choice for this resistor is $R_B = 1\ k\Omega$. This resistor should be omitted if the circuit is to be designed for the highest possible upper cutoff frequency.

Because the quiescent voltage across R_D is specified to be 7.5 V, we can solve for the drain bias resistor to obtain $R_D = 7.5V/1.5\ mA = 5\ k\Omega$. In addition, we have $R_{S1} + R_{S2} = 5\ k\Omega$ because the quiescent DC voltage across this sum resistance is also specified to be 7.5 V. The intrinsic source resistance is given by

$$r_s = \frac{1}{g_m} = \frac{1}{2\sqrt{\beta I_D}} = 707\ \Omega \tag{14.18}$$

The resistor R_{S1} can be determined from the midband small-signal voltage gain given by Eq. 14.10. In this equation, we will assume that $R_g << R_{G1}\|R_{G2}$. To obtain a voltage gain of -5, it follows from Eq. 14.10 that $R_{S1} = 245\ \Omega$. Finally, the expression for R_{S2} becomes $R_{S2} = 5k - 245 = 4.75\ k\Omega$.

The lower cutoff frequency of the CS amplifier is a function of the capacitors C_1, C_2, and C_S. Capacitors C_1 and C_2 set pole time constants whereas C_S sets both a pole and a zero time constant. The time constants are given by

$$\tau_1 = (R_g + R_{G1}\|R_{G2})C_1 \tag{14.19}$$

$$\tau_2 = (R_D + R_L)C_2 \tag{14.20}$$

$$\tau_{S1} = [R_{S2}\|(R_{S1} + r_s)]C_S \tag{14.21}$$

$$\tau_{S2} = R_{S2}C_S \tag{14.22}$$

where τ_{S1} is the pole time constant and τ_{S2} is the zero time constant. The approximate expression for the lower half-power or -3 dB cutoff frequency is given by

$$f_\ell = \frac{1}{2\pi}\sqrt{\frac{1}{\tau_1^2} + \frac{1}{\tau_2^2} + \frac{1}{\tau_{S1}^2} - \frac{1}{2\tau_{S2}^2}} \tag{14.23}$$

14.2.5 Voltage Controlled Attenuator

Eq. 14.1 for the drain current in the JFET when biased in the triode region can be rewritten

$$\frac{i_D}{v_{DS}} = 2\beta\left(v_{GS} - V_{TO} - \frac{v_{DS}}{2}\right) \tag{14.24}$$

If the substitution $v_{GS} = V_1 + v_{DS}/2$ is made, where V_1 is an externally applied DC voltage, this equation can be written

$$\frac{i_D}{v_{DS}} = 2\beta(V_1 - V_{TO}) \tag{14.25}$$

It follows from this result that the JFET looks like a resistor having the conductance value $2\beta(V_1 - V_{TO})$, i.e. it looks like a voltage controlled resistor whose value is controlled by the voltage V_1. This voltage controlled resistor can be used to design an electronic attenuator, i.e. an attenuator for which the gain is set by a voltage. Such circuits are used in remote volume controls and in audio compressors and limiters. A voltage controlled attenuator circuit which uses the JFET as a voltage variable resistor is shown in Fig. 14.7. For this circuit, it follows by superposition that

$$v_{GS} = (2) \times \frac{v_{DS}}{4} + \left(1 + \frac{1}{4}\right)V_C = \frac{v_{DS}}{2} + 1.25V_C$$

Therefore, $V_1 - V_{TO} = 1.25V_C$ for this circuit.

Relaxation Oscillator

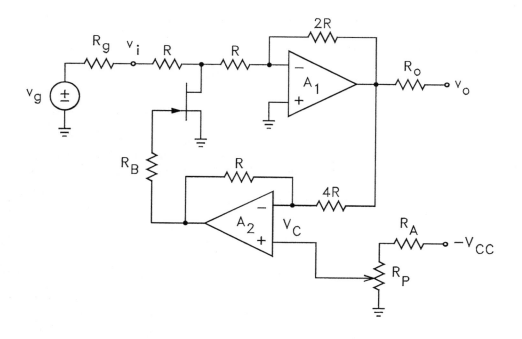

Figure 14.7: Circuit diagram of the JFET voltage controlled attenuator.

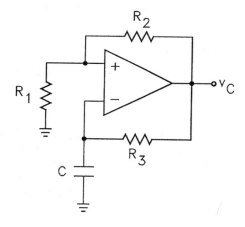

Figure 14.8: Circuit diagram of the relaxation oscillator.

14.2. THEORY

The circuit shown in Fig. 14.8 is called a relaxation oscillator. Another name for the circuit is astable multivibrator. It is introduced here because it is be used in the following section as the local oscillator for a JFET switching modulator. The relaxation oscillator is a feedback circuit which has both positive and negative feedback. When the power supply is first turned, the capacitor is initially uncharged. This makes the voltage at the inverting input to the op-amp zero. The positive feedback from the output to the non-inverting input will cause the output voltage to jump to either the positive saturation voltage or the negative saturation voltage of the op-amp. Let us assume that it jumps to the positive saturation voltage V_{SAT}. The capacitor voltage $v_C(t)$ will then increase toward this voltage according to the equation

$$v_C(t) = V_{SAT}\left[1 - e^{-t/R_3 C}\right] \qquad (14.26)$$

Let the voltage at the non-inverting input be denoted by V_{TH}. This voltage is given by

$$V_{TH} = \frac{R_1}{R_1 + R_2} V_{SAT} \qquad (14.27)$$

When the capacitor voltage becomes larger than V_{TH}, the op-amp output voltage will switch from the positive saturation voltage $+V_{SAT}$ to the negative saturation voltage $-V_{SAT}$. The capacitor then begins charging toward the negative saturation voltage according to the equation

$$v_C(t) = -(V_{SAT} + V_{TH})\left[1 - e^{-t/R_3 C}\right] + V_{TH} \qquad (14.28)$$

The capacitor charges according to this equation until its voltage reaches $-V_{TH}$ when the voltage at the inverting terminal of the op-amp is equal to the voltage at the noninverting terminal. This causes the op-amp output to switch to $+V_{SAT}$ and the process repeats. It follows that the circuit puts out a square wave with the period

$$T = 2R_3 C \; \ell n \left[1 + \frac{2R_1}{R_2}\right] \qquad (14.29)$$

The frequency of the square wave is given by $f = 1/T$.

The rise and fall times of the square wave put out by the circuit is a function of the op-amp slew rate. The slew rate must be high enough so that the time that it takes the op-amp to switch between the negative and positive saturation voltages is a small fraction of the period of the square wave.

14.2.6 Chopper Modulator

Fig. 14.9 gives the circuit diagram of a JFET chopper modulator. This circuit can be used to produce an AM modulated output signal. The input signal is applied from a source having an open-circuit output voltage v_G and an output resistance R_G. In order for the circuit to produce an AM modulated signal, the voltage v_G must have both a DC and an AC component as follows:

$$v_G = V_{DC} + v_m \cos \omega_m t \qquad (14.30)$$

The output voltage v_O is an amplitude modulated waveform if v_G is multiplied by a high-frequency carrier signal. That is, v_O is of the form

$$v_O = [V_{DC} + v_m \cos\omega_m t] \cos \omega_c t \qquad (14.31)$$

where ω_c is the radian frequency of the carrier signal.

In the circuit of Fig. 14.9, the carrier signal is a square wave rather than a sine wave. The carrier input voltage is v_C which is a square wave with a frequency f_c. The amplitude of this square wave must be large enough to turn the JFET completely off during its negative half cycle. Thus the JFET is switched between its cutoff state and its triode region at the frequency of the square wave. The waveform that appears across the resistor R_L is of the form

$$v_L(t) = [V_{DC} + v_m \cos \omega_m t] \frac{1 + s(t)}{2} \qquad (14.32)$$

where $s(t)$ is a square wave with the peak values $-1\,V$ and $+1\,V$. The diff amp in the figure converts this signal to a symmetrically modulated AM waveform.

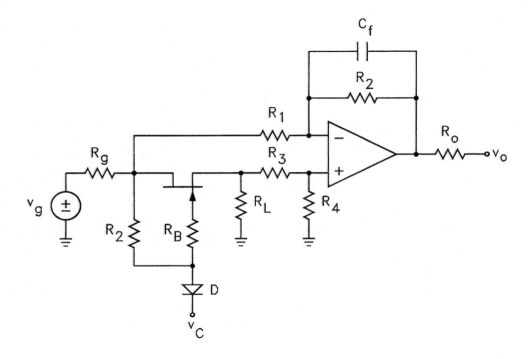

Figure 14.9: Circuit diagram of the JFET chopper modulator circuit.

14.3 Preliminary Derivations

Derive Eqs. 14.10, 14.17, and 14.29.

14.4 Preliminary Calculations

1. Determine the new drain current, drain voltage, and source voltage for the JFET in the circuit of the CS example in the Theory section if the JFET is replaced with one having the parameters $\beta = 1.875 \times 10^{-4}\ A/V^2$ and $V_{TO} = -4\ V$. Use the resistor values found in the example.

2. For the circuit in Fig. 14.6, determine the resistance values required to bias the circuit at $I_D = I_{DSS}/2$, $V_D = V_{CC}/2$, and $V_{DS} = V_{CC}$, where $V_{CC} = 15\ V$. The load resistance is $R_L = 100\ k\Omega$ and the generator resistance is $R_g = 50\ \Omega$. Assume that the JFET has parameters $\beta = 2 \times 10^{-4}\ A/V^2$, $V_{TO} - 3.5\ V$, and $\lambda \simeq 0$. The circuit is to be designed for a midband voltage gain of -5 and an input resistance of $800\ k\Omega$.

14.5 Preliminary SPICE Simulations

The JFET used in this experiment is the 2N5457 n-channel JFET. Typical SPICE parameters for this device are as follows: BETA $= 0.333M$ (β_O), VTO $= -3$ (V_{TO}), LAMBDA $= 0.01$ (λ), CGD $=$ 6P (c_{gdo}), CGS $=$ 6P (c_{gso}). Although these parameters are representative, the parameters of individual 2N5457 JFET's can be different.

1. Write the code for the SPICE input deck to simulate the CS amplifier example in the Theory section. Use the JFET parameters given above and the resistor values from the example. The SPICE analyses to be performed are the .OP, the .AC, and the .TRAN analyses. The frequency range for the .AC analysis is to be from 1 Hz to 10 MHz with at least 20 points per decade. The .TRAN analysis should use a sine wave for v_g with a DC level of zero, a frequency of 1 kHz, and at least 20 points per cycle. The sine-wave amplitude should be set to at least three values to investigate the clipping and large signal behavior. For the capacitors, use the values $C_1 = 0.1\ \mu F$, $C_2 = 1\ \mu F$, and $C_S = 100\ \mu F$. Perform a Fourier analysis in conjunction with the transient analysis and determine the input signal level for which the total harmonic distortion is 10%.

2. Write the code for the SPICE input deck to simulate the circuit shown in Fig. 14.7. Omit the potentiometer from the circuit. Use independent DC voltage sources for v_g and v_c. Use a dual or "nested" DC sweep to plot v_o versus v_g on the same plot for various values of v_c. The amplitude of the input voltage v_g should not exceed $0.5V_{TO}$. The resistor values in the circuit are $R_g = 50$, $R = 10\ k\Omega$, and $R_B = 1\ k\Omega$.

14.6 Experimental Procedures

14.6.1 Preparation

(a) Prepare the electronic breadboard to provide buses for the positive and negative power supply rails and the circuit ground. Each power supply rail should be decoupled with a 100 Ω 1/4 W resistor, and a 100 μF 25 V (or greater) capacitor. The resistors are connected in series with the external power supply leads and the capacitors are connected from power supply rail to ground on the circuit side of the resistors. The capacitors must be installed with proper polarity to prevent reverse polarity breakdown. **(b)** The transistor that specified for the experiment is the 2N5457 n-channel JFET. The pin layout for the 2N5457 is given in Fig. 14.1a.

14.6.2 Output Characteristics Curve Tracer

The following circuit will be used to obtain a plot of the drain current versus the drain-to-source voltage for various values of gate-to-source voltages for the N channel JFET. The plot will be made on the screen of the oscilloscope by using the oscilloscope in its XY mode. The drain-to-source voltage will be connected to the X axis of the oscilloscope (Channel 1) and a voltage proportional to the drain current will be connected to the Y axis. A differential amplifier is used to obtain a voltage across a resistor in series with the drain which is, therefore, proportional to the drain current. Both channels of the oscilloscope must be DC coupled. A DC voltmeter will be used to measure the gate-to-source voltage. **(a)** Assemble the circuit of Fig. 14.4. The element values are $R_A = 47\ k\Omega$, $R_P = 10\ k\Omega$, $R_{SA} = 1\ k\Omega$, $R_B = 1\ k\Omega$, $R_o = 100\ \Omega$, and $R = 470\ k\Omega$. The power supplies voltages are $+15\ V$ and $-15\ V$. The diode is a 1N4148 small signal fast recovery diode (the arrow points from the unbanded end to the banded end). **(b)** Set the function generator to produce a $10\ V$ peak-to-peak sine wave with a DC level of $5\ V$ and a frequency of 100 Hz. Prepare the oscilloscope for $X-Y$ operation. Connect v_{O2} to the Y-axis of the oscilloscope (channel 2) and v_{O1} to the X-axis (channel 1). Ground both channels of the oscilloscope and set the dot near the lower left portion of the CRT graticule. Unground both channels and set the coupling for both channels to DC. Set the volts per division for each channel to a value that permits the display to be easily sketched. If the oscilloscope has a bandwidth limit feature, turn it on. (Turn channel 1 off and press the channel 2 invert button on the oscilloscope.) Use the digital multimeter to measure the DC voltage from the wiper of the potentiometer to ground. **(c)** Obtain the sketches of v_{O2} versus v_{O1} for $v_{GS} = 0\ V$, $-0.5\ V$, $-1.0\ V$, $-1.5\ V$, etc., until the v_{O2} becomes negligibly small. The plots represent the drain current versus the drain-to-source voltage for constant values of gate-to-source voltage. Because the resistor R_{SA} is $1\ k\Omega$, the voltage v_{O2} corresponds to the drain current in mA. **(d)** For a drain-to-source voltage of $V_{DS} = 5\ V$, use the setup to determine the drain-to-source saturation current I_{DSS} and the gate-to-source voltage V_{GS} for the drain current $I_D = I_{DSS}/4$. **(e)** Use Eqs. 14.8 and 14.9 and the data obtained in the preceding step to determine β and V_{TO} for the JFET. Remember that V_{TO} is negative.

14.6.3 Transfer Characteristics Curve Tracer

(a) Assemble the circuit of Fig. 14.5. The element values are $R_P = 10\ k\Omega$, $R_{SA} = 1\ k\Omega$, $R_B = 1\ k\Omega$, $R_o = 100\ \Omega$, $R = 470\ k\Omega$, and $R_{BL} = 10\ k\Omega$. The power supply voltages are $+15\ V$ and $-15\ V$. The diode is a 1N4148 type. Set the function generator to produce a sine wave for v_g with a frequency of 100 Hz, a peak-to-peak value of $6\ V$, and a DC level of $-3\ V$. The output v_{O2} is connected to the vertical input of the oscilloscope and v_{O1} to the horizontal input. Ground both channels of the oscilloscope and set the dot near the lower right portion of the CRT graticule. Unground both channels and set both channels to DC coupling. Set the volts per division to a value that permits the display to be easily sketched. Use the digital multimeter to measure the DC voltage from the wiper of the pot to ground, i.e. the drain-to-source voltage. Sketch the display for the drain-to-source voltage equal to $2\ V$, $4\ V$, $6\ V$, and $8\ V$.

14.6.4 Common Source Amplifier Assembly

(a) The JFET measured in the preceding part is to be used in the CS amplifier in Fig. 14.6. Use the data obtained in the preceding part to calculate the resistor values for the amplifier. The design specifications are: midband small-signal voltage gain $v_o/v_g = -5$, midband small-signal input resistance $= 500\ k\Omega$, $I_D = I_{DSS}/2$, $V_{DS} = V_{CC}$, and $I_D R_D = V_{CC}/2$, where $V_{CC} = 15\ V$. The load and generator resistances are $R_L = 100\ k\Omega$ and $R_g = 50\ \Omega$. The design should follow the procedure presented in the Example in the Theory section. **(b)** Assemble the circuit with the nearest 5% resistor values. The capacitor values are specified to be $C_1 = C_2 = 0.1\ \mu F$ and $C_S = 100\ \mu F$. The electrolytic capacitor must be installed with the correct polarity to prevent reverse polarity breakdown. Use the digital multimeter to measure the voltage across resistor R_D. Use this measurement to determine the quiescent drain current I_D. If this value is not reasonably close to the design value, check the circuit for errors.

14.6.5 Common-Source Amplifier Gain Measurement

(a) Set the function generator to produce a sine wave with a DC level of $0\ V$, a peak-to-peak value of $1\ V$, and a frequency of 1 kHz. Use the dual channel oscilloscope to observe the circuit input and output signals. If the oscilloscope has a bandwidth limit feature, turn it on. Both channels should be DC coupled. Neither the input nor the output signals should have a DC offset. Measure the small-signal voltage gain and determine if it is reasonably close to the design value of -3. If it is not, review the calculations and examine the circuit for errors. (b) Vary the frequency of the input signal and determine the lower frequency f_ℓ and the upper frequency f_u at which the gain magnitude is reduced by the factor 0.707, i.e. it is decreased by 3 dB. Measure and record the small-signal voltage gain versus frequency from a lower frequency of $f_\ell/2$ to an upper frequency $2f_u$. The test frequencies should be chosen so that they are approximately equally spaced on a logarithmic scale. (Turn the bandwidth limit feature of the oscilloscope off to determine the upper frequency.) Due to the performance limitations of the function generator and/or oscilloscope combination it may not be possible to obtain the upper frequency at which the gain has been reduced by the factor 0.707. If this is the case, use the maximum frequency at which the function generator and oscilloscope combination permits a valid gain measurement to be made. (It is not necessary to take many data points in the midband range.)

14.6.6 Common-Source Amplifier Clipping Measurement

Set the frequency of the function generator to 1 kHz. Increase the input signal level until the output signal begins to distort (unequal positive and negative peaks and/or obvious distortion of the sine-wave shape). Continue to increase the input signal amplitude until clipping of either the negative or positive peak occurs. Record the amplitude of the input and sketch the output signal. Increase the amplitude of the input until both peaks show clipping. Record the amplitude of the input signal and sketch the output signal.

14.6.7 Voltage Controlled Attenuator

(a) Assemble the circuit shown in Fig. 14.7. The element values are $R = 10$ $k\Omega$, $R_A = 47$ $k\Omega$, $R_P = 10$ $k\Omega$, and $R_B = 1$ $k\Omega$. Because the $4R$ value of 40 $k\Omega$ is not a standard value, use two 20 $k\Omega$ resistors in series. (b) Set the function generator to produce a sine-wave output with a frequency of 1 kHz, a DC level of 0 V, and an amplitude of 0.5 V peak-to-peak. Use the dual channel oscilloscope to observe v_i and v_o as functions of time. Use the digital multimeter to measure the DC voltage v_c. Measure and record the signal v_o as v_c is varied from 0 V to the most negative value that can be obtained in steps of 0.2 V. (c) Prepare the oscilloscope for $X-Y$ operation. Connect the horizontal input to v_i and the vertical input to v_o. Use the oscilloscope to obtain the v_o versus v_i transfer characteristics of the circuit for several values of v_c. Sketch the characteristics and note any curvature of the curves. The curves for a perfectly linear attenuator would not exhibit curvature.

14.6.8 Relaxation Oscillator

(a) Assemble the circuit in Fig. 14.8. The element values are $C = 0.001$ μF, $R_1 = 10$ $k\Omega$, $R_2 = 10$ $k\Omega$, and $R_3 = 4.7$ $k\Omega$. Use a 741 type op-amp. Observe and sketch the output signal waveform using the oscilloscope. Measure the frequency of the signal. (b) Turn the power supply off, change the op-amp to a TL071 type, and repeat the preceding step. (c) Based on the results of the first two steps, decide which op-amp should be used in the next section which requires that $v_c(t)$ be a square wave.

14.6.9 Chopper Modulator

(a) Assemble the circuit of Fig. 14.9. The element values are $C_f = 100$ pF, $R_o = 100$ Ω, $R_L = 10$ $k\Omega$, $R_1 = 220$ $k\Omega$, $R_2 = 110$ $k\Omega$, $R_3 = 110$ $k\Omega$, $R_4 = 220$ $k\Omega$, and $R_B = 100$ $k\Omega$. Use a TL071 type for the op amp. The v_c input is the output of the relaxation oscillator assembled in the previous step. Connect the function generator to the v_g input and set the generator to produce a sine wave with a frequency of 1 kHz, a DC level of 2 V, and a peak-to-peak amplitude of 1 V. (b) Set the sweep time for the oscilloscope at 0.2 ms per division. Connect channel 1 of the oscilloscope to the circuit output v_o and channel 2 to the function generator output. Set the oscilloscope trigger source to channel 2. (c) Vary the amplitude and DC offset level of the function generator and observe the effect on v_o. Sketch several representative waveforms for v_o. Increase the sweep rate of the oscilloscope until the chopping action can be observed and sketch the waveform.

14.6.10 Curve Tracer

If a transistor curve tracer is available, obtain a plot of I_D versus V_{DS} for several values of V_{GS} in the range from $0\ V$ to $-5\ V$.

14.7 Laboratory Report

The laboratory report should include:

- all preliminary derivations, calculations, and SPICE simulations
- Sketches (drawn to scale) of the experimentally measured static terminal characteristics obtained in procedure parts 14.6.2 and 15.7.7
- calculation of β and V_{TO} based on this experimental data
- plot of the gain versus frequency for the common source amplifier examined in procedure part 14.6.5. The plot should be made by plotting the magnitude of the gain versus frequency on log-log graph paper or the dB gain versus frequency on linear-log graph paper
- sketch of the experimentally measured values of v_o versus v_i for various values of v_c for the voltage controlled attenuator
- An explanation of why the output of the relaxation oscillator depends on the choice of the op amp
- Representative sketches of the output of the chopper modulator should be included.
- plot of I_D versus V_{DS} for stepped values of V_{GS} obtained with a laboratory curve tracer. It should be contrasted with the results obtained in step 14.6.2

14.8 References

1. E. J. Angelo, *Electronics: BJTs, FETs, and Microcircuits*, McGraw-Hill, 1969.
2. W. Banzhaf, *Computer-Aided Circuit Analysis Using SPICE*, Prentice-Hall, 1989.
3. M. N. Horenstein, *Microelectronics Circuits and Devices*, Prentice-Hall, 1990.
4. P. Horwitz and W. Hill, *The Art of Electronics*, 2nd edition, Cambridge University Press, 1989.
5. P. Horowitz and I. Robinson, *Laboratory Manual for The Art of Electronics*, Cambridge University Press, 1981.
6. J. H. Krenz, *An Introduction to Electrical and Electronic Devices*, Prentice-Hall, 1987.
7. R. Mauro, *Engineering Electronics*, Prentice-Hall, 1989.
8. F. H. Mitchell and F. H. Mitchell, *Introduction to Electronic Design*, Prentice-Hall, 1988.
9. Motorola, *Small-Signal Semiconductors*, DL 126, Motorola, 1987.
10. C. J. Savant, M. S. Roden, and G. L. Carpenter, *Electronic Circuit Design*, Benjamin Cummings, 1987.
11. A. S. Sedra and K. C. Smith, *Microelectronics Circuits*, 4th edition, Oxford, 1998.
12. D. L. Schilling and C. Belove, *Electronic Circuits: Discrete and Integrated*, McGraw-Hill, 1968.
13. P. W. Tuinenga, *SPICE*, Prentice-Hall, 1988.

Chapter 15

The BJT Differential Amplifier

15.1 Object

The object of this experiment is to design a NPN BJT differential amplifier and to measure the small-signal differential and common-mode voltage gains, the common-mode rejection ratio, and the large signal clipping behavior.

15.2 Theory

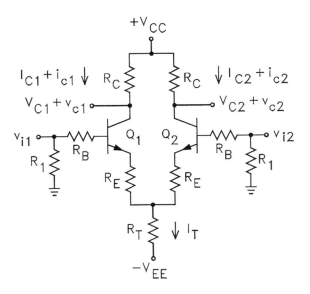

Figure 15.1: Differential Amplifier with resistive tail supply.

The differential amplifier or diff amp is one of the most important building blocks in analog circuit design. Fig. 15.1 shows the circuit diagram of a diff amp with a resistive tail supply. If the two BJTs are reasonably well matched and $v_{i1} = v_{i2} = 0$, the quiescent collector currents in the transistors are given by

$$I_{C1} = I_{C2} = \frac{V_{EE} - V_{BE}}{\frac{R_B}{\beta} + \frac{R_E + 2R_T}{\alpha}} \tag{15.1}$$

where V_{BE} is the quiescent base-to-emitter voltage, α is the emitter-to-collector current gain, and β is the base-to-collector current gain. The two current gains are related by $\alpha = \beta/(\beta + 1)$. The quiescent voltages at the collectors of the transistors are given by

$$V_{C1} = V_{C2} = V_{CC} - I_C R_C \tag{15.2}$$

15.2.1 Differential and Common-Mode Transconductances

When signals are applied to the inputs of the diff amp, the small-signal collector currents in the transistors can be written as differential and common-mode components given by

$$i'_{c1} = G_d(v_{i1} - v_{i2}) + G_{cm}\frac{v_{i1} + v_{i2}}{2} \tag{15.3}$$

$$i'_{c2} = G_d(v_{i2} - v_{i1}) + G_{cm}\frac{v_{i1} + v_{i2}}{2} \tag{15.4}$$

where G_d is the small-signal differential transconductance gain and G_{cm} is the small-signal common-mode transconductance gain. These are given by

$$G_d = \frac{\alpha}{2(r'_e + R_E)} \tag{15.5}$$

$$G_{cm} = \frac{\alpha}{r'_e + R_E + 2R_T} \tag{15.6}$$

where r'_e is the small-signal resistance seen looking into the emitter of Q_1 or Q_2 given by

$$r'_e = \frac{R_B + r_x}{1 + \beta} + r_e \tag{15.7}$$

In this equation, r_x is the BJT small-signal base spreading resistance, $r_e = V_T/I_E$ is the intrinsic emitter resistance, and $V_T = kT/q$ is the thermal voltage. At $27^0\ C$, V_T has the value $0.0259\ V$. Eqs. 15.3 through 15.6 assume that the small-signal collector-to-emitter resistance r_0 for each transistor is an open circuit.

15.2.2 Small-Signal Output Voltages

The small-signal voltages at the collectors of the two transistors are given by

$$v_{c1} = -i'_{c1} R_C \tag{15.8}$$

$$v_{c2} = -i'_{c2} R_C \tag{15.9}$$

It follows from Eqs. 15.3, 15.4, 15.8, and 15.9 that v_{c1} and v_{c2} are related to v_{i1} and v_{i2} by

$$v_{c1} = -\left(G_d + \frac{G_{cm}}{2}\right) R_C v_{i1} + \left(G_d - \frac{G_{cm}}{2}\right) R_C v_{i2} \tag{15.10}$$

$$v_{c2} = -\left(G_d + \frac{G_{cm}}{2}\right) R_C v_{i2} + \left(G_d - \frac{G_{cm}}{2}\right) R_C v_{i1} \tag{15.11}$$

15.3. JFET TAIL SUPPLY

In most cases, $G_{cm} \ll G_d$ so that these equations can be approximated by

$$v_{c1} \cong -G_d R_C (v_{i1} - v_{i2}) \tag{15.12}$$

$$v_{c2} \cong +G_d R_C (v_{i1} - v_{i2}) \tag{15.13}$$

Thus each output voltage is approximately proportional to the difference between the two input voltages.

15.2.3 Differential and Common-Mode Voltage Gains

The differential output voltage from the circuit is given by $v_{od} = v_{c1} - v_{c2}$. The differential voltage gain A_d is defined as the ratio v_{od}/v_{id} for the differential input signals $v_{i1} = -v_{i2} = v_{id}/2$. It follows that the differential voltage gain is given by

$$A_d = \frac{v_{od}}{v_{id}} = -2G_d R_C = \frac{-\alpha R_C}{r'_e + R_E} \tag{15.14}$$

The common-mode output voltage from the circuit is given by $v_{ocm} = (v_{c1} + v_{c2})/2$. The common-mode voltage gain A_{cm} is defined as the ratio v_{ocm}/v_{icm} for the common-mode input signals $v_{i1} = v_{i2} = v_{icm}$. It follows that the common-mode voltage gain is given by

$$A_{cm} = \frac{v_{ocm}}{v_{icm}} = -G_{cm} R_C = \frac{-\alpha R_C}{r'_e + R_E + 2R_T} \tag{15.15}$$

15.2.4 Common-Mode Rejection Ratio

A figure of merit for the diff amp is the common-mode rejection ratio or $CMRR$. This is defined as the ratio of the differential voltage gain to the common-mode voltage gain. A perfect diff amp would have a common-mode gain of zero so that its $CMRR$ would be infinite. The $CMRR$ is given by

$$CMRR = \frac{A_d}{A_{cm}} = \frac{2G_d}{G_{cm}} = 1 + \frac{2R_T}{r'_e + R_E} \tag{15.16}$$

If the output voltage is taken to be either v_{c1} or v_{c2} rather than the differential output voltage $v_{od} = v_{c1} - v_{c2}$, the $CMRR$ would be one-half the value given by this equation. The $CMRR$ is often expressed in decibels by the expression $20 \log_{10}(A_d/A_{cm})$.

15.3 JFET Tail Supply

Fig. 15.2 shows the diff amp with a N-channel JFET constant current source tail supply in place of the resistor R_T. The use of a constant current source in place of the resistor results in a significant improvement in the CMRR. The JFET drain current is given by

$$I_D = \beta(V_{GS} - V_{T0})^2 = I_{DSS}\left(1 - \frac{V_{GS}}{V_{T0}}\right)^2 \tag{15.17}$$

where β is the transconductance parameter, V_{T0} is the threshold voltage, and I_{DSS} is the drain-to-source saturation current. The gate-to-source voltage V_{GS} is given by

$$V_{GS} = -I_S R_S \tag{15.18}$$

Because the JFET gate current is zero, the drain current I_D is equal to the source current I_S.

The diff amp $CMRR$ with the JFET tail supply is given by

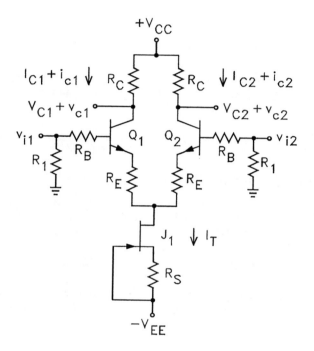

Figure 15.2: Differential amplifier with a JFET constant current tail supply.

$$CMRR = 1 + \frac{2r_{id}}{r'_e + R_E} \tag{15.19}$$

where r_{id} is the small-signal resistance seen looking into the JFET drain. This is given by

$$r_{id} = \frac{r_0 + r_s \| R_S}{1 - \dfrac{R_S}{R_S + r_s}} \tag{15.20}$$

In this equation, r_0 is the JFET small-signal drain-to-source resistance and r_s is the source intrinsic resistance. These are given by

$$r_s = \frac{1}{2\sqrt{\beta I_D}} \tag{15.21}$$

$$r_0 = \frac{1 + \lambda V_{DS}}{\lambda I_D} \tag{15.22}$$

where V_{DS} is the drain-to-source voltage and λ is the JFET channel length modulation factor.

15.4 Preliminary Derivations

1. Derive Eqs. 15.1 and 15.2 for the quiescent collector currents and voltages of the BJT's in Fig. 15.1. Note that the resistors labeled R_1 are short circuited when $v_{i1} = v_{i2} = 0$.

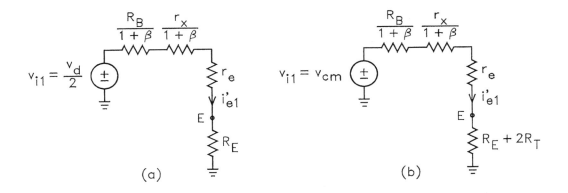

Figure 15.3: (a) Emitter equivalent circuit for Q_1 for a differential input signal. (b) Emitter equivalent circuit for Q_1 for a common-mode input signal.

2. For the differential input signals $v_{i1} = -v_{i2} = v_{id}/2$ to the circuit in Fig. 15.1, it follows by symmetry that the small-signal voltage at the junction of R_T and the two R'_Es is zero. Therefore, this node can be grounded in the small-signal circuit. This decouples the two transistors so that the small-signal emitter equivalent circuit for Q_1 is the circuit of Fig. 15.3a. Use this circuit and the relation $i'_c = \alpha i'_e$ to derive Eq. 15.5 for the small-signal differential transconductance gain of the diff amp.

3. If R_T in Fig. 15.1 is replaced by two parallel resistors, each having the value $2R_T$, it follows by symmetry for the common-mode input signals $v_{i1} = v_{i2} = v_{icm}$ that all of i'_{e1} flows through one of the resistors and all of i'_{e2} flows through the other. This decouples the two transistors so that the small-signal emitter equivalent circuit for Q_1 is the circuit of Fig. 15.3b. Use this circuit and the relation $i'_c = \alpha i'_e$ to derive Eq. 15.5 for the small-signal common-mode transconductance gain of the diff amp.

4. (a) Derive Eqs. 15.12 and 15.13 for the differential and common-mode voltage gains of the diff-amp. (b) Derive Eqs. 15.14 and 15.15 for v_{c1} and v_{c2}.

15.5 Preliminary Calculations

1. (a) For the diff-amp in Fig. 15.1, it is given that $R_1 = 10\ k\Omega$, $R_B = 1\ k\Omega$, $V_{CC} = +15\ V$, and $-V_{EE} = -15\ V$. Each transistor in the diff-amp is to be biased at a quiescent collector current of 1 mA and a quiescent collector voltage of $V_{CC}/2$. The circuit is to be designed for a differential gain $A_d = -20$ (26 dB). For the BJT's, assume that $V_{BE} = 0.65\ V$, $\alpha = 0.99$, and $r_x = 10\ \Omega$. Use Eqs. 15.1, 15.2, and 15.12 to calculate the required values of the resistors R_C, R_E, and R_T. Round off the resistor values to the nearest 5% standard value. (b) Calculate the $CMRR$ of the diff-amp. Express it both as a linear number and in dB.

2. (a) The JFET in Fig. 15.2 has the parameters $I_{DSS} = 3$ mA and $V_{T0} = -3\ V$. Calculate the value of R_S which biases the two BJT's at a quiescent collector current of 1 mA. Use the BJT parameters given in the preceding part. (b) Calculate the values of the JFET transconductance parameter β. Calculate the source intrinsic resistance r_s. If $\lambda = 0.01$, calculate the JFET drain-to-source resistance r_0. Calculate the $CMRR$ of the diff-amp with the JFET tail supply. Use numerical values from the preceding part.

15.6 Preliminary SPICE Simulations

1. (a) Using the resistor values and power supply voltages in part 1 of the preliminary calculations, perform a SPICE simulation to verify the quiescent bias current and voltages for the diff-amp of Fig. 15.1. The appropriate SPICE analysis is the ".OP" analysis. The ".OUT" file contains the bias calculations. For the BJT's, use the SPICE model parameters IS = 1.26E − 14, BF = 99, and RB = 10. The input voltages v_{i1} and v_{i2} should be zeroed for the simulation. (b) Repeat the SPICE simulation of the preceding step for the diff-amp circuit of Fig. 2. Use the resistor values and power supply voltages from the preceding step and the value of R_S calculated in part 2 of the preliminary calculations. For the JFET, use the SPICE model parameters BETA= β, VTO = −3, and LAMBDA= 0.01, where the value of β is calculated in part 2 of the preliminary calculation.

2. Using the numerical values from the preceding part, perform a SPICE simulation to verify the small-signal DC differential gain of the circuits of Figs. 15.1 and 15.2. The SPICE ".TF" analysis is the appropriate analysis. The diff-amp inputs must be driven differentially by connecting an independent voltage source VIN from v_{i1} to v_{i2}. The output variable for the ".TF" analysis is the small-signal differential voltage $v_{c1} - v_{c2}$. The form of the ".TF" command line is ".TF V(N1,N2) VIN", where N1 is the node number for v_{c1} and N2 is the node number for v_{c2}. The ".TF" command also calculates the small-signal input and output resistances. The data calculated by the ".TF" command is in the SPICE ".OUT" file.

3. With appropriate modifications to the procedure, repeat the preceding part to obtain the small-signal DC common-mode gains of the circuits in Figs. 15.1 and 15.2. The two inputs must be connected in shunt and the common input node driven by the source VIN. The common-mode output voltage is $v_{ocm} = (v_{c1} + v_{c2})/2$. This can be realized in SPICE by connecting the collectors of Q_1 and Q_2 in shunt. (Use a zero voltage source for a short circuit.) From the SPICE simulation data, calculate the $CMRR$'s for both diff-amp circuits and compare these to the values predicted in the preliminary calculations.

4. (a) Write the code for the SPICE input deck to perform a ".DC" sweep on the diff-amps of Figs. 15.1 and 15.2. For the analysis, connect a differential voltage source VIN between the v_{i1} and v_{i2} inputs. Sweep the value of VIN over the range from −1 V to +1 V with 20 points per volt. (b) Use the PROBE graphics feature of PSpice to display the two output voltages v_{c1} and v_{c2} as a function of VIN. From the simulations, determine the output clipping levels for both outputs of each diff-amp. What is the maximum and minimum values of VIN if clipping is to be avoided? (c) Calculate the slopes of the graphs of v_{c1} and v_{c2} versus VIN. How do these relate to the small-signal voltage gains of the circuit?

15.7 Experimental Procedures

15.7.1 Preparation

Prepare the electronic breadboard to provide buses for the positive and negative power supply rails and the circuit ground. Each power supply rail should be decoupled with a 100 Ω, 1/4 W resistor, and a 100 μF, 25 V (or greater) capacitor. The resistors are connected in series with the external power supply leads and the capacitors are connected from power supply rails to ground on the circuit side of the resistors. The capacitors must be installed with the proper polarity to prevent reverse polarity breakdown.

15.7.2 Transistor Matching

Use the transistor curve tracer to select two 2N3904 NPN BJTs that have loosely matched output characteristics. A pin diagram is give in Fig. 15.4a. (Perfectly matched transistors are not a requirement for this experiment.)

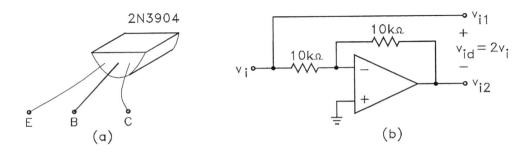

Figure 15.4: (a) Pin diagram for the 2N3904. (b) Op-amp inverter circuit used to generate differential input signals to the diff-amp.

15.7.3 Assembly

(a) Assemble the diff-amp circuit of Fig. 15.1 on the breadboard. (b) Ground the two inputs and connect the two outputs to the inputs of the dual channel oscilloscope with the oscilloscope inputs DC coupled. (c) Set the power supply voltages to $+15\ V$ and $-15\ V$. (d) Use the digital multimeter to measure V_{BE} for each transistor and verify that it is approximately equal to the 0.65 V value assumed for the calculations. A voltage in the range of 0.6 V to 0.7 V is normal. (Observe the DC collector voltages on the oscilloscope while doing this. If the DC voltages change, the multimeter is perturbing the bias currents. It may be necessary to measure V_B and V_E and subtract the two to obtain V_{BE}.) (e) Measure the DC voltage across R_T and the two R'_Cs and calculate the currents through these resistors to verify that the voltages are correct.

15.7.4 Function Generator

(a) Connect the output of the function generator to the oscilloscope. Set the generator for a 1 kHz sine wave. Verify that the DC offset of the generator signal is zero. (b) Set the function generator for the smallest peak-to-peak output signal that it can produce. If this is greater than 0.1 V, a $L-pad$ attenuator must be used to reduce the generator signal to a level which will not overload the diff-amp. The output impedance of the $L-pad$ should be less than or equal to 100 Ω.

15.7.5 Single Ended Input Small-Signal Gain

(a) With v_{i2} grounded, connect the output of the function generator to v_{i1}. (Use the $L-pad$ attenuator if required.) Connect the two outputs of the diff-amp to the inputs of the dual channel oscilloscope with its inputs AC coupled. Measure the small-signal gain from v_{i1} to v_{c1} and from v_{i1} to v_{c2}. (b) Set the dual channel oscilloscope to measure the differential output voltage $v_{od} = v_{c1} - v_{c2}$. Measure the gain from v_{i1} to v_{od}. (c) Repeat the measurements with v_{i1} grounded and the function generator connected to v_{i2}.

15.7.6 Differential Input Small-Signal Gain

(a) Connect a 741 op amp as shown in Fig. 15.4b to generate differential input signals for the diff amp. The power supply voltages for the op amp are $\pm 15\ V$. If the $L-pad$ attenuator is required, it should be between the function generator and the v_i input to the op amp circuit. The generator level should be set to obtain approximately the same peak output voltages from the diff amp obtained in the preceding part. (b) Measure the small-signal gain from the differential input voltage $v_{id} = v_{i1} - v_{i2}$ to v_{c1} and from v_{id} to v_{c2}. (Note that $v_{id} = 2v_i$.) (c) Set the dual channel oscilloscope to measure the differential output voltage $v_{od} = v_{c1} - v_{c2}$. Measure the gain from v_{id} to v_{od}. Calculate the differential gain $A_d = v_{od}/v_{id}$ of the diff-amp.

15.7.7 Transfer Characteristic

(a) Connect v_i to the horizontal input of the oscilloscope and set the oscilloscope for XY mode with the two y inputs DC coupled. (It may be necessary to set the oscilloscope to the chop sweep mode.) While observing the voltage transfer characteristics from v_i to v_{c1} and v_{c2}, increase the function generator output level until the curves just begin to exhibit limiting on both sides. Record the curves. (b) Change the oscilloscope to a time sweep and record the clipped sine-wave waveforms. Notice any differences between the positive and negative clipped peaks. (c) Repeat steps (a) and (b) with the function generator output adjusted to obtain hard limiting.

15.7.8 Common Mode Gain

(a) Change the oscilloscope coupling to AC. Adjust the function generator output level to approximately the value used in part 15.7.6. Disconnect the v_{i2} input from the op amp and connect it in shunt with v_{i1} so that the diff amp is driven with a common-mode signal. Connect a short-circuit jumper wire between the v_{c1} and v_{c2} outputs. The output voltage is now the common-mode output voltage v_{ocm}. Use the dual channel oscilloscope to measure v_{icm} and v_{ocm}. (The output signal for a common-mode input is small. If a distorted waveform is obtained, the oscilloscope input preamps are being overloaded.) (b) Calculate the common-mode voltage gain $A_{cm} = v_{ocm}/v_{icm}$. Calculate the common-mode rejection ratio $CMRR$.

15.7.9 Current Source Design

(a) Using a 2N5457 JFET and a $1\ k\Omega$ potentiometer for R_P, assemble the circuit of Fig. 15.5a on the breadboard. (A $10\ k\Omega$ potentiometer is acceptable as a substitute.) The pin diagram for the 2N5457 is given in Fig. 15.5b. Set the multimeter in series with the JFET drain to measure current. With $V_{EE} = -15\ V$, set $R_P = 0$. Record the value of the drain current as the JFET drain-to-source saturation current I_{DSS}. Adjust R_P to obtain a drain current of $I_{DSS}/4$. Remove R_P from the circuit and use the multimeter to measure its resistance. Calculate and record the value of the JFET threshold voltage given by $V_{T0} = -I_{DSS}R_S/2$. Calculate and record the value of the JFET transconductance parameter β. (b) Put R_P back the circuit. Adjust its value to obtain a drain current $I_D = 2\ mA$. Remove R_P and use the multimeter to measure its value. (c) Replace R_P with the nearest value 5% value resistor and re-measure I_D. If the current is not within $\pm 5\%$ of 2 mA, change the resistor to the next higher or lower 5% value until the correct current is obtained. (If the JFET has a drain-to-source saturation current I_{DSS} less than 2 mA, it will not be possible to obtain a current of 2 mA. In this case, connect the JFET gate to its source with no resistor in series with the source lead. In this case, the drain current is I_{DSS}.)

15.8. LABORATORY REPORT

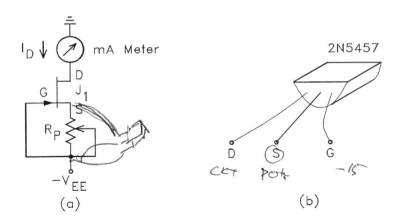

Figure 15.5: (a) Circuit used to deterrmine the JFET source resistor. (b) Pin diagram for the 2N5457.

15.7.10 Differential Amplifier with JFET Current Source Tail Supply

(a) Assemble the circuit of Fig. 15.2 on the breadboard using the value of R_S obtained in the preceding part.
(b) Verify that the currents are the desired values. If the currents are not correct, adjust R_S as required.
(c) Repeat the measurements of parts 15.7.5 through 15.7.7 above.

15.7.11 Square Wave Response

Before disassembly of the circuit, observe the response of the diff amp to square waves and sine waves of different amplitudes and frequencies.

15.8 Laboratory Report

The lab report should include:

- all preliminary derivations, calculations, and SPICE simulations. For SPICE simulations of the circuit in Fig. 15.2, use the use the values for the JFET model parameters β and V_{T0} obtained in part 15.7.9 of the experimental procedures. For the JFET channel length modulation factor, use the value $\lambda = 0.01$

- a comparison of the theoretically predicted, the simulated, and the experimentally observed values should be made and sources of error identified

- an explanation for the observed clipping behavior. How would the clipped waveforms be altered if only one input were driven while the other was grounded?

15.9 References

1. E. J. Angelo, *Electronics: BJT's, FET's, and Microcircuits*, McGraw-Hill, 1969.
2. W. Banzhaf, *Computer-Aided Circuit Analysis Using SPICE*, Prentice-Hall, 1989.
3. M. N. Horenstein, *Microelectronics Circuits and Devices*, Prentice-Hall, 1990.
4. P. Horowitz & W. Hill, *The Art of Electronics*, 2nd edition, Cambridge University Press, 1989.

5. P. Horowitz & I. Robinson, *Laboratory Manual for The Art of Electronics,* Cambridge University Press, 1981.
6. J. H. Krenz, *An Introduction to Electrical and Electronic Devices,* Prentice-Hall, 1987.
7. R. Mauro, *Engineering Electronics,* Prentice-Hall, 1989.
8. F. H. Mitchell & F. H. Mitchell, *Introduction to Electronic Design,* Prentice-Hall, 1988.
9. Motorola, Inc., *Small-Signal Semiconductors,* DL 126, Motorola, 1987.
10. C. J. Savant, M. S. Roden, & G. L. Carpenter, *Electronic Circuit Design,* Benjamin Cummings, 1987.
11. A. S. Sedra & K. C. Smith, *Microelectronics Circuits,* 4th edition, Oxford, 1998.
12. D. L. Schilling & C. Belove, *Electronic Circuits: Discrete and Integrated,* McGraw-Hill, 1968.
13. P. W. Tuinenga, *SPICE,* Prentice-Hall, 1988.

Chapter 16

MOSFET Amplifier

16.1 Object

The object of this experiment is to design and evaluate a single-stage common-source MOSFET amplifier with a resistive load and a differential MOSFET amplifier with a resistive load and an active load.

16.2 Device Equations

Whereas the JFET has a diode junction between the gate and the channel, the metal-oxide semiconductor FET or MOSFET differs primarily in that it has an oxide insulating layer separating the gate and the channel. The circuit symbols are shown in Fig. 16.1. Each device has gate (G), drain (D), and source (S) terminals. Four of the symbols show an additional terminal called the body (B) which is not normally used as an input or an output. It connects to the drain-source channel through a diode junction. In discrete MOSFETs, the body lead is connected internally to the source. When this is the case, it is omitted on the symbol as shown in four of the MOSFET symbols. In integrated-circuit MOSFETs, the body usually connects to a dc power supply rail which reverse biases the body-channel junction. In the latter case, the so-called "body effect" must be accounted for when analyzing the circuit.

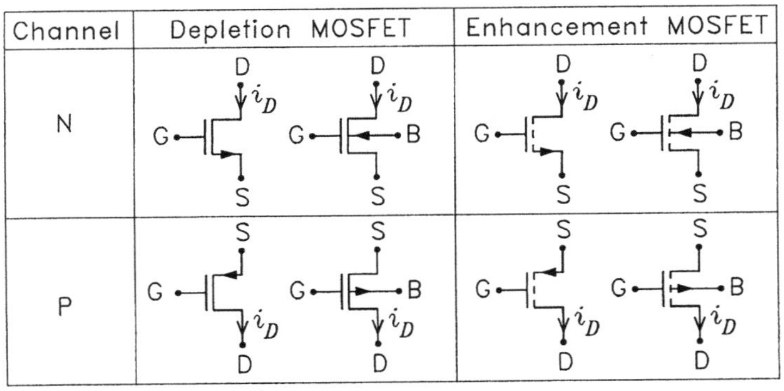

Figure 16.1: MOSFET symbols.

The discussion here applies to the n-channel MOSFET. The equations apply to the p-channel device if the subscripts for the voltage between any two of the device terminals are reversed, e.g. v_{GS} becomes v_{SG}.

16.5. SMALL-SIGNAL EQUIVALENT CIRCUITS

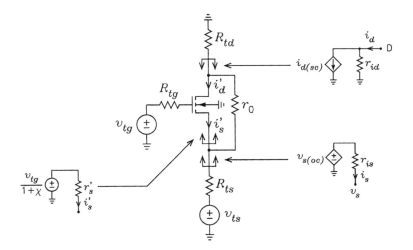

Figure 16.11: Summary of the small-signal equivalent circuits. Set $\chi = 0$ if the body is connected to the source.

where
$$V_{GG} = \frac{V^+ R_2 + V^- R_1}{R_1 + R_2} \tag{16.40}$$

The drain current can be obtained by solving these equations simultaneously. The solution is

$$I_D = \frac{1}{4KR_S^2} \left[\sqrt{1 + 4KR_S(V_{GG} - V^- - V_{TO})} - 1 \right]^2$$

If a desired drain current is specified, the gate and source bias networks can be designed for the desired current. The gate resistors should not be too small to prevent loading of the input signal source. C_1, C_2, and C_3 set the lower cutoff frequency of the amplifier. These must be chosen to have a low impedance at the desired signal frequencies.

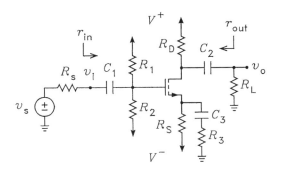

Figure 16.12: Common-source amplifier.

For a small-signal ac analysis, set $V^+ = V^- = 0$ and replace all capacitors with short circuits. The small-signal mid-band output voltage can be obtained from the equivalent circuit of Fig. 16.9 with $R_{td} = R_D \| R_L$

The output voltage is given by
$$v_o = -i_{d(sc)} r_{id} \| R_D \| R_L \tag{16.41}$$

To solve for the voltage gain, Eq. (16.29) is used for $i_{d(sc)}$ with the values

$$\chi = 0 \quad r'_s = r_s \quad v_{tg} = v_g \frac{R_1 \| R_2}{R_s + R_1 \| R_2} \quad R_{ts} = R_S \| R_3 \quad v_{ts} = 0 \tag{16.42}$$

It follows that the gain is given by

$$\frac{v_o}{v_g} = \frac{R_1 \| R_2}{R_s + R_1 \| R_2} \frac{-r_{id} \| R_D \| R_L}{r_s (1 + R_S \| R_3 / r_0) + R_S \| R_3} \tag{16.43}$$

where r_{id} is given by

$$r_{id} = r_0 \left(1 + \frac{R_S \| R_3}{r_s}\right) + R_S \| R_3 \tag{16.44}$$

To solve for the gain v_o/v_i, simply set $R_s = 0$ in Eq. (16.43). The output resistance seen by R_L and the input resistance seen by the source, respectively, are given by

$$r_{\text{out}} = r_{id} \| R_D \quad r_{\text{in}} = R_1 \| R_2 \tag{16.45}$$

The lower cutoff frequency f_ℓ can be estimated from the equation

$$f_\ell \simeq \sqrt{f_1^2 + f_2^2 + f_{Sp}^2 - 2f_{Sz}^2} \tag{16.46}$$

where

$$f_1 = \frac{1}{2\pi (R_s + r_{\text{in}}) C_1} \quad f_2 = \frac{1}{2\pi (r_{\text{out}} + R_L) C_2}$$
$$f_{Sp} = \frac{1}{2\pi (r_{iso} \| R_S + R_3) C_3} \quad f_{Sz} = \frac{1}{2\pi (R_S + R_3) C_3} \tag{16.47}$$

and

$$r_{iso} = \frac{r_0 + R_D \| R_L}{1 + r_0 / r_s} \tag{16.48}$$

16.5.6 Common-Drain Amplifier

Figure 16.13 shows the circuit diagram of a common-drain amplifier. The bias solution is the same as for the common source amplifier in Fig. 16.12. To calculate the small-signal midband voltage gain, set $V^+ = V^- = 0$ and replace capacitors with short circuits. The equivalent circuit of Fig. 16.10 with $R_{ts} = R_S \| R_L$ can be used to write

$$v_o = v_{s(oc)} \frac{R_S \| R_L}{r_{is} + R_S \| R_L} \tag{16.49}$$

where by Eq. (16.37)

$$v_{s(oc)} = v_s \frac{R_1 \| R_2}{R_s + R_1 \| R_2} \frac{r_0}{r_0 + r_s} \quad r_{is} = \frac{r_0}{1 + r_0 / r_s} = r_0 \| r_s \tag{16.50}$$

It follows that the voltage gain is given by

$$\frac{v_o}{v_g} = \frac{R_1 \| R_2}{R_s + R_1 \| R_2} \frac{r_0}{r_0 + r_s} \frac{R_S \| R_L}{r_{is} + R_S \| R_L} \tag{16.51}$$

16.5. SMALL-SIGNAL EQUIVALENT CIRCUITS

To solve for v_o/v_i, simply set $R_g = 0$ in this equation. The output resistance seen by R_L and the input resistance seen by the source, respectively, are given by

$$r_{\text{out}} = R_S \| r_{is} \qquad r_{\text{in}} = R_1 \| R_2 \tag{16.52}$$

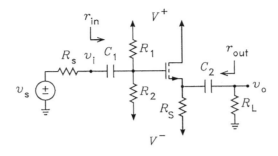

Figure 16.13: Common-drain amplifier.

The lower cutoff frequency f_ℓ can be estimated from the equation

$$f_\ell \simeq \sqrt{f_1^2 + f_2^2} \tag{16.53}$$

where

$$f_1 = \frac{1}{2\pi \left(R_s + R_1 \| R_2\right) C_1} \qquad f_2 = \frac{1}{2\pi \left(r_{is} \| R_S + R_L\right) C_2} \tag{16.54}$$

16.5.7 Common-Gate Amplifier

Figure 16.14 shows the circuit diagram of a common-gate amplifier. The bias solution is the same as for the common source amplifier in Fig. 16.12. The small-signal midband output voltage can be obtained from the equivalent circuit of Fig. 16.9 with $R_{td} = R_D \| R_L$, where it is assumed that $V^+ = V^- = 0$ and the three capacitors are signal short circuits. The output voltage is given by

$$v_o = -i_{d(sc)} r_{id} \| R_D \| R_L \tag{16.55}$$

To solve for the voltage gain, Eq. (16.29) is used for $i_{d(sc)}$ with the values

$$\chi = 0 \qquad r'_s = r_s \qquad v_{tg} = 0 \qquad R_{ts} = R_S \| R_s \qquad v_{ts} = v_s \frac{R_S}{R_s + R_S} \tag{16.56}$$

The gain is given by

$$\frac{v_o}{v_s} = \frac{R_S}{R_s + R_S} \frac{-r_{id} \| R_D \| R_L}{r_s \left(1 + R_S \| R_s / r_0\right) + R_S \| R_s} \tag{16.57}$$

where r_{id} is given by

$$r_{id} = r_0 \left(1 + \frac{R_S \| R_s}{r_s}\right) + R_S \| R_s \tag{16.58}$$

To solve for v_o/v_i, simply set $R_s = 0$ in Eq. (16.57). The output resistance seen by R_L and the input resistance seen by the source, respectively, are given by

$$r_{\text{out}} = r_{id} \| R_D \qquad r_{\text{in}} = R_S \| r_{is} \tag{16.59}$$

where
$$r_{is} = \frac{r_0 + R_D \| R_L}{1 + r_0/r_s} \qquad (16.60)$$

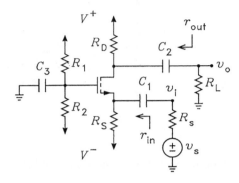

Figure 16.14: Common-gate amplifier.

The lower cutoff frequency f_ℓ can be estimated from the equation

$$f_\ell \simeq \sqrt{f_1^2 + f_2^2} \qquad (16.61)$$

where

$$f_1 = \frac{1}{2\pi (R_s + r_{\text{in}}) C_1} \qquad f_2 = \frac{1}{2\pi (r_{\text{out}} + R_L) C_2} \qquad (16.62)$$

16.5.8 MOSFET Differential Amplifier with Resistive Load

Biasing

Figure 16.15 shows the circuit diagram of a MOSFET differential amplifier with resistive loads. The current source I_{TO} sets the dc source current in each device. The resistor R_T models the output resistance of the source. If M_1 and M_2 are matched, the current I_T divides equally between the sources of the two MOSFETs so that

$$I_{D1} = I_{D2} = \frac{I_T}{2} \simeq \frac{I_{T0}}{2} \qquad (16.63)$$

where the approximation holds for R_T large. Resistors R_{G1} and R_{G2} provide a ground reference for the gates when no signal source is connected to the inputs. Because the gate current is approximately zero, these resistors can be very large. In the following, it is assumed that $R_{G1} = R_{G2}$ and $R_{D1} = R_{D2}$.

16.5. SMALL-SIGNAL EQUIVALENT CIRCUITS

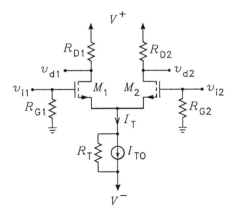

Figure 16.15: Differential amplifier.

Small-Signal Gain

To solve for the gain, set $V^+ = V^- = 0$ and $I_{TO} = 0$. Let the input voltages be expressed by their differential and common-mode components as follows:

$$v_{i1} = v_{icm} + \frac{v_{id}}{2} \qquad v_{i1} = v_{icm} - \frac{v_{id}}{2} \qquad (16.64)$$

where v_{icm} is the common-mode component and v_{id} is the differential component. These are given by

$$v_{icm} = \frac{v_{i1} + v_{i2}}{2} \qquad v_{id} = v_{i1} - v_{i2} \qquad (16.65)$$

Superposition of v_{icm} and v_{id} can be used to solve for the small-signal drain currents. With $v_{icm} = 0$, we have

$$v_{i1} = \frac{v_{id}}{2} \qquad v_{i2} = -\frac{v_{id}}{2} \qquad (16.66)$$

With equal but opposite sign voltages applied to the diff-amp inputs, it follows by symmetry that the small-signal voltage is zero at the node common to the MOSFET sources and there is no signal current in R_T. The equivalent circuit of Fig. 16.9 can be used to write

$$v_{o1} = -i_{d1(sc)} r_{id1} \| R_{D1} \qquad v_{o2} = -i_{d2(sc)} r_{id2} \| R_{D2} \qquad (16.67)$$

where, by Eq. (16.29), $i_{d1(sc)}$ and $i_{d2(sc)}$ are given by

$$i_{d1(sc)} = \frac{v_{id}}{2} \frac{1}{r'_{s1}} = \frac{v_{id}}{2} g_{m1} \qquad i_{d2(sc)} = -\frac{v_{id}}{2} \frac{1}{r'_{s2}} = -\frac{v_{id}}{2} g_{m1}$$

where $v_{ts} = 0$, $\chi = 0$, $r'_s = r_s$, and $R_{ts} = 0$.

With $v_{id} = 0$, we have

$$v_{i1} = v_{icm} \qquad v_{i2} = v_{icm} \qquad (16.68)$$

With equal voltages applied to the diff-amp inputs, it follows by symmetry that $i_{s1(sc)} = i_{s2(sc)}$. Thus if R_T is split into two parallel resistors of value $2R_T$ each, all of $i_{s1(sc)}$ would flow through one and all of $i_{s2(sc)}$ would flow through the other. The equivalent circuit of Fig. 16.9 with $v_{ts} = 0$, $\chi = 0$, $r'_s = r_s$, and $R_{ts} = 2R_T$ connected from the source node to ground can be used to write

$$i_{d1(sc)} = i_{s1(sc)} = \frac{v_{icm}}{r_s + 2R_T} \qquad i_{d2(sc)} = i_{s2(sc)} = \frac{v_{icm}}{r_s + 2R_T} \qquad (16.69)$$

Adding the differential and common-mode components, we obtain

$$i_{d1(sc)} = i_{s1(sc)} = \frac{v_{icm}}{r_s + 2R_T} + \frac{v_{id}}{2r_s} = G_{m(cm)}\frac{v_{i1} + v_{i2}}{2} + G_{m(d)}(v_{i1} - v_{i2}) \quad (16.70)$$

$$i'_{d2} = i'_{s2} = \frac{v_{icm}}{r_s + 2R_T} - \frac{v_{id}}{2r_s} = G_{m(cm)}\frac{v_{i1} + v_{i2}}{2} - G_{m(d)}(v_{i1} - v_{i2}) \quad (16.71)$$

where the common-mode and differential transconductances are given by

$$G_{m(cm)} = \frac{1}{r_s + 2R_{TO}} \qquad G_{m(d)} = \frac{1}{2r_s} \quad (16.72)$$

To obtain an approximate solution, let us assume that $r_{id1} = r_{id2} = \infty$. The small-signal drain voltages are given by $v_{d1} = -i_{d1(sc)}R_D$ and $v_{d2} = -i_{d2(sc)}R_D$. When the relations for the drain currents are used, it follows that

$$v_{d1} = -\left(\frac{G_{m(cm)}}{2} + G_{m(d)}\right)R_D v_{i1} - \left(\frac{G_{m(cm)}}{2} - G_{m(d)}\right)R_D v_{i2} \quad (16.73)$$

$$v_{d2} = -\left(\frac{G_{m(cm)}}{2} + G_{m(d)}\right)R_D v_{i2} - \left(\frac{G_{m(cm)}}{2} - G_{m(d)}\right)R_D v_{i1} \quad (16.74)$$

If $2R_T \gg r_s$, it follows that $G_{m(cm)} \ll G_{m(d)}$. In this case, the drain voltages are approximately given by

$$v_{d1} = -G_{m(d)}R_D(v_{i1} - v_{i2}) = -\frac{R_D}{2r_s}(v_{i1} - v_{i2}) \quad (16.75)$$

$$v_{d2} = +G_{m(d)}R_D(v_{i1} - v_{i2}) = \frac{R_D}{2r_s}(v_{i1} - v_{i2}) \quad (16.76)$$

Thus the output voltages are proportional to the difference between the input voltages. This is the reason that the circuit is called a differential amplifier.

Differential and Common-Mode Voltage Gains

The differential output voltage from the circuit is given by $v_{od} = v_{d1} - v_{d2}$. The differential voltage gain $A_{v(d)}$ is defined as the ratio v_{od}/v_{id} for the differential input signals $v_{i1} = -v_{i2} = v_{id}/2$. It follows that the differential voltage gain is given by

$$A_{v(d)} = \frac{v_{od}}{v_{id}} = -2G_{m(d)}R_D = -\frac{R_D}{r_s} \quad (16.77)$$

The common-mode output voltage is given by $v_{o(cm)} = (v_{d1} + v_{d2})/2$. The common-mode voltage gain $A_{v(cm)}$ is defined as the ratio v_{ocm}/v_{icm} for the common mode input signal $v_{i1} = v_{i2} = v_{icm}$. It follows that the common-mode voltage gain is given by

$$A_{v(cm)} = \frac{v_{ocm}}{v_{icm}} = -G_{m(cm)}R_D = \frac{-R_D}{r_s + 2R_T} \quad (16.78)$$

Common-Mode Rejection Ratio

A figure of merit for the differential amplifier is the common-mode rejection ratio or $CMRR$. This is defined as the ratio of the differential voltage gain to the common-mode voltage gain. A perfect differential amplifier would have a common-mode gain of zero so that its $CMRR$ would be infinite. The $CMRR$ is given by

$$CMRR = \frac{A_{v(d)}}{A_{v(cm)}} = \frac{2G_{m(d)}}{G_{m(cm)}} = 1 + \frac{2R_T}{r_s} \quad (16.79)$$

If the output voltage is taken to be either v_{d1} or v_{d2} rather than the differential output voltage $v_{o(d)} = v_{d1} - v_{d2}$, the $CMRR$ would be one-half the value given by this equation. The $CMRR$ is often expressed in decibels by the expression $20\log_{10}(A_{v(d)}/A_{v(cm)})$.

16.5.9 MOSFET Differential Amplifier with Active Load

Figure 16.16 shows the MOSFET differential amplifier connected to a current mirror load. An op amp is used to convert the output current i_o into a voltage. The op amp also sets the dc voltage at the drains of M_2 and M_4 via the bias voltage V_B connected to the non-inverting input of the op amp.

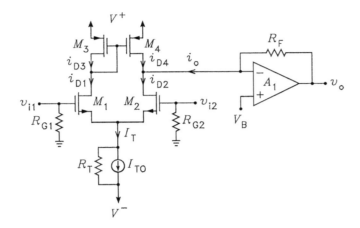

Figure 16.16: Differential amplifier with active load.

Because the inverting input to the op amp is a virtual ac ground, the small-signal output voltage is given by
$$v_o = R_F i_o \tag{16.80}$$
where the current i_o is given by
$$i_o = i_{d2} - i_{d4} = i_{d2} - i_{d3} = i_{d2} - i_{d1} \tag{16.81}$$

If we assume that the $CMRR$ is large, we can write $i_{d1} - i_{d2} = 2G_{m(d)}(v_{i1} - v_{i2})$, where $G_{m(d)}$ is given by Eq. 16.72. Thus v_o can be written
$$v_o = -2G_{m(d)} R_F (v_{i1} - v_{i2}) = -\frac{R_F}{r_s}(v_{i1} - v_{i2}) \tag{16.82}$$

16.6 Preliminary Derivations

1. Derive the expressions for the differential-mode and common-mode voltage gains of the MOSFET differential amplifier with a resistive tail supply, R_T.

2. Derive Eq. 16.82.

16.7 Preliminary Calculations

1. Design a common-source MOSFET amplifier to meet the specifications given by the laboratory instructor. Use $V^+ = 15$ V. Assume that the MOSFET has the parameters: $K = 2 \times 10^{-4}$ A/V^2, $V_{T0} = 2$ V, $\lambda = 0$, and $\gamma = 0$.

2. Design a MOSFET differential amplifier with resistive load having the small-signal differential gain specified by the laboratory instructor. Assume that an ideal current source is available for biasing and that

the positive power supply voltage and device parameters are the same as for problem 1. The negative power supply voltage is $V^- = -V^+$.

3. Design a MOSFET differential amplifier having the small-signal differential voltage gain specified by the laboratory instructor. Assume that the device parameters are the same as for problem 1.

16.8 Preliminary SPICE Simulations

1. Perform a SPICE simulation for the circuit designed in problem 1 of the Preliminary Calculations section. The simulations to be performed are `.op`, `.tran`, and `.ac`. Determine the quiescent operating point, the small-signal voltage gain, and the positive and negative clipping levels. The SPICE parameters are: KP=0.4m, VTO=2, GAMMA=0, and LAMBDA=0. Perform a harmonic analysis in conjunction with the `.tran` analysis.

2. Perform a SPICE simulation for the circuit designed in problem 2 of the Preliminary Calculations section. The simulations to be performed are `.op`, `.tran`, and `.ac`. Determine the quiescent operating point, the small-signal single-ended, differential, and common-mode voltage gain, and the positive and negative clipping levels. The SPICE parameters are: KP=0.4m, VTO=2, GAMMA=0, and LAMBDA=0 for each of the n channel enhancement mode MOSFETs. Perform a harmonic analysis in conjunction with the `.tran` analysis.

3. Perform a SPICE simulation for the circuit designed in problem 3 of the Preliminary Calculations section. The simulations to be performed are `.op`, `.tran`, and `.ac`. Determine the quiescent operating point, the small-signal single-ended, differential, and common-mode voltage gain, and the positive and negative clipping levels. The SPICE parameters are: KP=0.4m, VTO=2, GAMMA=0, and LAMBDA=0 for each of the n channel enhancement mode MOSFETs and KP=0.4m, VTO=-2, GAMMA=0, and LAMBDA=0 for each of the p channel enhancement mode MOSFETs. Perform a harmonic analysis in conjunction with the `.tran` analysis.

16.9 Experimental Procedures

16.9.1 Preparation

Prepare the electronic breadboard to provide buses for the positive and negative power supply rails and the circuit ground. Each power supply rail should be decoupled with a 100 Ω resistor, 25 V (or greater) capacitor. The resistors are connected in series with the external power supply leads and the capacitors are connected from the power supply rails to ground on the circuit side of the resistor. The capacitors must be installed with the proper polarity to prevent reverse polarity breakdown.

16.9.2 Parameter Measurement

The MOSFET devices that will be used in this experiment are contained in the MC14007/CD4007 IC. This IC has 3 n channel and 3 p channel enhancement mode MOSFETs. The pinouts are shown in Fig. 16.17. Normally, pin 14 is required to be the most positive pin and pin 7 is required to be the most negative pin.

Use the laboratory transistor curve tracer to obtain a plot of the output characteristics for one of the n channel and p channel enhancement mode MOSFET. Set the DC offset on the curve tracer to zero and do not connect pins 14 or 7 to anything.

From the output characteristics ascertain the transistor parameters K and V_{TO}. If they different significantly from those used in the preliminary calculations, repeat these calculations using the measured values for these transistor parameters.

16.9. EXPERIMENTAL PROCEDURES

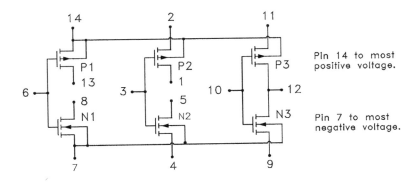

Figure 16.17: Pin connections for the MC14007/CD4007.

16.9.3 Common-Source Amplifier

Assemble the common-source MOSFET amplifier shown in Fig. 16.18 or the circuit specified by the laboratory instructor. Be certain to connect pin 14 to the positive power supply rail. Use the value for R_D that was calculated for the design value of the small-signal gain with a power supply voltage $V^+ = 15$ V. Adjust the 10 kΩ pot until the DC drain current has the desired value; measure the current by measuring the DC voltage across the resistor R_D and dividing this voltage by the value of R_D. (The coupling capacitor on the drain of the transistor can be dispensed with if the oscilloscope is AC coupled.)

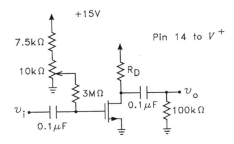

Figure 16.18: Common-source amplifier.

Set the function generator to produce a sine wave with a frequency of 1 kHz, a DC level of zero, and an peak-to-peak value of 1 V. Measure the small-signal voltage gain. Vary the frequency of the function generator and determine the upper and lower break frequencies. Vary the amplitude of the sine wave produced by the function generator and determine the positive and negative clipping values. Also note the output level at which noticeably distortion occurs.

Switch the input to a 1 kHz square wave and observe the output. The amplitude of the square wave should be sufficiently small so that the amplifier does not clip.

16.9.4 MOSFET Differential Amplifier with Resistive Load

Assemble the circuit shown in Fig. 16.19 using the component values calculated in the preliminary calculations section. Be certain to connect pin 14 of the IC to the positive power supply rail. (The design of the JFET current source used for the tail supply is found in Exp 15 on the BJT differential amplifier.)

17.2. THEORY

17.2.4 Series-Shunt Feedback

An example series-shunt feedback amplifier is shown in Fig. 17.4, where the bias sources and networks are omitted for simplicity. It is assumed that the dc solution to the circuit is known. The feedback network is in the form of a voltage divider and consists of resistors R_{F1} and R_{F2}. Because the input to the feedback network connects to the v_o node, the amplifier is said to employ shunt or voltage sampling. The output of the feedback network connects to the emitter of Q_1. The Thévenin voltage and resistance looking out of the emitter of Q_1 is

$$v_{te1} = v_o \frac{R_{F2}}{R_{F1} + R_{F2}} \qquad R_{te1} = R_{F1} \| R_{F2}$$

The collector current in Q_1 is proportional to the error voltage $v_e = v_i - v_{te1}$, i.e. to a series combination of two voltages. For this reason, the amplifier is said to employ series or voltage summing at the input. The type of feedback is then described by giving the type of input summing followed by the type of output sampling. Thus the circuit is a series-shunt feedback amplifier.

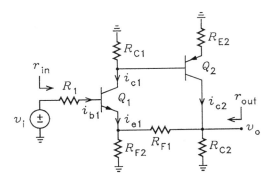

Figure 17.4: Series-shunt amplifier circuit.

In order for the amplifier to have negative feedback, the voltage gain from the emitter of Q_1 to the collector of Q_2 must be inverting. When the feedback signal is applied to its emitter, Q_1 is a common-base stage which has a non-inverting voltage gain. Q_2 is a common-emitter stage which has an inverting gain. Thus the amplifier has an inverting voltage gain from the emitter of Q_1 to the collector of Q_2 so that the feedback is negative.

The object is to solve for the voltage gain, the input resistance and the output resistance of the amplifier. To do this, first draw the equivalent circuit with feedback removed. This is shown in Fig. 17.5. The circuit seen looking out of the emitter of Q_1 is replaced with a Thévenin equivalent circuit with respect to v_o. A current-controlled current source with value i_{e1} connects to the junction between R_{F1} and R_{F2} at the output. In order to calculate the output resistance r_{out} as part of the analysis, the external current source i_t is added in shunt with the v_o node. For the analysis, the collector output resistance of each transistor will be neglected, i.e. it will be assumed that $r_{ic} = \infty$. This is equivalent to assuming that $r_0 = \infty$ in the small-signal BJT model. To solve for the collector current in each transistor, we use the equations $i_c = G_m (v_{tb} - v_{te})$ for the NPN BJT and $i_c = -G_m (v_{tb} - v_{te})$ for the PNP BJT, where G_m is the effective transconductance, v_{tb} is the Thévenin equivalent voltage seen looking out of the base, and v_{te} is the Thévenin equivalent voltage seen looking out of the emitter.

Superposition can be used to write the following equations for the circuit in Fig. 17.5:

$$v_o = (i_{c2} + i_t) R_{C2} \| (R_{F2} + R_{F1}) + i_{e1} \frac{R_{F2}}{R_{F2} + R_{F1} + R_{C2}} R_{C2} \qquad (17.13)$$

$$i_{c1} = G_{m1} v_e \qquad v_e = v_i - v_o \frac{R_{F2}}{R_{F1} + R_{F2}} \qquad (17.14)$$

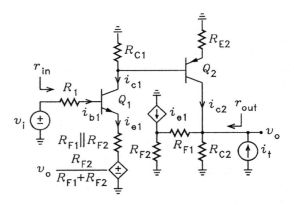

Figure 17.5: Series-shunt amplifier with feedback removed.

$$i_{c2} = -G_{m2}v_{tb2} = -G_{m2}\left(-i_{c1}R_{C1}\right) \qquad i_{b1} = \frac{i_{c1}}{\beta_1} \qquad i_{e1} = \frac{i_{c1}}{\alpha_1} \qquad (17.15)$$

where v_e is the error voltage and G_{m1} and G_{m2} are transconductances given by

$$G_{m1} = \frac{\alpha_1}{r'_{e1} + R_{F1}\|R_{F2}} \qquad G_{m2} = \frac{\alpha_2}{r'_{e2} + R_{E2}} \qquad (17.16)$$

In these equations, r'_{e1} and r'_{e2} are the small-signal resistances seen looking into the emitters of Q_1 and Q_2. They are given by

$$r'_{e1} = \frac{R_1 + r_{x1}}{1 + \beta_1} + r_{e1} \qquad r'_{e2} = \frac{R_{C1} + r_{x2}}{1 + \beta_2} + r_{e2} \qquad (17.17)$$

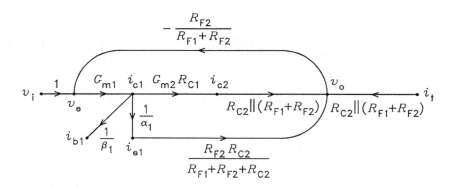

Figure 17.6: Flow graph for the series-shunt amplifier.

The flow graph for the circuit equations is given in Fig. 17.6. There are two touching loops in the graph.

17.2. THEORY

The determinant is given by

$$\Delta = 1 - \left(\frac{-R_{F2}}{R_{F1} + R_{F2}}\right) G_{m1} \left[G_{m2} R_{C1} \times R_{C2} \| (R_{F1} + R_{F2}) \right. \tag{17.18}$$
$$\left. + \frac{1}{\alpha_1} \times \frac{R_{F2} R_{C2}}{R_{F2} + R_{F1} + R_{C2}} \right]$$

There are two forward paths from v_1 to v_o and each path touches both loops. Thus $\Delta_k = 1$ for each path. It follows from Mason's formula that the voltage gain is given by

$$\frac{v_o}{v_i} = \frac{1}{\Delta} \times G_{m1} \left[G_{m2} R_{C1} \times R_{C2} \| (R_{F1} + R_{F2}) + \frac{1}{\alpha_1} \times \frac{R_{F2} R_{C2}}{R_{F2} + R_{F1} + R_{C2}} \right] \tag{17.19}$$

When this expression is compared to Eq. (17.2), it follows that A and b are given by

$$A = G_{m1} \left[G_{m2} R_{C1} \times R_{C2} \| (R_{F1} + R_{F2}) + \frac{1}{\alpha_1} \times \frac{R_{F2} R_{C2}}{R_{F2} + R_{F1} + R_{C2}} \right] \tag{17.20}$$

$$b = \frac{R_{F2}}{R_{F1} + R_{F2}} \tag{17.21}$$

The input resistance to the circuit is given by the reciprocal of the gain from v_i to i_{b1}. The output resistance is given by the gain from i_t to v_o. Mason's formula can be used to write these as follows:

$$r_{\text{in}} = \left(\frac{i_{b1}}{v_i}\right)^{-1} = \left[\frac{1}{\Delta} \times G_{m1} \times \frac{1}{\beta_1}\right]^{-1} = \Delta \times \frac{\beta_1}{G_{m1}} \tag{17.22}$$

$$r_{\text{out}} = \frac{v_o}{i_t} = \frac{1}{\Delta} \times R_{C2} \| (R_{F1} + R_{F2}) \tag{17.23}$$

Because Δ corresponds to the amount of feedback, it follows that the voltage gain is divided by the amount of feedback, the input resistance is multiplied by the amount of feedback, and the output resistance is divided by the amount of feedback. We can conclude that the series-shunt topology is suited for amplifiers where the voltage gain is to be set by the feedback and a high input resistance and a low output resistance are desired.

An alternate expression for G_{m1} is

$$G_{m1} = \frac{\beta_1}{R_1 + r_{ib1}} \tag{17.24}$$

where r_{ib1} is the small-signal resistance seen looking into the base of Q_1 given by

$$r_{ib1} = r_{x1} + (1 + \beta_1)(r_{e1} + R_{F1} \| R_{F2}) \tag{17.25}$$

When Eq. (17.24) is used in Eq. (17.22), it follows that the input resistance can be written

$$r_{\text{in}} = \Delta \times (R_1 + r_{ib1}) \tag{17.26}$$

Eq. (17.3) can be used to write the approximate voltage gain as

$$\frac{v_o}{v_i} \simeq \frac{1}{b} = 1 + \frac{R_{F1}}{R_{F2}} \tag{17.27}$$

The amount of feedback corresponds to the determinant Δ. It follows that the voltage gain is divided by the amount of feedback, the output resistance is divided by the amount of feedback, and the input resistance is multiplied by the amount of feedback. It follows that the series-shunt amplifier is used in applications where a high input resistance and a low output resistance are desired.

17.2.5 Shunt-Shunt Feedback

Fig. 17.7 shows the circuit diagram of an example shunt-shunt feedback amplifier, where the bias circuits have been omitted for simplicity. The feedback network consists of the resistor R_F which connects between the output and input nodes. The source is modeled by a Norton equivalent circuit. This is necessary in order to obtain a gain in the form of Eq. (17.2). Because R_F connects to the output node, the amplifier is said to have voltage or shunt sampling. Because the current fed back through R_F to the input node combines in parallel with the source current, the circuit is said to have current or shunt summing. Thus the amplifier is said to have shunt-shunt feedback. In order for the feedback to be negative, the voltage gain from v_{b1} to v_o must be inverting. Q_1 is a common-emitter stage which has an inverting gain. Q_2 is a common-collector stage which has a non-inverting gain. Thus the overall voltage gain is inverting so that the feedback is negative.

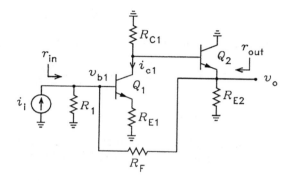

Figure 17.7: Shunt-shunt amplifier example.

Figure 17.8 shows the equivalent circuit with feedback removed. The feedback network at the input is modeled by a Norton equivalent circuit with respect to v_o. The feedback network at the output is modeled by a Thévenin equivalent circuit with respect to v_{b1}. The external current source i_t is added to the circuit so that the output resistance r_{out} can be calculated as part of the analysis. As in the series-shunt example, we assume that $r_0 = \infty$ for each transistor. The small-signal Thévenin equivalent circuit seen looking into the emitter of Q_2 consists of the voltage source $v_{tb2} = -i_{c1} R_{C1}$ in series with the resistance r'_{e2}. Using superposition, we can write the following equations for the circuit:

$$\begin{aligned} v_o &= v_{tb2} \times \frac{R_{E2} \| R_F}{r'_{e2} + R_{E2} \| R_F} + v_{b1} \frac{R_{E2} \| r'_{e2}}{R_F + R_{E2} \| r'_{e2}} + i_t \left(r'_{e2} \| R_{E2} \| R_F \right) \\ &= -i_{c1} R_{C1} \frac{R_{E2} \| R_F}{r'_{e2} + R_{E2} \| R_F} + v_{b1} \frac{R_{E2} \| r'_{e2}}{R_F + R_{E2} \| r'_{e2}} + i_t \left(r'_{e2} \| R_{E2} \| R_F \right) \end{aligned} \quad (17.28)$$

$$i_{c1} = G_{m1} v_{b1} \quad (17.29)$$

$$v_{b1} = i_e \left(R_1 \| R_F \| r_{ib1} \right) \qquad i_e = i_i + \frac{v_o}{R_F} \quad (17.30)$$

where i_e is the error current, i.e. the total Norton current delivered to the v_{b1} node, and G_{m1}, r_{ib1}, and r'_{e2} are given by

$$G_{m1} = \frac{\alpha_1}{r'_{e1} + R_{E1}} \qquad r'_{e1} = \frac{r_{x1}}{1 + \beta_1} + r_{e1} \quad (17.31)$$

$$r_{ib1} = r_{x1} + (1 + \beta_1)(r_{e1} + R_{E1}) \quad (17.32)$$

$$r'_{e2} = \frac{R_{C1} + r_{x2}}{1 + \beta_2} + r_{e2} \quad (17.33)$$

17.2. THEORY

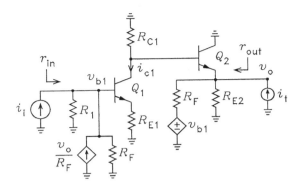

Figure 17.8: Shunt-shunt amplifier with feedback removed.

Because v_{tb1} is taken to be v_{b1} in Eq. (17.29), it follows that R_{tb1} is calculated with $v_{b1} = 0$.

The flow graph for the circuit equations is shown in Fig. 17.9. The determinant of the flow graph is given by

$$\Delta = 1 - \frac{R_1 \| R_F \| r_{ib1}}{R_F} \left[G_{m1} (-R_{C1}) \frac{R_{E2} \| R_F}{r'_{e2} + R_{E2} \| R_F} + \frac{r'_{e2} \| R_{E2}}{R_F + r'_{e2} \| R_{E2}} \right] \qquad (17.34)$$

This corresponds to the amount of feedback. From the flow graph, the transconductance gain, input resistance, and output resistance can be written by inspection

$$\frac{v_o}{i_i} = \frac{R_1 \| R_F \| r_{ib1}}{\Delta} \left[G_{m1} (-R_{C1}) \frac{R_{E2} \| R_F}{r'_{e2} + R_{E2} \| R_F} + \frac{r'_{e2} \| R_{E2}}{R_F + r'_{e2} \| R_{E2}} \right] \qquad (17.35)$$

$$r_{in} = \frac{v_{b1}}{i_i} = \frac{R_1 \| R_F \| r_{ib1}}{\Delta} \qquad (17.36)$$

$$r_{out} = \frac{v_o}{i_t} = \frac{r'_{e2} \| R_{E2} \| R_F}{\Delta} \qquad (17.37)$$

It can be seen from these expressions that the transresistance gain, the input resistance, and the output resistance are all decreased by a factor equal to the amount of feedback.

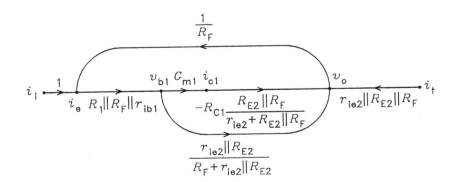

Figure 17.9: Flow graph for the shunt-shunt amplifier.

It follows from Eqs. (17.34) and (17.35) that A and b for the amplifier are given by

$$A = (R_1 \| R_F \| r_{ib1}) \left[G_{m1}(-R_{C1}) \frac{R_{E2} \| R_F}{r'_{e2} + R_{E2} \| R_F} + \frac{r'_{e2} \| R_{E2}}{R_F + r'_{e2} \| R_{E2}} \right] \quad (17.38)$$

$$b = \frac{-1}{R_F} \quad (17.39)$$

The expression for A has two terms. In all cases of interest, the first term in the brackets in Eq. (17.38) dominates so that A is negative. Because b is also negative, it follows that the product bA is positive. This product must always be positive for an amplifier with negative feedback. Eq. (17.3) can be used to write the approximate transresistance gain as

$$\frac{v_o}{i_i} \simeq \frac{1}{b} = -R_F \quad (17.40)$$

17.2.6 Series-Series Feedback

Figure 17.10 shows the circuit diagram of a series-series feedback amplifier, where the bias circuits have been omitted for simplicity. The feedback network consists of resistors R_{F1} and R_{F2}. This network samples the voltage at the collector of Q_2 and feeds a voltage back into the emitter of Q_1. The output voltage from the circuit is the voltage at the emitter of Q_1. Thus the feedback network does not sample an output voltage. Instead, it samples a voltage which is proportional to the output current, i.e. it samples a voltage proportional to $i_{c2} = \alpha_2 i_o$. Because i_o is the current through the load resistor R_{E2}, it follows that the feedback is proportional to the load current. This is called series sampling. The input summing is identical to that for the circuit of Fig. 17.4 which is series summing. Therefore, the circuit is a series-series feedback amplifier.

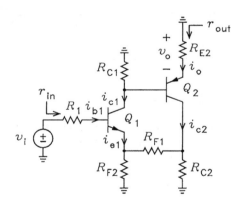

Figure 17.10: Series-series feedback amplifier.

To analyze the circuit, first remove the feedback. The circuit with feedback removed is shown in Fig. 17.11. An external voltage source v_t is included in series with R_{E2} in order to calculate the output resistance r_{out}. Notice that r_{out} is labeled looking into R_{E2} from signal ground. To simplify the analysis, it will be assumed that $r_0 = \infty$ for both BJTs. The Thévenin equivalent circuit seen looking into the emitter of Q_2 consists of the voltage source $v_{tb2} = -i_{c1} R_{C1}$ in series with the resistor r'_{e2}. To solve for i_o/v_i, r_{in}, and r_{out}, we can write the following equations:

$$i_o = i_{e2} = \frac{v_t - v_{tb2}}{r'_{e2} + R_{E2}} = \frac{v_t + i_{c1} R_{C1}}{r'_{e2} + R_{E2}} \quad (17.41)$$

17.4. PRELIMINARY DERIVATIONS

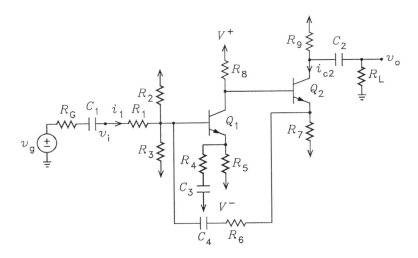

Figure 17.22: Shunt-series amplifier.

Also, $R_3 = (-10 + 15)/0.05\text{m} = 100$ kΩ. If we assume that the open-loop gain is infinite, the input resistance is R_1 and the current gain is $i_{c2}/i_1 = -(1 + R_6/R_7)$. Thus, we pick $R_1 = 10$ kΩ and $(1 + R_6/1.25\text{k}) = 10$. Solution for R_6 yields $R_6 = 21.1$ kΩ. We will pick the standard value $R_6 = 22$ kΩ. To complete the design, we pick $C_1 = C_2 = 10$ μF, $C_3 = 100$ μF, $C_4 = 10$ μF, and $R_4 = 100$ Ω.

17.4 Preliminary Derivations

17.4.1 Series-Shunt

1. Derive Eqs. 17.18 through 17.23.

17.4.2 Shunt-Shunt

1. Derive Eqs. 17.34 through 17.37.

17.4.3 Series-Series

1. Derive Eqs. 17.45 through 17.48

17.4.4 Shunt-Series

1. Derive Eqs. 17.57 through 17.59.

17.5 Preliminary Calculations

17.5.1 Series-Shunt

Design the circuit shown in Fig. 17.19 using the parameters specified by the laboratory instructor. Recommended ranges are 0.5 mA to several mA for the dc collector currents and 0 V for the dc bias voltage at the collector of Q_2. Recommended ranges for the midband small-signal voltage gain are from 5 to 20 (V/V).

The input resistance should be 10 kΩ or larger. Use ±15 V for the power supply voltages. Assume that $C_1 = C_2 = 10$ μF and $C_3 = 100$ μF. Use 100 kΩ for the load resistor R_L.

17.5.2 Shunt-Shunt

Design the circuit shown in Fig. 17.20 using the parameters specified by the laboratory instructor. Recommended ranges are 0.5 mA to several mA for the dc collector currents and 0 V for the dc bias voltage for the emitter of transistor Q_2. Recommended ranges for the midband small-signal transresistance gain are from 5 to 20 kΩ. Use ±15 V for the power supply voltages. Assume that $C_1 = C_2 = 10$ μF and $C_3 == C_4 = 100$ μF. Use 100 kΩ for the load resistor R_L. The resistor R_1 is included so that the input current can be sampled. A suggested value is $R_1 = 10$ kΩ.

17.5.3 Series-Series

Design the circuit shown in Fig. 17.21 using the parameters specified by the laboratory instructor. Recommended ranges are 0.5 mA to several mA for the quiescent collector currents and 0 V for the dc bias voltage at the collector of Q_2. Recommended ranges for the midband small-signal transconductance gain are from 5×10^{-3} S to 20×10^{-5} S. Use ±15 V power supply voltages. Assume that $C_1 = C_2 = 10$ μF and $C_3 = 100$ μF. Use $R_L = 100$ kΩ.

17.5.4 Shunt-Series

Design the circuit shown in Fig. 17.22 using the parameters specified by the laboratory instructor. Recommended ranges are 0.5 mA to several mA for the quiescent collector currents and 0 V for the dc bias voltage for the emitter of Q_2. Recommended ranges for the midband small-signal current gain are from 5 to 20 (amps/amp). Use ±15 V for the power supply voltages. Assume that $C_1 = C_2 = 10$ μF and $C_3 = C_4 = 100$ μF.

17.6 Preliminary SPICE Simulations

17.6.1 Series-Shunt

Simulate the circuit shown in Fig. 17.19 using the values determined in the Preliminary Calculations. Perform .OP, .AC, and .TRAN analyses. Check to see if either transistor is in either the cutoff or saturation states. If they are, then the rest of the SPICE analysis is totally meaningless and an error of some sort has been made.

17.6.2 Shunt-Shunt

Simulate the circuit shown in Fig. 17.20 using the values determined in the Preliminary Calculations. Perform .OP, .AC, and .TRAN analyses. Check to see if either transistor is in either the cutoff or saturation states. If they are, then the rest of the SPICE analysis is totally meaningless and an error of some sort has been made.

17.6.3 Series-Series

Simulate the circuit shown in Fig. 17.21 using the values determined in the Preliminary Calculations. Perform .OP, .AC, and .TRAN. Check to see if either transistor is in the cutoff or saturation states. If they are, then the rest of the SPICE analysis is totally meaningless and an error of some sort has been made.

17.7. PROCEDURE

17.6.4 Shunt-Series

Simulate the circuit shown in Fig. 17.22 using the values determined in the Preliminary Calculations. Perform .OP, .AC, and .TRAN. Check to see if either transistor is in either the cutoff or saturation states. If they are, then the rest of the SPICE analysis is totally meaningless and an error of some sort has been made.

17.7 Procedure

17.7.1 Preparation

Prepare the electronic breadboard to provide busses for the positive and negative power supply rails and the circuit ground. Each power supply rail should be decoupled with a 100 Ω, 1/4 W resistor and a 100 μF, 25 V (or greater) capacitor. The resistors are connected in series with the external power supply leads and the capacitors are connected from power supply rail to ground on the circuit side of the resistors. The capacitors must be installed with the proper polarity to prevent reverse polarity breakdown.

17.7.2 Series Shunt

Bias

Assemble the circuit in Fig. 17.19 with the exception of the function generator, capacitors, and the load resistor. Measure the dc collector-to-emitter voltage of each transistor to make sure that neither transistor is in either the cutoff or saturation states. (If they are the circuit is worthless as an amplifier.) Measure the dc collector current in each transistor and the dc voltage at the collector of Q_2. If the currents are not within 20% of the design value or if the dc collector voltage of transistor Q_2 is more than 1 V in magnitude, check the circuit for errors. Only when the circuit is properly biased is the rest of the experiment meaningful.

Gain

Assemble the entire circuit shown in Fig. 17.19. Make certain that the polarity of the electrolytic capacitors are correct.

Set the function generator to produce a sine wave with an appropriate amplitude (small enough so that the output is not distorted), a dc level of zero, and a frequency of 5 kHz. Measure the small signal midband voltage gain, A_v, at this frequency. Determine the value of the open-loop gain A using the formula

$$A_v = \frac{A}{1 + bA} \quad (17.71)$$

where A is the open-loop gain, A_v is the closed-loop midband small signal voltage gain, and b is the feedback factor. For this circuit the feedback factor is

$$b = \frac{R_{F2}}{R_{F1} + R_{F2}} \quad (17.72)$$

Note that both A and b as well as A_v have units of volts/volts or are dimensionless since they are voltage ratios.

Place an emitter bypass capacitor from the emitter of Q_2 to ground. This should be a large electrolytic capacitor such as 100 μF. Since it must be inserted with the proper polarity, first measure the dc voltage at the emitter of Q_2 to determine the polarity. Remove the capacitor.

Transient Response

Apply a 100 Hz square wave to the input and observe and sketch the output. From the sketch estimate the peak overshoot.

Frequency Response

Determine and plot the gain versus frequency as the frequency of the source varies from 10 Hz to 1 MHz. From the plot estimate the phase margin.

17.7.3 Shunt Shunt

Bias

Assemble the circuit in Fig. 17.20 with the exception of the function generator, capacitors, and the load resistor. Measure the dc collector-to-emitter voltage of each transistor to make sure that neither transistor is in either the cutoff or saturation states. (If they are the circuit is worthless as an amplifier.) Measure the dc collector current in each transistor and the dc voltage at the emitter of Q_2. If the currents are not within 20% of the design value or if the dc emitter voltage of transistor Q_2 is more than 1 V in magnitude, check the circuit for errors. Only when the circuit is properly biased is the rest of the experiment meaningful.

Gain

Assemble the entire circuit shown in Fig. 17.20. Make certain that the polarity of the electrolytic capacitors are correct.

Set the function generator to produce a sine wave with an appropriate amplitude (small enough so that the output is not distorted), a dc level of zero, and a frequency of 5 kHz. Measure the small signal midband voltage gain, $A_v = V_o/V_i$, at this frequency. Determine the value of the open-loop transresistance gain A using the formula

$$A_r = A_v R_2 = \frac{A}{1 + bA} \qquad (17.73)$$

where A is the open-loop transresistance gain, R_2 is the Norton equivalent resistance seen looking out of the base of Q_1 with the feedback removed, and b is the feedback factor. For this circuit the feedback factor is

$$b = -\frac{1}{R_F} \qquad (17.74)$$

Note that A_v is dimensionless or volts/volt, A_r and A have units of ohms, and b has units of mhos. Also note that since this is an inverting amplifier that A should be negative.

Place an emitter bypass capacitor from the emitter of Q_1 to ground and observe the effect. This should be a large electrolytic capacitor such as 100 μF which is polarized and must be inserted with the proper polarity. First measure the dc voltage at the emitter to determine the polarity. Remove the capacitor.

Transient Response

Apply a 100 Hz square wave to the input and observe and sketch the output. From the sketch estimate the peak overshoot.

Frequency Response

Determine and plot the gain versus frequency as the frequency of the source varies from 10 Hz to 1 MHz. From the plot estimate the phase margin.

17.7.4 Series-Series

Bias

Assemble the circuit in Fig. 17.21 with the exception of the function generator, capacitors, and the load resistor. Measure the dc collector-to-emitter voltage of each transistor to make sure that neither transistor is in either the cutoff or saturation states. (If they are the circuit is worthless as an amplifier.) Measure

17.7. PROCEDURE

the dc collector current in each transistor and the dc voltage at the collector of Q_2. If the currents are not within 20% of the design value, check the circuit for errors. Only when the circuit is properly biased is the rest of the experiment meaningful.

Gain

Assemble the entire circuit shown in Fig. 17.21. Make certain that the polarity of the electrolytic capacitors are correct.

Set the function generator to produce a sine wave with an appropriate amplitude (small enough so that the output is not distorted), a dc level of zero, and a frequency of 5 kHz. Measure the small signal midband transconductance gain, $A_m = I_o/V_i$, at this frequency. Determine the value of the open-loop transconductance gain A using the formula

$$A_m = \frac{A_v}{R_{E2}} = \frac{A}{1 + bA} \qquad (17.75)$$

where A is the open-loop transconductance gain and b is the feedback factor. For this circuit the feedback factor is

$$b = \alpha_2 \times \frac{R_{C2} R_{F2}}{R_{C2} + R_{F1} + R_{F2}} \qquad (17.76)$$

Note that A_m and A have dimensions of mhos or siemens and b has units of ohms

Transient Response

Apply a 100 Hz square wave to the input and observe and sketch the output. From the sketch estimate the peak overshoot.

Frequency Response

Determine and plot the gain versus frequency as the frequency of the source varies from 10 Hz to 1 MHz. From the plot estimate the phase margin.

17.7.5 Shunt-Series Feedback Amplifier

Bias

Assemble the circuit in Fig. 17.22 with the exception of the function generator, capacitors, and the load resistor. Measure the dc collector-to-emitter voltage of each transistor to make sure that neither transistor is in either the cutoff or saturation states. (If they are the circuit is worthless as an amplifier.) Measure the dc collector current in each transistor and the dc voltage at the emitter of Q_2. If the currents are not within 20% of the design value, check the circuit for errors. Only when the circuit is properly biased is the rest of the experiment meaningful.

Gain

Assemble the entire circuit shown in Fig. 17.22. Make certain that the polarity of the electrolytic capacitors are correct.

Set the function generator to produce a sine wave with an appropriate amplitude (small enough so that the output is not distorted), a dc level of zero, and a frequency of 5 kHz. Measure the small signal midband current gain, $A_i = I_o/I_i$, at this frequency. Determine the value of the open-loop current gain A using the formula

$$A_i = A_v \frac{R_{C2}}{R_2} = \frac{A}{1 + bA} \qquad (17.77)$$

where A is the open-loop current gain and b is the feedback factor. For this circuit the feedback factor is

$$b = -\frac{1}{\alpha_2}\left(1 + \frac{R_F}{R_{E2}}\right) \tag{17.78}$$

Note that A_i and b are dimensionless or amps/amp. Also note that since this is an inverting amplifier that A should be negative.

Transient Response

Apply a 100 Hz square wave to the input and observe and sketch the output. From the sketch estimate the peak overshoot.

Frequency Response

Determine and plot the gain versus frequency as the frequency of the source varies from 10 Hz to 1 MHz. From the plot estimate the phase margin.

17.8 Laboratory Report

Turn in all sketches and calculations.

For each circuit compute A from the measured voltage gain and the value of b computed from the feedback network. Determine Ab and ascertain whether or not the condition $Ab \gg 1$.

Answer any supplementary questions posed by the laboratory instructor.

17.9 References

1. E. J. Angelo, *Electronics: BJTs, FETs, and Microcircuits,* McGraw-Hill, 1969.
2. W. Banzhaf, *Computer-Aided Circuit Analysis Using SPICE,* Prentice-Hall, 1989.
3. S. G. Burns and P. R. Bond, *Principles of Electronic Circuits,* West, 1987.
4. M. N. Horenstein, *Microelectronics Circuits and Devices,* Prentice-Hall, 1990.
5. P. Horwitz and W. Hill, *The Art of Electronics,* 2nd edition, Cambridge University Press, 1989.
6. P. Horowitz and I. Robinson, *Laboratory Manual for The Art of Electronics,* Cambridge University Press, 1981.
7. J. H. Krenz, *An Introduction to Electrical and Electronic Devices,* Prentice-Hall, 1987.
8. Leach, W. M., Jr., *Introduction to Electroacoustics and Audio Amplifier Design,* Dubuque, Iowa, Kendall/Hunt, 1998, ISBN 0-7872-5410-X.
9. R. Mauro, *Engineering Electronics,* Prentice-Hall, 1989.
10. F. H. Mitchell and F. H. Mitchell, *Introduction to Electronic Design,* Prentice-Hall, 1988.
11. Motorola, *Small-Signal Semiconductors,* DL 126, Motorola, 1987.
12. R. B. Northrop, *Analog Electronic Circuits,* Addison Wesley, 1990.
13. C. J. Savant, M. S. Roden, and G. L. Carpenter, *Electronic Circuit Design,* Benjamin Cummings, 1987.
14. A. S. Sedra and K. C. Smith, *Microelectronic Circuits,* 4th edition, Oxford, 1998.
15. D. L. Schilling and C. Belove, *Electronic Circuits: Discrete and Integrated,* McGraw-Hill, 1968.
16. P. W. Tuinenga, *SPICE,* Prentice-Hall, 1988.

Chapter 18

Linear Op-Amp Oscillators

18.1 Object

The object of this experiment is to design, simulate, assemble, and evaluate three linear op-amp oscillators.

18.2 Theory

The oscillator is a fundamental building block in electronic circuits and systems. Oscillators are found in the circuits which generate clock signals in digital systems and carrier signals in communications systems as well as in the signal generators and function generators which generate electronic test signals in the laboratory.

This experiment considers a class of electronic oscillators known as linear op-amp oscillators. These circuits employ op amps that are operated in the linear or active mode to directly produce sinusoidal output signals. The linear op-amp oscillator can be used to produce sinusoidal waveforms with frequencies varying from the infrasonic to a few hundred kilohertz. The lower frequency limitation is due to the size of practical circuit components while the upper limitation is due to slew rate limitations of the op amps. One of the problems with linear oscillators is the stabilization of the output signal amplitude. This is accomplished in this experiment with nonlinear diode limiting circuits.

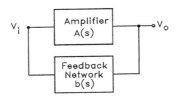

Figure 18.1: Block diagram of a general oscillator circuit showing the amplifier and feedback network.

A linear oscillator consists of an amplifier and a feedback network. The basic structure is shown in Fig. 18.1. It is assumed here that the amplifier can be modeled as a voltage controlled voltage source having the voltage gain transfer function $A(s)$, where s is the complex frequency. A feedback network with the voltage transfer function $b(s)$ samples the amplifier output voltage and feeds part of this output back into the input of the amplifier. The output voltage V_o of the amplifier is related to the input voltage V_i by

$$V_o = A(s)V_i \tag{18.1}$$

The voltage fed back to the input of the amplifier is given by

$$V_i = b(s)V_o \tag{18.2}$$

It follows from these two equations that the output voltage from the amplifier must satisfy the relation

$$V_o = b(s)A(s)V_o \tag{18.3}$$

For the condition $V_o \neq 0$, Eq. 18.3 can be satisfied only if

$$b(s)A(s) = 1 \tag{18.4}$$

If there is a value of $s = \sigma + j\omega$ for which this condition is satisfied, the circuit can have an output voltage that has the complex signal representation

$$v_O(t) = V_{01}e^{st} = V_{01}e^{(\sigma + j\omega)t} \tag{18.5}$$

where V_{01} is a constant. If $\sigma < 0$, $v_O(t)$ damps out with time. If $\sigma > 0$, $v_O(t)$ increases with time. If $\sigma = 0$, $v_O(t)$ is a constant amplitude sinusoidal signal. It follows, therefore, that the condition for constant amplitude sinusoidal oscillations at a single frequency is that there exist only one value of ω such that the loop-gain transfer function satisfies $b(j\omega)A(j\omega) = 1$. The value of ω for which this is satisfied is the frequency of oscillation.

Because oscillators have an output signal but no input signal, they might appear to violate some of the more celebrated laws of physics. This isn't the case because power must be supplied to the circuits in order for them to produce a signal. This power comes from the dc power supplies. Without this power, the active devices in the oscillator would not amplify. Conservation of energy requires the average power supplied by an oscillator to its load be less than the dc input power to the oscillator circuit. The power that is lost is dissipated as heat in the resistors and active devices in the circuit.

Three different linear sinusoidal oscillators are described in the following. These are the Wien-bridge oscillator, the phase shift oscillator, and the quadrature oscillator.

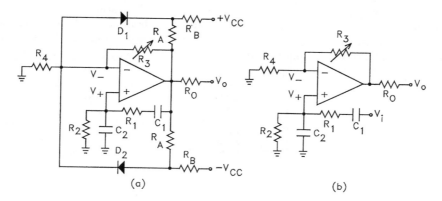

Figure 18.2: (a) Wien-bridge oscillator with diode limiter circuit. (b) Circuit for determining the loop-gain of the Wien-bridge oscillator.

18.2.1 The Wien-Bridge Oscillator

The circuit shown in Fig. 18.2a is an op-amp realization of what is known as the Wien-bridge oscillator. The op amp and resistors R_3 and R_4 form the amplifier part of the oscillator. The voltage gain of the amplifier

18.2. THEORY

is given by $A(s) = V_o/V_{i+} = 1 + R_3/R_4$. The oscillator feedback network connects V_o to V_{i+}. The voltage gain of this network is $b(s) = V_{i+}/V_o$. The circuit is redrawn in Fig. 18.2b with the feedback loop broken. The diode limiter network which stabilizes the amplitude of the output voltage is omitted. The loop-gain transfer function is $b(s)A(s) = V_o/V_i$. It can be shown that

$$b(s)A(s) = \frac{\left(1 + \frac{R_3}{R_4}\right) R_2 C_1 s}{R_1 R_2 C_1 C_2 s^2 + (R_1 C_1 + R_2 C_2 + R_2 C_1)s + 1} \tag{18.6}$$

For $s = j\omega$, it follows that this expression has a positive real value at the radian frequency

$$\omega = \omega_0 = \frac{1}{\sqrt{R_1 R_2 C_1 C_2}} \tag{18.7}$$

The circuit will oscillate at this frequency provided that $b(j\omega_0)A(j\omega_0) = 1$. In a design, the components R_1, R_2, C_1, and C_2 can be selected to obtain a particular frequency of oscillation while the loop gain can be made unity by selecting the appropriate ratio R_3/R_4.

In practice, it is impossible to obtain components that are precise enough to set the loop-gain magnitude to exactly unity at the desired frequency of oscillation. If the gain magnitude is slightly larger than unity, the oscillations will grow in amplitude until the op amp clips. If the gain magnitude is slightly less than unity, the amplitude will decay until it eventually drops below the ambient noise level, if oscillations occur in the first place. A practical oscillator circuit, therefore, requires that the gain magnitude be made slightly larger than unity to initiate oscillations and then reduced to unity when a desired output level has been achieved. The diodes in the circuit of Fig. 18.2a accomplish the gain reduction.

The circuit of Fig. 18.2a is a practical Wien-bridge oscillator with diode limiters that set the peak output voltage. The magnitude of the open-loop gain is set to slightly larger than unity by adjustment of R_3 which sets the ratio R_3/R_4. Oscillations are then guaranteed to occur and the amplitude will increase until the diodes become forward biased on the positive and negative peaks of the output signal. If it is assumed that $R_1 = R_2$ and $C_1 = C_2$, the magnitude of the peak output voltage which just turns on the diodes satisfies the equation

$$\left[\frac{R_B}{R_A + R_B} - \frac{1}{3}\right] \times |V_o| = \frac{R_A}{R_A + R_B} V_{CC} + V_\gamma \tag{18.8}$$

where V_γ is the threshold or cut-in voltage for the diodes.

18.2.2 The Phase-Shift Oscillator

The circuit shown in Fig. 18.3 is an op-amp realization of what is known as a phase shift oscillator. Three cascaded high-pass RC networks are used to obtain a phase shift of $+180°$ at some frequency. The resistor R connected to the inverting terminal of the op-amp is both the input resistor for the op-amp gain stage and the final R for the three cascaded RC networks. If the diode limiting circuit is not considered, it can be shown that the loop gain transfer function is given by

$$b(s)A(s) = -\frac{R_F}{R} \frac{(RCs)^3}{(RCs)^3 + 6(RCs)^2 + 5(RCs) + 1} \tag{18.9}$$

For $s = j\omega$, it follows that this expression has a positive real value at the radian frequency

$$\omega = \omega_0 = \frac{1}{RC\sqrt{6}} \tag{18.10}$$

The circuit will oscillate at this frequency provided that $b(j\omega_0)A(j\omega_0) = 1$. In a design, the components R and C can be selected to obtain a particular frequency of oscillation while the loop gain can be made unity by selecting the appropriate ratio R_F/R.

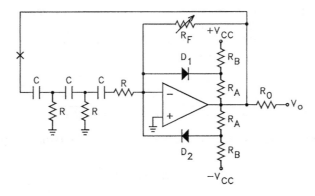

Figure 18.3: Phase-shift oscillator with diode limiter circuit.

The diode limiting network for the phase shift oscillator performs the same function as that for the Wien-bridge oscillator. The magnitude of the peak output voltage which just turns on the diodes is given by

$$|V_o| = \frac{R_A}{R_B}(V_{CC} + V_\gamma) + V_\gamma \qquad (18.11)$$

where V_γ is the threshold or cut-in voltage for the diodes.

18.2.3 The Quadrature Oscillator

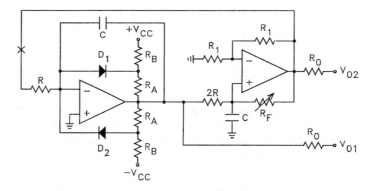

Figure 18.4: Quadrature oscillator with diode limiter circuit.

The circuit shown in Fig. 18.4 is known as a quadrature oscillator. It has two sinusoidal outputs that are in phase quadrature, i.e. they are 90° out of phase. If the diode limiting circuit is not considered, the oscillator consists of an inverting integrator in cascade with a non-inverting integrator. The latter is realized by a first-order filter which uses an op-amp negative impedance converter to realize a negative resistance in parallel with the capacitor connected to ground.

The loop-gain transfer function of the quadrature oscillator circuit is given by

$$b(s)A(s) = \frac{-1}{(RCs)^2 + \left(\frac{1}{2} - \frac{R}{R_F}\right)RCs} \qquad (18.12)$$

For $s = j\omega$, this transfer function has a positive real value when $R/R_F = 0.5$. When this condition is satisfied, it follows that the circuit will oscillate at the frequency where the loop-gain magnitude is unity. The radian frequency is given by

$$\omega = \omega_0 = \frac{1}{RC} \qquad (18.13)$$

To ensure oscillations in a practical quadrature oscillator, the ratio R/R_F is made slightly larger than 0.5 to cause the loop-gain magnitude to be slightly larger than unity. The peak level of the output signal is then controlled with the diode limiting circuit. The limiting circuit shown in Fig. 18.4 is identical to the one in the phase shift oscillator circuit of Fig. 18.3 so that the magnitude of the peak output voltage is given by Eq. 18.11.

18.3 Preliminary Derivations

Derive Eqs. 18.6 through 18.13 for the loop-gain transfer function, the frequency of oscillation, and the peak limited output voltage for the Wien-bridge, the phase shift, and the quadrature oscillators.

18.4 Preliminary Calculations

1. Design a Wien-bridge oscillator, a phase-shift oscillator, and a quadrature oscillator having the frequency of oscillation specified by the laboratory instructor. Choose standard values for the capacitors and calculate the values of the required resistors. Capacitor values should be less than $1\,\mu\text{F}$ to avoid using electrolytic capacitors in the feedback networks. For best results with op-amp circuits, it is suggested that all resistors be in the range $1\,\text{k}\Omega$ to $100\,\text{k}\Omega$.

2. Design diode limiting networks for each oscillator in the preceding part. The network is to be designed to limit the peak output voltage from each oscillator to 5 V. The limiting networks are to be designed so that the quiescent current through each network is 0.5 mA. Assume power supply voltages of $+15\,\text{V}$ and $-15\,\text{V}$ and a diode threshold voltage $V_\gamma = 0.6\,\text{V}$.

18.5 Preliminary SPICE Simulations

1. For the resistor and capacitor values calculated in the previous part, write the SPICE code required to calculate the ac gain and phase (".AC" command line) of the loop gain as a function of frequency for each of the oscillator circuits. These simulations should verify that the loop-gain transfer function has a gain of unity and a phase of zero at the desired frequency of oscillation. For the Wien-bridge oscillator, the open-loop gain circuit is given in Fig. 18.2b. For the phase-shift and quadrature oscillators the loop should be broken at the point indicated by an "X" in Figs. 18.3 and 18.4. The limiting circuits are to be omitted for this analysis.

2. Modify the SPICE code for each open-loop oscillator to include the diode limiter circuits. For the diodes, use the SPICE model parameters IS = 9.32N and N = 2. Change the voltage generator in each circuit to a sinusoidal time-dependent function (SIN) having a frequency equal to the frequency of oscillation. Perform transient analyses (".TRAN" command line) on each circuit to verify that the diode limiter circuits clip the output signal at the desired $5\,V$ level.

18.6 Experimental Procedures

18.6.1 Preparation

Prepare the electronic breadboard to provide buses for the positive and negative power supply rails and the circuit ground. Each power supply rail should be decoupled with a $100\,\Omega$, $1/4\,\text{W}$ resistor and a $100\,\mu\text{F}$, $25\,\text{V}$ (or greater) capacitor. The resistors are connected in series with the external power supply leads and the capacitors are connected from power supply rail to ground on the circuit side of the resistors. The capacitors must be installed with the proper polarity to prevent reverse polarity breakdown.

18.6.2 Component Measurement

Assemble the resistor and capacitor values calculated in the preliminary calculations section. Use the digital multimeter to measure the actual value of the resistors. Use the capacitor meter to measure the actual value of the capacitors. The measured values of these elements should be used in subsequent calculations.

18.6.3 Wien-Bridge Oscillator

Assemble the open loop circuit for the Wien bridge oscillator shown in Fig. 18.2b. The power supply voltages are $+15\,\text{V}$ and $-15\,\text{V}$. Use 741 op amps. Resistor R_3 should be a standard value in series with a $10\,\text{k}\Omega$ potentiometer so that the value of R_3 can be adjusted at least 20% above and 20% below the design value. Use $100\,\Omega$ for R_o.

Open-Loop Measurements

(a) Connect a function generator to the input of the circuit and use a dual channel oscilloscope to monitor the input and output voltages. Trigger the oscilloscope from the channel connected to the input. Set the function generator to produce a 1 kHz sine wave with a amplitude of 1 V and a dc level of zero. (b) Vary the frequency of the function generator until the input and output are in phase. Use the frequency counter to measure this frequency.

Closed-Loop Oscillator

(a) Adjust resistor R_3 (by varying the $10\,\text{k}\Omega$ potentiometer) until the measured gain of the circuit is unity. (b) Turn off the power supply, disconnect R_3 from the circuit, and measure its resistance with the digital multimeter. Replace R_3 in the circuit. (c) Disconnect the function generator from the circuit and connect the output of the circuit directly to the input. Turn the power supply on and observe the output. There should be a sinusoid at the same frequency as the function generator setting obtained in step for the open loop circuit. If no output is visible, slowly increase the value of R_3 until the circuit oscillates. Use the frequency counter to measure the frequency of oscillation. Sketch the waveform.

Limiters

(a) Turn the power supply off and modify the circuit to include the diode limiter circuit of Fig. 18.2a. The diodes are 1N4148 or equivalent. Initially use the value for R_3 experimentally obtained in the part for the closed loop circuit. Turn the power supply on and observe the voltage at the nodes common to R_A, R_B, and the diodes to determine when the limiting diodes just turn on. Measure the amplitude and frequency of the output sinusoid. Sketch the output waveform.

18.6.4 Phase Shift Oscillator

Repeat steps in section 18.6.3 for the phase-shift oscillator. The open-loop gain circuit can be obtained by breaking the circuit in Fig. 18.3 at the point indicated on the circuit diagram with an "X".

18.6.5 Quadrature Oscillator

Assemble the quadrature oscillator. Adjust the potentiometer so that it is not oscillating. Then slowly adjust the potentiometer until the circuit is just over the threshold of oscillations and the two output voltages are stable and undistorted. Measure the amplitude and frequency of both outputs. Sketch the waveforms. Set the oscilloscope for XY operation and obtain a plot of v_{O1} versus v_{O2}. Sketch the waveform.

18.7 Laboratory Report

The laboratory report should include:

- all preliminary derivations, calculations, and SPICE simulations
- a comparison of the theoretically predicted, the simulated, and the experimentally observed values with sources of error identified
- sketches of all the waveforms observed on the oscilloscope

18.8 References

1. E. J. Angelo, *Electronics: BJT's, FET's, and Microcircuits*, McGraw-Hill, 1969.
2. W. Banzhaf, *Computer-Aided Circuit Analysis Using SPICE*, Prentice-Hall, 1989.
3. M. N. Horenstein, *Microelectronics Circuits and Devices*, Prentice-Hall, 1990.
4. P. Horowitz & W. Hill, *The Art of Electronics*, 2nd edition, Cambridge University Press, 1989.
5. P. Horowitz & I. Robinson, *Laboratory Manual for The Art of Electronics*, Cambridge University Press, 1981.
6. J. H. Krenz, *An Introduction to Electrical and Electronic Devices*, Prentice-Hall, 1987.
7. R. Mauro, *Engineering Electronics*, Prentice-Hall, 1989.
8. F. H. Mitchell & F. H. Mitchell, *Introduction to Electronic Design*, Prentice-Hall, 1988.
9. Motorola, Inc., *Small-Signal Semiconductors*, DL 126, Motorola, 1987.
10. C. J. Savant, M. S. Roden, & G. L. Carpenter, *Electronic Circuit Design*, Benjamin Cummings, 1987.
11. A. S. Sedra & K. C. Smith, *Microelectronics Circuits*, 4th edition, Oxford, 1998.
12. D. L. Schilling & C. Belove, *Electronic Circuits: Discrete and Integrated*, McGraw-Hill, 1968.
13. P. W. Tuinenga, *SPICE*, Prentice-Hall, 1988.

Chapter 19

Switched-Capacitor Filters

19.1 Object

The object of this experiment is to design, implement, and experimentally examine some elementary switched-capacitor filters. Second-order low-pass, high-pass, and band-pass filters will be implemented with a universal second-order state-variable switched-capacitor filter.

19.2 Theory

Active filter design with op amps is a robust mature discipline. All of the classical filter types that were implemented in bygone eras with resistors, capacitors, and inductors may now be implemented solely with resistors, capacitors, and op amps. The accuracy of these filters is limited by only the precision of the components and the properties of the physical op amps employed. If discrete resistors and capacitors are used, variations within the manufacturer's stated tolerance may produce unacceptable error in the design unless extremely expensive components are employed.

The resistors and capacitors required for filter design may be fabricated on monolithic integrated circuits along with the op amps but they usually require a large amount of area and are subject to temperature drift and other annoying effects such as parasitic capacitance. Resistors fabricated on integrated circuits are usually restricted to values less than $10\,\text{k}\Omega$ while the upper limit for capacitors is approximately $100\,\text{pF}$. Also, it is quite difficult to obtain precise values of passive components fabricated on integrated circuits. Such dedicated analog filter integrated circuits are available but they are rather expensive.

The problem of component variation may be overcome with switched-capacitor filters. These use small integrated circuit capacitors whose terminals are switched by a high frequency clock signal using MOSFET switches to simulate large values of resistance. The MOSFETs are fabricated on the same integrated circuit while the clock may be external or also resident on the integrated circuit.

switched-capacitor filters are not a panacea. They are digital circuits and are, therefore, subject to aliasing. The Nyquist criterion requires that the waveform be sampled at a rate at least twice its bandwidth to prevent aliasing. Normally the clock frequency is picked to be large compared to the critical frequencies of the filter (50 to 100 times larger) to prevent aliasing. Also, the output is a discrete rather than continuous waveform. To minimize both of these defects it is customary to precede the digital switched-capacitor filter with an anti-aliasing analog low-pass filter to limit the bandwidth and to follow the digital filter with an analog deglitching filter. The break frequencies of these analog filters are not crucial so this use of analog filters at the input and output is not a major impediment.

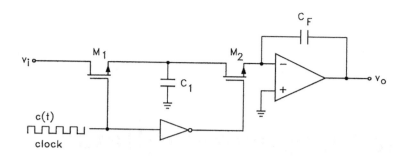

Figure 19.1: Switched-capacitor integrator.

19.2.1 Switched-Capacitor Integrator

A switched-capacitor integrator is shown in Fig. 19.1. The clock signal $c(t)$ with frequency f_c and period T_c ($f_c = 1/T_c$) is applied to both the gate input of MOSFET M_1 and the digital inverter. The signal applied to the gate of MOSFET M_2 is the complement of the clock. Hence, excepts for the switching transient, one MOSFET is on while the other is off.

When the clock is high MOSFET M_1 is on and M_2 is off. Capacitor C_1 has a charge $\Delta q = C_1 v_i$ placed on it by the input to the filter. If the clock frequency is large compared to the bandwidth of v_i the input may be considered to be constant during the sampling interval $(T_c/2)$. During the next clock half cycle M_1 is off and M_2 is on which places the top node of the capacitor C_1 at the virtual ground of the op amp which causes the charge on it to be transferred to C_F. The average current flowing into capacitor C_1 is

$$i(t) = \frac{\Delta q}{T_c} = \frac{C_1}{T_c} v_i = C_1 f_c v_i \tag{19.1}$$

which means that it is equivalent to a resistor $R_{eq} = 1/C_1 f_c$. The output of the op amp is then given by

$$v_o = -\frac{1}{C_F} \int_{-\infty}^{t} i(u)\, du = -\frac{C_1 f_c}{C_F} \int_{-\infty}^{t} v_i(u)\, du = -\frac{1}{C_F R_{eq}} \int_{-\infty}^{t} v_i(u)\, du \tag{19.2}$$

which makes this circuit an integrator. Integrators are the heart of the state variable filter which means that any of the classical filters may be realized with this switched-capacitor arrangement. Other more elaborate topologies are also employed in switched-capacitor filters but the circuit in Fig. 19.1 illustrates the basic principle.

Since charge is transferred in spurts from capacitor C_1 to capacitor C_F this makes the output voltage discrete rather than continuous. The voltage increments are reduced to acceptable values by picking the clock frequency to be large which is also required to prevent aliasing. A deglitching analog low-pass filter cascaded with the output may also be used to smooth the output voltage.

Since the output of the switched-capacitor integrator depends on the ratio of two capacitances, this can easily be fabricated on an integrated circuit. Although precise values of components are difficult to control, maintaining ratios is relatively simple.

Anent the equivalent resistance being set as $R_{eq} = 1/C_1 f_c$, this makes controlling the critical frequencies of the filter elementary. The system clock sets the critical frequencies. Therefore, such filters may be easily electronically tuned.

19.2. THEORY

When a MOSFET is on the drain to source resistance is not zero but has a certain value know as the on-resistance R_{on} which is normally several hundred ohms. This means that the charging and discharging of capacitor C_1 is not instantaneous but limited by an RC time constant $\tau_{on} = R_{on}C_1$. This sets the upper limit for the system clock. Proper operation requires that the period of the system clock be large compared to the charging time constant. With current MOS technology this makes operation of switched-capacitor filters above a few hundred kilo-hertzs impossible.

19.2.2 Second Order Filter Categories

Because switched-capacitor filters are digital circuits, the appropriate mathematical artifice to analyze them is the z transform. However, classical frequency domain analysis is sufficiently accurate and more amendable to a mathematical tractable analysis. This discussion will be limited to second order filters because that is what is implemented in the device that will be ultimately employed in this experiment.

Low-Pass Filter

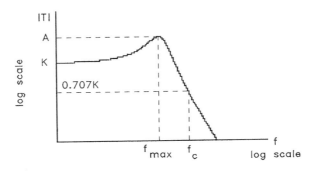

Figure 19.2: Second-order low-pass filter.

The complex transfer function for a second-order low-pass filter is

$$T(s) = K \frac{1}{(s/\omega_0)^2 + (1/Q)(s/\omega_0) + 1} \tag{19.3}$$

where K is the DC gain of the filter, Q is the quality factor, and ω_0 is the resonant frequency of the filter. The magnitude of the complex transfer function is plotted in Fig. 19.2. The minus 3 dB, half-power, critical, or cutoff frequency, f_c, is the frequency at which the gain is reduced to $K/\sqrt{2}$ and is given by

$$f_c = f_0 \sqrt{\left(1 - \frac{1}{2Q^2}\right) + \sqrt{\left(1 - \frac{1}{2Q^2}\right)^2 + 1}} \tag{19.4}$$

which is a frequency larger than f_0. The maximum value of the magnitude of the complex transfer function A occurs at the frequency $f_{\max}$ where

$$A = \frac{KQ}{\sqrt{1 - 1/4Q^2}} \tag{19.5}$$

$$f_{\max} = f_0 \sqrt{1 - 1/2Q^2} \tag{19.6}$$

which is a frequency smaller than f_0.

The proper selection of Q will produce any of the classical filter configurations such as Butterworth, Chebyshev, and Bessel. The choice of Q is given in Table 1.

Filter Type	Pass-Band Ripple	Q	f_c/f_0
Bessel	—	0.577	0.786
Butterworth	—	0.707	1.000
Chebyshev	$0.1\ dB$	0.767	1.078
Chebyshev	$0.2\ dB$	0.797	1.111
Chebyshev	$0.3\ dB$	0.821	1.136
Chebyshev	$0.5\ dB$	0.864	1.176
Chebyshev	$1\ dB$	0.957	1.246
Chebyshev	$2\ dB$	1.129	1.333
Chebyshev	$3\ dB$	1.305	1.389

Table 19.1: Second-Order Low-Pass Filter Parameters

High-Pass Filter

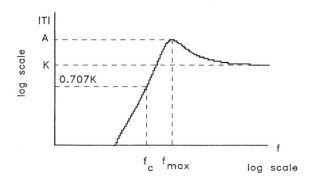

Figure 19.3: Second-order high-pass filter.

The complex transfer function for a second-order high-pass filter is

$$T(s) = K \frac{(s/\omega_0)^2}{(s/\omega_0)^2 + (1/Q)(s/\omega_0) + 1} \tag{19.7}$$

where K is the high-frequency gain of the filter, Q is the quality factor, and ω_0 is the resonant frequency of the filter. The magnitude of the complex transfer function is plotted in Fig. 19.3. The minus 3 dB, half-power, critical, or cutoff frequency, f_c, is the frequency at which the gain is reduced to $K/\sqrt{2}$ and is given by

$$f_c = \frac{f_0}{\sqrt{(1 - 1/2Q^2) + \sqrt{(1 - 1/2Q^2)^2 + 1}}} \tag{19.8}$$

which is a frequency smaller than f_0. The maximum value of the magnitude of the complex transfer function A occurs at the frequency $f_{\max}$ where

$$A = \frac{KQ}{\sqrt{1 - 1/4Q^2}} \tag{19.9}$$

19.2. THEORY

and

$$f_{max} = \frac{f_0}{\sqrt{1 - 1/4Q^2}} \quad (19.10)$$

which is a frequency larger than f_0.

Band-Pass Filter

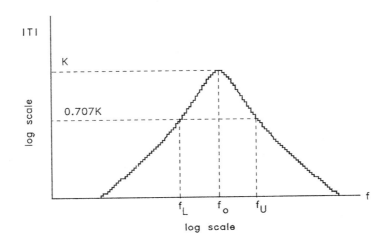

Figure 19.4: Band-pass filter.

The complex transfer function for the second-order band-pass filter is

$$T(s) = K \frac{(1/Q)(s/\omega_0)}{(s/\omega_0)^2 + (1/Q)(s/\omega_0) + 1} \quad (19.11)$$

where f_0 is the resonant frequency and also the center frequency, Q is the quality factor, and K is the gain at the center frequency. The plot of the magnitude of the complex transfer function as a function of frequency is shown in Fig. 19.4. This circuit has both an upper, f_U, and lower, f_L, half-power frequencies given by

$$f_U = f_0 \left(\frac{1}{2Q} + \sqrt{\left(\frac{1}{2Q}\right)^2 + 1} \right) \quad (19.12)$$

and

$$f_L = f_0 \left(-\frac{1}{2Q} + \sqrt{\left(\frac{1}{2Q}\right)^2 + 1} \right) \quad (19.13)$$

The difference between these two half-power frequencies is known as the half-power bandwidth, Δf, which is given by

$$\Delta f = f_U - f_L = \frac{f_0}{Q} \quad (19.14)$$

which reveals that
$$Q = \frac{f_0}{\Delta f} \tag{19.15}$$
which is the reason why Q is called the quality factor, i.e. the larger Q is the sharper the peak in the filter becomes. The center frequency is also the geometric mean of the half-power frequencies, i.e.
$$f_0 = \sqrt{f_U f_L} \tag{19.16}$$

Notch Filter

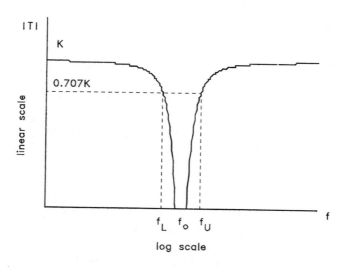

Figure 19.5: Notch filter.

The complex transfer function of the second order notch filter is given by
$$T(s) = K \frac{(s/\omega_0)^2 + 1}{(s/\omega_0)^2 + (1/Q)(s/\omega_0) + 1} \tag{19.17}$$
where ω_0 is the notch frequency in radians/sec, Q is the quality factor, and K is both the high and low frequency gain. A plot of the magnitude of the complex transfer function for the notch filter is shown in Fig. 19.5.

All of the equations for the upper and lower half-power frequencies as well as the quality factor for the band-pass filter are equally applicable for the notch filter. This is because the transfer function of the notch filter, $T_N(s)$, is related to that of the band-pass filter, $T_B(s)$, by
$$T_N(s) = K - T_B(s) \tag{19.18}$$
Thus, either filter may be obtained by summing the other with a gain constant.

All-Pass Filter

The transfer function of the second-order all-pass filter is given by

$$T(s) = K \frac{(s/\omega_0)^2 - (1/Q)(s/\omega_0) + 1}{(s/\omega_0)^2 + (1/Q)(s/\omega_0) + 1} \tag{19.19}$$

which has a constant amplitude and a phase that varies with frequency. The magnitude of the complex transfer function is K at all frequencies whereas the angle of the complex transfer function is given by

$$\phi(j\omega) = \angle T(j\omega) = -2\arctan\left[\frac{(1/Q)(\omega/\omega_0)}{1 - (\omega/\omega_0)^2}\right] \tag{19.20}$$

Biquadratic Filter

The second order biquadratic filter has the complex transfer function

$$T(s) = K \frac{(s/\omega_z)^2 + (1/Q_z)(s/\omega_z) + 1}{(s/\omega_z)^2 + (1/Q_p)(s/\omega_p) + 1} \tag{19.21}$$

which has a DC gain of K and a high frequency gain of $K(\omega_p/\omega_z)^2$. A special case of the biquadratic filter is obtained by letting the quality factor for the numerator become infinite (which puts the zeroes on the imaginary axis as is the case for the notch filter) – these are known as elliptic filters. If $\omega_p > \omega_z$ it is a high-pass elliptic filter and if $\omega_p < \omega_z$ it is a low-pass elliptic filter (and, of course, if $\omega_p = \omega_z$ it is a notch filter). Elliptic filters feature the fastest transition from the pass to the stop band but, unlike the garden variety high- and low-pass filters, do not have a gain that is a monotonic function of frequency.

19.3 Devices

A plethora of switched-capacitor devices are available as off-the-shelf integrated circuits from a variety of manufacturers. Prominent among these are the devices from Motorola which are primarily intended for telecommunications applications such as the MC145432 which features a six pole notch filter operated by a highly stable crystal oscillator internal clock or the MC145414 which is a fifth-order elliptic filter. Some filters are totally digital in that both the clock frequency and the filter parameters are directly set by input digital codes.

The device that will be employed in this experiment in the MF10 Universal Monolithic Dual switched-capacitor Filter because it is the cheapest switched-capacitor device available and it offers a degree of flexibility to the user. A functional block diagram is shown in Fig. 19.6. The square symbols indicate that these are pin numbers on the integrated circuit. It has two identical filters known and A and B. Separate clocks are provided for each of the two filters. The circuit components shown in this figure are internal to the integrated circuit. Each of the symbols show as a triangle with an integral symbol inside are switched-capacitor integrators of the type discussed in 19.2.1. Because there are two integrators and op amp summers, this makes this a second order state variable filter.

The symbol with a one plus and two minus signs in a circle with an X is a summer with two inverting and one non-inverting inputs. One of the inverting inputs is a pin on the IC while the other is controlled by the S_{AB} input on pin 6; the position of this switch determines whether the feedback is closed around the input op amp or the triple input summer.

The level shift input on pin 9 determines the type of clock being used. This pin is grounded for TTL clock signals or $\pm 5\ V$ CMOS clock signals. The same type of clock must be applied to both filter A and B but different frequency clocks may be used at the two inputs. The $50/100/CL$ input on pin 12 determines whether the critical frequencies are obtained as the clock frequency divided by 50 or 100 or whether the filter is turned off; if this pin is connected to the positive power supply the clock frequency is divided by 50, if it

CHAPTER 19. SWITCHED-CAPACITOR FILTERS

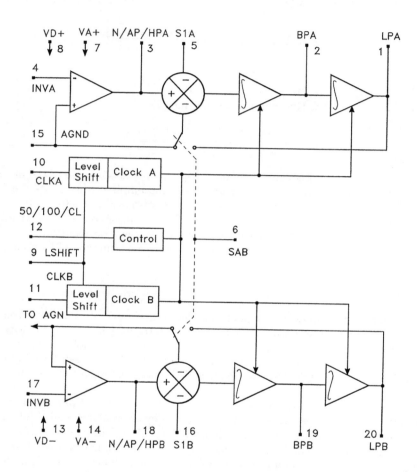

Figure 19.6: MF10 switched-capacitor filter.

19.3. DEVICES

is grounded the clock frequency is divided by 100, and if it is connected to the negative supply filtration is inhibited.

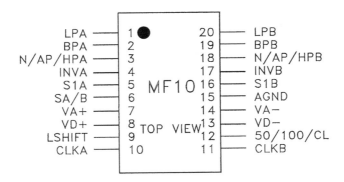

Figure 19.7: Pinouts for MF10

The pinouts for the MF10 are given in Fig. 19.7. A more complete description of this device is given in the manufacturer's data sheet in the appendix to this experiment. The notation used in the manufacturer's data sheet differs slightly from that used in this experiment.

The manufacturer defines several modes or circuit topologies for the MF10. The one that will be used in this experiment is Mode 3a shown in Fig. 19.8. The resistors $R1$, $R2$, $R3$, and $R4$ are external resistors that are connected to the pins shown. There are three outputs low-, high-, and band-pass at pins 1, 3, and 2 respectively. One external op amp summer with resistors R_L, R_U, and R_g is used to sum the high- and low-pass outputs to produce a notch output. The input is v_i.

The resonant frequency f_0 is given by

$$f_0 = \frac{f_{CLK}}{100} \times \sqrt{\frac{R2}{R4}} \quad \text{or} \quad \frac{f_{CLK}}{50} \times \sqrt{\frac{R2}{R4}} \tag{19.22}$$

depending on whether the clock is being divided by 100 or 50; this is center frequency of both the band-pass and notch filter. The quality factor is given by

$$Q = \sqrt{\frac{R2}{R4}} \times \frac{R3}{R2} \tag{19.23}$$

The DC gain of the low-pass filter is

$$K_{LP} = -\frac{R4}{R1} \tag{19.24}$$

the high-frequency gain of the high-pass filter is

$$K_{HP} = -\frac{R2}{R1} \tag{19.25}$$

and the gain at the center frequency of the band-pass filter is

$$K_{BP} = -\frac{R3}{R1} \tag{19.26}$$

The gain of the notch filter at the notch frequency f_0 is given by

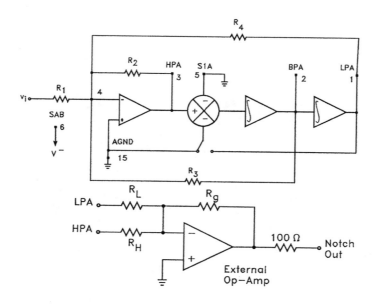

Figure 19.8: Mode 3A for MF10

$$A_n = \left| Q \left(\frac{R_g}{R_L} K_{LP} - \frac{R_g}{R_H} K_{HP} \right) \right| \tag{19.27}$$

which would normally be picked to be zero. The gain of the notch filter at DC is

$$K_N = \frac{R_g}{R_L} \times K_{LP} \tag{19.28}$$

while the gain at the frequency which is half of the clock frequency is

$$K_{\frac{CLK}{2}} = -\frac{R_g}{R_H} \times K_{HP} \tag{19.29}$$

19.4 Preliminary Derivations

1. Derive Eqns. 19.4, 19.5, and 19.6.
   2. Derive Eqn. 19.20.

19.5 Preliminary Calculations

Compute appropriate values for the circuits components required to produce filters with the parameters specified by the laboratory instructor.

19.6 Preliminary SPICE Simulations

Simulate the transfer function for the circuits that were designed in the preliminary calculations section. Use the Analog Behavioral Modeling feature of SPICE to achieve this.

19.7. EXPERIMENTAL PROCEDURES

Example. Use the Analog Behavioral Modeling feature of SPICE to plot the transfer function of a second order Chebyshev Low-Pass Filter with a DC gain of unity, a cut off frequency of 10 kHz, and 3 dB of ripple in the pass band.

Solution. The appropriate equation is Eqn. 19.3. Table 19.1 discloses that $Q = 1.305$ and $f_c/f_0 = 1.389$ where f_c is the cutoff frequency given as 10 kHz. The *PSpice* input deck is given by

```
Analog Behavioral Modeling Example
.param pi = 3.1415926
.param two_pi = {2*pi}
.param q = 1.305
.param wc = {two_pi*10k}
.param w0 = {wc/1.389}
.param k = 1
vi in 0 ac 1
ri in 0 1
e out 0 laplace {v(in)} =
+{k/(((s/w0)*((s/w0)+1/q)+1))}
rout out 0 1
.ac dec 30 1k 100k
.probe
.end
```

There are three nodes in this circuit: `in`, `out`, and `0`. Note that any mathematical operation must be enclosed by the squiggly brackets "`{}`". Some of the conventions of programming languages such as BASIC and FORTRAN must be followed, namely left brackets and right brackets must match in number. Unfortunately, there is no code for raising a variable to a power, e.g. s^2 is written as "`s*s`". The user plots `vdb(out)`.

19.7 Experimental Procedures

19.7.1 Preparation

Prepare the electronic breadboard to provide buses for the positive and negative power supply rails and the circuit ground. Three power supplies will be used in this experiment: +5 V, and −5 V. Each power supply rail should be decoupled with a 100 Ω, 25 V (or greater) capacitor. The resistors are connected in series with the external power supply leads and the capacitors are connected from the power supply rails to ground on the circuit side of the resistor. The capacitor must be installed with the proper polarity to prevent reverse polarity breakdown.

19.7.2 Clock

The clock that will be used for the experiment in the MC1555/555 timer shown in Fig. 19.9. The power supply voltage for this IC will be +5 V so that it will produce a TTL clock. The clock frequency is

$$f_{CLK} = \frac{1.44}{(R_1 + 2R_2)C_T} \tag{19.30}$$

and the duty cycle is

$$Duty\ Cycle = \frac{R_1 + R_2}{R_1 + 2R_2} \tag{19.31}$$

Design the timer so that the clock frequency is either $f_0 \times 100$ or $f_0 \times 50$. Pick the duty cycle to be close to 50%. The lower of the two frequencies is the most judicious choice; although the higher clock frequency

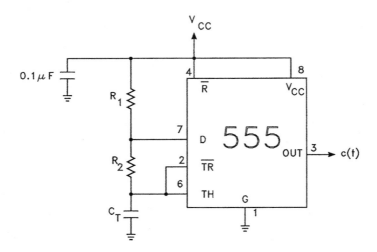

Figure 19.9: 555/MC1555 Timer

would produces fewer glitches in the output of the filter, if the timer were operated at this high frequency on an untidy breadboard, the waveform might be distorted and degrade the performance of the filter.

19.7.3 Second-Order Filter

Assemble the second-order state variable filter shown in Fig. 19.8 using the component values computed in the preliminary calculations section with the power off. This is the mode of the filter labeled 3a by the manufacturer.

Pins 5, 9, and 15 should be connected to ground.

Pins 7, 8, and 12 should be connected to the positive power supply, +5 V. (Connecting pin 12 to the positive power supply sets the filter to the $f_{CLK}/50$ mode. If the $f_{CLK}/100$ mode is desired, pin 12 should be grounded.)

Pins 6, 13, and 14 should be connected to the negative power supply, −5 V.

Standard connections should be made for the external op amp. An TL071 would be a better choice than a 741 since the 741 requires power supply voltages of at least ± 8 V which would incinerate the MF10.

The clock input is pin 3 of the MC1555 timer which is connected to pin 10 of the MF10 for filter A.

Turn the power on and measure the frequency response of the filter. Data should be taken of the four outputs as the frequency of the input is varied one decade above and below f_0.

Increase the frequency of the input until it is approximately equal to the clock frequency. Explain why the low- and band-pass outputs are nonzero in the laboratory report.

19.7.4 Fourth-Order Filter

Assemble the fourth-order filter with the characteristics specified by the laboratory instructor. This is obtained by cascading the two second-order filters found in the MF10. Repeat the measurement of the frequency response.

19.8 Laboratory Report

The laboratory report should include:

- plots of the experimental measured frequency response of all of the filters that were designed
- SPICE simulations for the frequency response of the filters that were designed
- an explanation of why the output of the low- and band-pass filters were not zero when the input signal was a sine wave with a frequency near the clock frequency

19.9 References

1. P. E. Allen and E. Sánchez-Sinencio, *switched-capacitor Circuits*, Van Nostrand Reinhold, 1984.
2. W. M. Leach, Jr., *Introduction to Electroacoustics and Audio Amplifier Design*, Kendall/Hunt Publishing Company, Dubuque, Iowa, 1998, ISBN 0-7872-5410-X.
3. C. Chen, *Active Filter Design*, Hayden, 1982.
4. P. Horowitz and W. Hill, *The Art of Electronics*, Second Edition, Cambridge University Press, 1989.
5. Z. H. Meiksin and P. C. Thackray, *Electronic Design with Off-the-Shelf Integrated Circuits*, Parker, 1980.
6. J. M. McMenamin, *Linear Integrated Circuits: Operation and Applications*, Prentice-Hall, 1985.
7. Motorola, *Telecommunications Device Data*, DL 136, Rev. 2, Motorola, 1989.
8. National Semiconductor, *Linear Databook 2*, National Semiconductor, 1987.
9. S. Soclof, *Design and Applications of Analog Integrated Circuits*, Prentice-Hall, 1991.

Chapter 20

A Voltage Regulator

20.1 Object

The object of this experiment is to assemble and test a voltage regulator circuit. The first version of the circuit has no feedback. The final version has feedback error correction and foldback current limiting.

20.2 Theory

20.2.1 Block Diagram

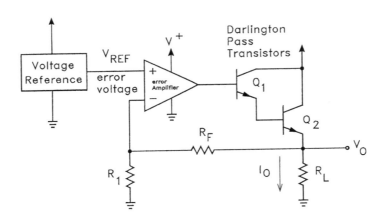

Figure 20.1: Block diagram of a regulated dc power supply.

Fig. 20.1 shows a block diagram of a simple feedback power supply regulator. The resistor R_L is known as the load. The goal is to maintain the output or load voltage V_O a constant, known as the regulated voltage, as the load resistor and consequently the load current, I_O, change. The voltage V^+ is assumed to be a nonregulated voltage. This is usually the output of a half or full wave rectifier cascaded with rectifier filter.

341

The regulated dc power supply is a negative feedback amplifier operating at a frequency of 0 Hz or dc. The input is the reference voltage V_{REF} which is compared with a portion of the output voltage V_O supplied by the feedback network consisting of resistors R_F and R_1. The error voltage is amplified and used to control the output current I_O supplied by the series pass transistors Q_1 and Q_2 to the load R_L.

Transistors Q_1 and Q_2 form a Darlington pair. This configuration means that the emitter current of Q_1 is the base current of transistor Q_2. This is equivalent to a single transistor with a overall current gain of $\beta_1\beta_2$. Since the transistor current gain β is large (typically $50 \to 200$) this means that the overall current gain can be immense. Since these two transistors are being operated in the emitter follower configuration the voltage gain from the base of Q_1 to the emitter of Q_2 is approximately unity. Since the amount of power dissipated in the pass transistors can be large one or both may be mounted on heat sinks to prevent thermal damage.

When the load current increases, which tends to cause the output voltage to drop, the error voltage increases which causes the error amplifier to increase the current supplied to the base of pass transistor Q_1 which tends to increase the load current supplied by transistor Q_2. Conversely when the output current drops the output voltage increases which causes the error voltage to become smaller which decreases the base current of Q_1 and decreases the load current. Therefore, the negative feedback amplifier compensates for or counteracts changes in the output current. This regulates the output voltage.

If the error amplifier is assumed to be ideal (no current flowing into either the inverting or noninverting inputs and a dc gain of infinity), the output voltage is given by

$$V_O = V_{REF}\left(1 + \frac{R_F}{R_1}\right) \tag{20.1}$$

This is the result expected for a noninverting amplifier with input V_{REF} and output V_O. The resistors R_F and R_1 are chosen to be large compared to the load so that most of the emitter current of Q_2 flows into the load and not into the feedback network.

The voltage reference is chosen to be relatively independent of changes in temperature and the unregulated voltage V^+. It may be a reverse biased Zener diode, an integrated circuit band gap reference, or the output of a "V_{BE}" multiplier. Unless it is very stable the output voltage will not be well regulated.

20.2.2 Differential Amplifier

The circuit shown in Fig. 20.2 is a regulated dc power supply implemented with discrete BJT transistors. The circuit consists of a V_{BE} multiplier voltage reference (Q_3) connected to the reference or noninverting input of an error amplifier (the differential amplifier consisting of transistors Q_4 and Q_5 with resistive load R_4). A JFET J_1 current source is used to bias the differential amplifier. A series pass output stage (Q_1 and Q_2) connects between the output of the error amplifier and the load. The output voltage is connected directly to the feedback or inverting input of the error amplifier using unity feedback, i.e. the feedback network has a gain of one. The reference voltage is the collector-to-emitter voltage across Q_3. If base current is neglected, the voltage across Q_3 is given by

$$V_{CE3} = V_{BE3}\left(1 + \frac{R_2}{R_3}\right) \tag{20.2}$$

The input to the error amplifier is the difference voltage $V_{CE3} - V_O$. Because the amplifier is part of a negative feedback loop, it adjusts V_O so as to make the error voltage approach zero. In this case, the output voltage is given by $V_O = V_{CE3}$.

20.2.3 BJT Equivalent Circuits

Subsequent derivations will require the base and emitter equivalent circuits. These are the equivalent circuits seen looking into the base and emitter terminals of a BJT.

The base and emitter equivalent circuits for a BJT are shown in Fig. 20.3. The voltage v_{tb} and resistor R_{tb} are the Thévenin equivalent circuit seen looking out the base terminal back into the base circuit. Similarly

20.2. THEORY

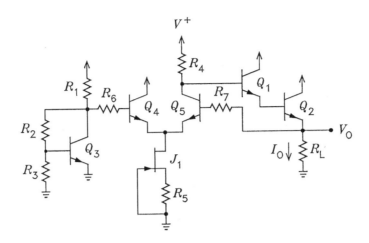

Figure 20.2: BJT differential amplifier voltage regulator with V_{BE} voltage reference.

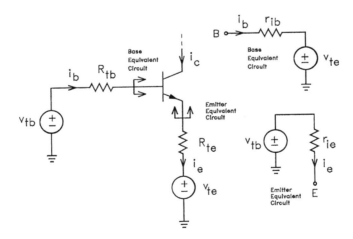

Figure 20.3: BJT equivalent circuits.

the voltage v_{te} and resistor R_{te} are the Thévenin equivalent circuit seen looking out the emitter terminal back into the emitter circuit.

The base equivalent circuit seen looking into the base terminal consists of the resistor r_{ib} in series with the voltage v_{te} where

$$r_{ib} = r_x + (1+\beta)(r_e + R_{te}) \tag{20.3}$$

where r_x is the base spreading resistance of the transistor and r_e is the intrinsic emitter resistance $r_e = V_T/I_E$ where $V_T = kT/q$ is the thermal voltage. The base spreading resistance is a parameter that must be determined from experimental measurements but is usually in the range of a tens to hundreds of Ohms. The intrinsic emitter resistance is also usually small, viz. it has a value of approximately 26 Ω at a temperature of 300° K when the emitter current is 1 mA. So the resistance seen looking into the base terminal consists of what's seen looking out of the emitter terminal multiplied by the current gain of the transistor (since $\beta + 1 \approx \beta$).

The emitter equivalent circuit seen looking into the emitter of a BJT transistor consists of the voltage v_{tb} in series with a resistor r_{ie} where

$$r_{ie} = \frac{R_{tb} + r_x}{1+\beta} + r_e \tag{20.4}$$

This discloses that the resistance seen looking into the emitter of a BJT consists of the resistance seen looking out of the base divided by the current gain of the transistor. For this reason the base is called a high impedance terminal and the emitter a low impedance terminal.

20.2.4 Small-Signal Output Impedance

A perfectly regulated dc power supply is modeled as an ideal voltage source. An actual regulated dc power supply is modeled as an ideal voltage source in series with a incremental output resistance. The smaller this resistance the more efficacious the power supply.

A plot of the output load voltage V_O versus the load current can be modeled as a straight line with a slope of $-r_{out}$ where r_{out} is the small signal incremental resistance given by

$$r_{out} = -\left.\frac{\partial V_O}{\partial I_O}\right|_{I_O=I_Q} \tag{20.5}$$

where I_Q is a quiescent or nominal operating point. It is assumed in the subsequent derivation that the transistor small signal output impedance is infinite which means that the Early voltage is assumed to be infinite.

The small signal equivalent circuit shown in Fig. 20.4 will be used to determine r_{out} for the regulated dc power supply shown in Fig. 20.2. The collector current for Q_5 is a controlled source so it is represented with the rectangular symbol for a controlled current source. The load resistor has been replaced with a test current source i_t. The small signal incremental output resistance is

$$r_{out} = \frac{v_O}{i_t} \tag{20.6}$$

The equivalent circuit seen looking out of the base of Q_1 is given by

$$v_{tb1} = -i_{c5} R_4 \tag{20.7}$$

and

$$R_{tb1} = R_4 \tag{20.8}$$

The small signal equivalent circuit seen looking out of the base of Q_2 is the small signal equivalent circuit seen looking into the emitter of Q_1 and is given by

$$v_{tb2} = v_{tb1} = -i_{c5} R_4 \tag{20.9}$$

20.2. THEORY

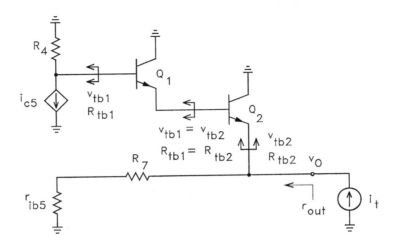

Figure 20.4: Output resistance equivalent circuit.

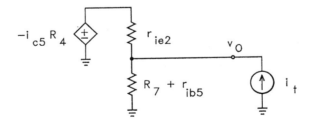

Figure 20.5: Equivalent circuit for emitter of Q_2.

and
$$R_{tb2} = r_{ie1} = \frac{R_4 + r_{x1}}{1 + \beta_1} + \frac{V_T}{I_{E1}} \tag{20.10}$$

The next step in the derivation of the output resistance is to replace Q_2 with its small signal emitter equivalent circuit as shown in Fig. 20.5. For this circuit

$$r_{ie2} = \frac{\frac{R_4 + r_{x1}}{1 + \beta_1} + \frac{V_T}{I_{E1}} + r_{x2}}{1 + \beta_2} + \frac{V_T}{I_{E2}} \tag{20.11}$$

Since Q_1 and Q_2 are in series

$$I_{E1} = I_{B2} = \frac{I_{E2}}{1 + \beta_2} \tag{20.12}$$

which makes Eq. 20.11

$$r_{ie2} = \frac{R_4 + r_{x1}}{(1 + \beta_1)(1 + \beta_2)} + \frac{r_{x2}}{(1 + \beta_2)} + \frac{2V_T}{I_{E2}} \tag{20.13}$$

If the feedback is disabled this would be r_{out}. Since the value of the current gains of the transistors are normally large the expression is approximately $r_{ie2} \approx 2V_T/I_{E2}$. With feedback the output is

$$v_O = i_t r_{ie2} \| (R_7 + r_{ib5}) - i_{c5} R_4 \frac{R_7 + r_{ib5}}{r_{ie2} + R_7 + r_{ib5}} \tag{20.14}$$

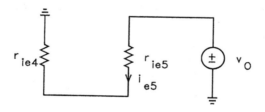

Figure 20.6: Emitter equivalent circuit for differential amplifier.

Assuming that the JFET current source is an ideal current source, the emitter equivalent circuit for the differential amplifier is shown in Fig. 20.6. The emitter equivalent resistors for the differential amplifier are given by

$$r_{ie4} = \frac{R_6 + r_{x4}}{1 + \beta_4} + \frac{V_T}{I_{E4}} \tag{20.15}$$

and

$$r_{ie5} = \frac{R_7 + r_{x5}}{1 + \beta_5} + \frac{V_T}{I_{E5}} \tag{20.16}$$

assuming that the reference voltage (V_{ce3}) is a perfect voltage source. The expression for i_{e5} is

$$i_{e5} = \frac{v_O}{r_{ie4} + r_{ie5}} \tag{20.17}$$

which yields

$$i_{c5} = \alpha_5 i_{e5} = \frac{\alpha_5 v_O}{r_{ie4} + r_{ie5}} \tag{20.18}$$

20.2. THEORY

Substituting Eq. 20.18 into Eq. 20.14 yields

$$v_o = i_t r_{ie2} \| (R_7 + r_{ib5}) - \frac{\alpha_5 v_O}{r_{ie4} + r_{ie5}} R_4 \frac{R_7 + r_{ib5}}{r_{ie2} + R_7 + r_{ib5}} \qquad (20.19)$$

which may be solved for v_O to yield

$$r_{out} = \frac{v_O}{i_t} = \frac{r_{ie2} \| (R_7 + r_{ib5})}{1 + \frac{\alpha_5 R_4}{r_{ie4} + r_{ie5}} \frac{R_7 + r_{ib5}}{r_{ie2} + R_7 + r_{ib5}}} \qquad (20.20)$$

as the expression for the output resistance where

$$r_{ib5} = r_{x5} + (1 + \beta_5) \left(\frac{V_T}{I_{E5}} + r_{ie4} \right) \qquad (20.21)$$

Simpler expressions for these parameters may be obtained by invoking the large β approximation which yields

$$\beta \to \infty \qquad (20.22)$$

$$\alpha \to 1 \qquad (20.23)$$

$$r_{ie2} \to \frac{2V_T}{I_{E2}} \qquad (20.24)$$

$$r_{ie4} \to \frac{V_T}{I_{E4}} \qquad (20.25)$$

$$r_{ie5} \to \frac{V_T}{I_{E5}} \qquad (20.26)$$

$$r_{ib5} \to \infty \qquad (20.27)$$

$$r_{out} \to \frac{\frac{2V_T}{I_{E2}}}{1 + \frac{R_4}{V_T \left(\frac{1}{I_{E4}} + \frac{1}{I_{E5}} \right)}} \qquad (20.28)$$

If the transistors in the differential amplifier are symmetrically biased, $I_{E4} = I_{E5} = I_{D1}/2$ where I_{D1} is the drain current in JFET J_1. The approximate expression for the output resistance becomes

$$r_{out} = \frac{\frac{2V_T}{I_{E2}}}{1 + \frac{R_4 I_{D1}}{4V_T}} \qquad (20.29)$$

The numerator of Eq. 20.29 is the output resistance with no feedback and the denominator is $1 + A_{OL}$ where A_{OL} is the open loop gain of the amplifier.

As a numerical example if the amplifier were biased at $I_{E2} = 2\,\text{mA}$, $I_{D1} = 1\,\text{mA}$, and $R_4 = 18\,\text{k}\Omega$ the output resistance without feedback is $25.9\,\Omega$ and with feedback it is only $0.148\,\Omega$. Even further reductions may be obtained by increasing the open loop gain of the amplifier.

In general the output resistance of the voltage regulator is given by

$$r_{out(FB)} = r_{out(nFB)} \frac{A_{CL}}{A_{OL} + 1} \qquad (20.30)$$

where $r_{out(FB)}$ is the output resistance with feedback, $r_{out(nFB)}$ is the output resistance with no feedback, A_{OL} is the open loop gain, and A_{CL} is the closed loop gain of the amplifier. The open loop gain can be made immense by replacing the resistive load on the differential amplifier with an active load which results in even smaller values for the output resistance.

20.2.5 Current Limit

The series pass transistors can be damaged due to excessive power dissipation. Since the current is much larger in Q_2 the power dissipation in Q_1 may be neglected. The power dissipated in Q_2 is given by

$$P_D = (V^+ - V_O) I_O \qquad (20.31)$$

If a short circuit is placed across the output the power dissipated is $P_D = V^+ I_O$. To prevent damage to Q_2 it is desirable to limit the maximum current that can be supplied to the load.

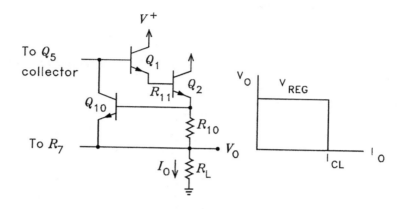

Figure 20.7: Simple current limit circuit.

A simple current limiting scheme is shown in Fig. 20.7. The resistor R_{10} is used to provide a voltage drop across the base-to-emitter terminal of transistor Q_{10}. When the output current is below the current limit value the voltage drop across is too small to turn Q_{10} on and it acts as an open circuit and the voltage regulator performs as described in previous sections. However, once the output current reaches the current limit value Q_{10} turns on and removes base current from Q_1 which prevents the output current from increasing.

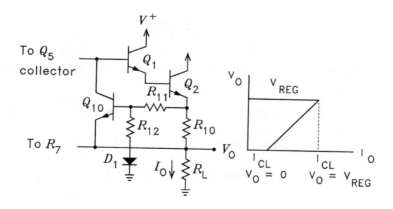

Figure 20.8: Foldback current limit circuit.

20.3. PRELIMINARY DERIVATIONS

An improved current limiting scheme can be obtained using the circuit shown in Fig. 20.8. The current limiting transistor responds to a combination of the load voltage and current. There are different current limits for when the output is the regulated value and when it is shorted. The diode causes the current limit transistor Q_{10} to turn on faster when the output is shorted. This scheme is called foldback current limiting.

For the circuit shown in Fig. 20.8 the current limit value is given by

$$I_{CL} = \frac{R_{10} + R_{11} + R_{12}}{R_{10}R_{12}} V_{ON} + \left(1 + \frac{R_{12}}{R_{11}}\right)(V_O - V_\gamma)\frac{1}{R_{12}} \qquad (20.32)$$

where V_{ON} is the base-to-emitter voltage at which transistor Q_{10} turns on and V_γ is the on voltage for the diode. Both of these on voltages are in the neighborhood of $0.5 \to 0.7\,V$. In deriving this equation it was assumed that $R_{11} + R_{12}$ is very large compared to R_{10} so that most of the load current flows through R_{10}.

20.3 Preliminary Derivations

1. Derive Eq. 20.32.

20.4 Preliminary Calculations

1. For the circuit shown in Fig. 20.2 if $V^+ = 18\,V$, calculate R_1, R_2, and R_3 so that $V_{CE3} = 10\,V$, $I_{C3} = 1\,mA$, and the current through R_2 is $0.1\,mA$. Assume $V_{BE3} = 0.65\,V$ and neglect I_{B3}, I_{B4}, and I_{B5}.

2. For the circuit shown in Fig. 20.2, calculate R_4 for $I_{D1} = 1.5\,mA$, $I_{C5} = I_{D1}/2$, $I_{B1} = 0$, and $V_{C5} = V_O + 2 \times 0.65\,V$, where $V_O = 10\,V$.

3. For the circuit shown in Fig. 20.2, determine R_5 to bias the JFET at $I_{D1} = 1.5\,mA$. Assume that the JFET has a pinch off voltage of $3\,V$ and an $I_{DSS} = 3\,mA$.

4. For the circuit shown in Fig. 20.8 determine values for R_{10}, R_{11}, and R_{12} so that the power that is dissipated in Q_2 when the load current limits at both the regulated voltage of $10\,V$ and when the output is shorted is $900\,mW$. Pick $R_{11} + R_{12} = 10R_{10}$ so that most of the load current flows through R_{10}. Assume that the on voltage for the current limit transistor Q_{10} and the diode D_1 is $0.5\,V$.

20.5 Preliminary SPICE Simulations

1. Perform a SPICE simulation of the circuit in Fig. 20.2 using the component values calculated in the Preliminary Calculations. Assume that the load resistor $R_L = 10\,k\Omega$. Use the SPICE parameters for the JFET given in the Preliminary Calculations. For the BJTs, assume that they have identical parameters: BF=200, IS=12.61f, and VA=200. Assume that the diode has parameters: N=1.89, BV=100, and IS=1.7n.

2. Replace the load resistor R_L with a voltage source with a value of $10\,V$. Perform a dc sweep of this voltage source from $0 \to 11\,V$. Plot the load voltage versus the load current. This should illustrate foldback current limiting.

20.6 Experimental Procedures

20.6.1 No Feedback

Fig. 20.9 shows the circuit with the feedback input to the error amplifier connected to the voltage reference. This is the first form of the circuit to be assembled and evaluated for the laboratory. Because the circuit has no feedback, the power supply is unregulated in this form.

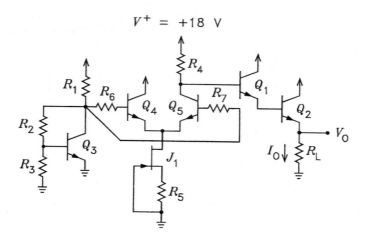

Figure 20.9: Circuit with no feedback.

V_{BE} Multiplier

Assemble the V_{BE} multiplier and make any adjustments required to obtain the specified bias current and voltage.

Current Source

Determine R_5 for a JFET drain current of 1.5 mA. The easiest way to do this is to connect the JFET in series with a dc ammeter to the laboratory bench dc power supply with a potentiometer connected as a variable resistor for R_5. Adjust R_5 for $I_{D1} = 1.5$ mA, then measure the value of the resistance of the potentiometer with an ohmmeter. Use the closest 5% fixed resistor. The laboratory bench dc power supply voltage should be set to a value equal to the expected drain voltage on the JFET. This voltage should be close to $10 - 0.65 = 9.35$ V.

Assembly

Assemble the circuit of Fig. 20.9 with $R_6 = R_7 = 1\,\text{k}\Omega$ and the value of R_4 obtained in the Preliminary Calculations. For the initial test, use $R_L = 10\,\text{k}\Omega$. Measure V_O. If it is not the specified value of 10 V, adjust R_4 until $V_O = 10$ V. There will probably be some temperature drift in V_O. This is normal because the circuit has no feedback.

Load Voltage versus Load Current Data

When the circuit is operational, measure V_O versus I_O for at least 5 values of I_O in the range $1\,\text{mA} \leq I_O \leq 50\,\text{mA}$. Observe V_O on an oscilloscope during the measurements. If the circuit oscillates, add a 1 μF capacitor in parallel with R_L. It may be necessary to use this capacitor in the subsequent steps.

Output Resistance

Calculate the output resistance r_{out} from the slope of the V_O versus I_O curve. Most pocket calculators have a routine that will calculate the slope by making a least squares fit of a straight line to the data. It should

20.6.2 Feedback

Connect the feedback input to the V_O node as shown in Fig. 20.2. Repeat the measurements described above to calculate r_{out}. By what factor has the output resistance decreased?

Current Mirror Active Load

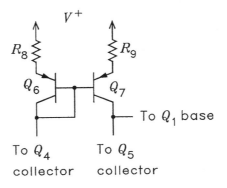

Figure 20.10: Current mirror active load.

Shown in Fig. 20.10 is a current mirror active load for the differential amplifier. Add this to the circuit and repeat the measurements for r_{out}. Use $R_8 = R_9 = 100\,\Omega$. This should greatly increase the open loop gain of the amplifier and decrease the output resistance.

Foldback Current Limiter

Fig. 20.8 shows a foldback current limiter protection circuit for the series pass stage. Add this to the circuit shown in Fig. 20.2 using the values for the resistors calculated in the Preliminary Calculations. Experimentally verify the current limit values. If necessary, adjust the resistors in the limiter circuit to meet the specifications.

20.7 Laboratory Report

Turn in all the calculations made, data taken and simulations performed. Plot the load voltage versus load current for the circuit with and without feedback. Compare the calculated, experimental, and simulation values for the output resistance and current limits.

Did the circuit oscillate for any values of load resistance?

Discuss any unusual results.

20.8 References

1. E. J. Angelo, *Electronics: BJT's, FET's, and Microcircuits*, McGraw-Hill, 1969.
2. W. Banzhaf, *Computer-Aided Circuit Analysis Using SPICE*, Prentice-Hall, 1989.
3. M. N. Horenstein, *Microelectronics Circuits and Devices*, Prentice-Hall, 1990.
4. P. Horwitz & W. Hill, *The Art of Electronics*, 2nd edition, Cambridge University Press, 1989.
5. P. Horowitz & I. Robinson, *Laboratory Manual for The Art of Electronics*, Cambridge University Press, 1981.
6. J. H. Krenz, *An Introduction to Electrical and Electronic Devices*, Prentice-Hall, 1987.
7. R. Mauro, *Engineering Electronics*, Prentice-Hall, 1989.
8. F. H. Mitchell & F. H. Mitchell, *Introduction to Electronic Design*, Prentice-Hall, 1988.
9. Motorola, Inc., *Small-Signal Semiconductors*, DL 126, Motorola, 1987.
10. C. J. Savant, M. S. Roden, & G. L. Carpenter, *Electronic Circuit Design*, Benjamin Cummings, 1987.
11. A. S. Sedra & K. C. Smith, *Microelectronics Circuits*, 4th edition, Holt, Oxford, 1998.
12. D. L. Schilling & C. Belove, *Electronic Circuits: Discrete and Integrated*, McGraw-Hill, 1968.
13. P. W. Tuinenga, *SPICE*, Prentice-Hall, 1988.

Chapter 21

The Operational Amplifier

21.1 Object

The object of this experiment is to assemble and evaluate a discrete operational-amplifier circuit. The general theory of the op amp is covered.

21.2 Notation

The notation used for variables in the equations in this experiment correspond to the conventions summarized in the following examples:

dc value – I_E total value, i.e. dc plus small-signal – v_{BE}
small-signal value – i_e phasor value – V_{be}

Although phasor notation is used in equations involving transfer functions, figure labels follow the dc and small-signal notations.

21.3 Op-Amp Model

The general purpose operational amplifier, or op amp, is an amplifier having two inputs and one output. The circuit is designed so that the output voltage is proportional to the difference between the voltages at the inputs. In general, the op amp can be modeled as a three-stage amplifier as shown in Fig. 21.1. The input stage is a differential amplifier (Q_1 and Q_2) with a current mirror load (Q_3 through Q_5). The second stage is a high-gain stage having an inverting gain. A capacitor connects the output of this stage to its input. This capacitor is called the compensating capacitor and it sets the bandwidth of the circuit. The output stage is a unity-gain stage which provides the current gain to drive low impedance loads.

21.4 Voltage-Gain Transfer Function

In the circuit of Fig. 21.1, let $V_{i1} = V_{id}/2$ and $V_{i2} = -V_{id}/2$, where V_{id} is the differential input voltage. For the diff-amp transistors Q_1 and Q_2, let us assume that the devices are matched, $I_{E1} = I_{E2} = I_Q/2$, $V_A = \infty$, and $\beta = \infty$. In this case, the small-signal Thévenin equivalent circuit seen looking into the emitter of Q_1 is the intrinsic emitter resistance r_e in series with $V_{id}/2$ and circuit seen looking into the emitter of Q_2 is r_e in series with $-V_{id}/2$. Thus we can write

$$I_{c1} = -I_{c2} = \frac{V_{id}}{2(R_E + r_e)} \tag{21.1}$$

CHAPTER 21. THE OPERATIONAL AMPLIFIER

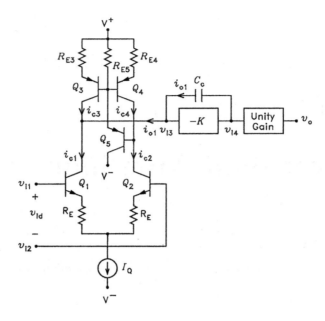

Figure 21.1: Model of the operational amplifier.

where r_e is given by

$$r_e = \frac{V_T}{I_E} = \frac{2V_T}{I_Q} \tag{21.2}$$

For the current mirror transistors Q_3, Q_4, and Q_5, let us assume that the devices are matched, $V_A = \infty$, and $\beta = \infty$. By the current mirror action, $I_{c3} = I_{c4}$. But, when base currents are neglected, $I_{c4} = I_{c2}$. Thus we have $I_{c3} = I_{c2}$. It follows that the output current from the diff amp stage can be written

$$I_{o1} = I_{c1} - I_{c2} = I_{c1} - (-I_{c1}) = 2I_{c1} = \frac{V_{id}}{R_E + r_e} \tag{21.3}$$

To calculate the output voltage, we assume that the second-stage input current is negligible so that all of I_{o1} flows through the compensating capacitor C_c. We also assume that the voltage gain of the output stage is unity. Thus we can write

$$V_o = V_{i4} = V_{i3} + \frac{I_{o1}}{C_c s} = \frac{V_o}{-K} + \frac{I_{o1}}{C_c s} \tag{21.4}$$

Solution for V_o yields

$$V_o = \frac{I_{o1}}{(1 + 1/K)C_c s} \simeq \frac{I_{o1}}{C_c s} = \frac{V_{id}}{(R_E + r_e)C_c s} \tag{21.5}$$

where the approximation assumes that $K \gg 1$. Thus we can define the transfer function

$$G(s) = \frac{V_o}{V_{id}} = \frac{1}{(R_E + r_e)C_c s} \tag{21.6}$$

21.5 Gain-Bandwidth Product

The Bode plot for $|G(j\omega)|$ is a straight line with a slope of -1 dec/dec or -20 dB/dec. The plot is shown in Fig. 21.2(a). The frequency at which $|G(j\omega)| = 1$ is called the unity-gain frequency or the gain-bandwidth

21.6. SLEW RATE

product. It is labeled ω_x in the figure and is given by

$$\omega_x = 2\pi f_x = \frac{1}{(R_E + r_e)\, C_c} = \frac{1}{(R_E + 2V_T/I_Q)\, C_c} \qquad (21.7)$$

For most general purpose IC op amps, f_x lies in the range $1\,\text{MHz} \leq f_x \leq 10\,\text{MHz}$. An alternate expression for $G(s)$ is

$$G(s) = \frac{\omega_x}{s} \qquad (21.8)$$

(a) Asymptotic open-loop gain. (b) Asymptotic closed-loop gain.

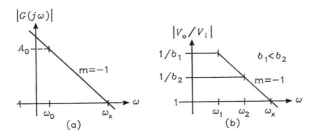

Figure 21.2: (a) Asymptotic open-loop gain. (b) Asymptotic closed-loop gain.

The circuit model used above predicts that $|G(j\omega)| \to \infty$ as $\omega \to 0$. In a physical op amp, the gain must remain finite as $\omega \to 0$. Fig. 21.2(a) shows the asymptotic low-frequency gain shelving at the value A_0 below the frequency ω_0, where ω_0 is a pole in the transfer function. Its value is determined by the output resistance of the diff amp and current mirror stages and by the input resistance to the second gain stage. These are assumed to be infinite in the derivation of Eq. (21.6), whereas they are finite in a physical op amp. In a general purpose IC op amp, typical values for A_0 and f_0 are $A_0 = 2 \times 10^5$ (106 dB) and $f_0 = 5$ Hz. If $f_0 \ll f_x$, it follows from Fig. 21.2(a) that $f_x = A_0 f_0$. This equation illustrates why f_x is called the gain-bandwidth product. When the pole at ω_0 is included in the transfer function, $G(s)$ can be written in the forms

$$G(s) = \frac{A_0}{1 + s/\omega_0} = \frac{A_0}{1 + sA_0/\omega_x} = \frac{\omega_x/\omega_0}{1 + s/\omega_0} \qquad (21.9)$$

The pole ω_0 is called the first or dominant pole in the op amp transfer function. In physical op amps, the transfer function has more than one pole. In order for the circuit to be stable when feedback is added, i.e. not to oscillate, the second and higher order poles must occur at frequencies above ω_x. The compensating capacitor C_c is chosen to obtain this condition. With a proper value of C_c, the higher order poles can usually be neglected in the analysis of the circuit.

Example 1 *An op amp is to be designed for $f_x = 5\,\text{MHz}$ and $I_Q = 0.5\,\text{mA}$. If $R_E = 0$, calculate the required value for C_c.*

Solution. $C_c = 1/(2\pi f_x r_e) = I_Q/(4\pi f_x V_T) = 307\,\text{pF}$, where it is assumed that $V_T = 0.0259\,\text{V}$.

21.6 Slew Rate

The op amp slew rate is the maximum value of the time derivative of its output voltage. In general, the positive and negative slew rates can be different. The circuit model of Fig. 21.1 predicts that the two are equal so that we can write

$$-SR \leq \frac{dv_O}{dt} \leq +SR \qquad (21.10)$$

where SR is the slew rate. To solve for this, we use Eq. (21.5) to write

$$sV_o = \frac{I_{o1}}{C_c} \qquad (21.11)$$

The s operator in a phasor equation becomes the d/dt operator in a time-domain equation. Thus we can write

$$\frac{dv_O}{dt} = \frac{i_{O1}}{C_c} \qquad (21.12)$$

It follows that the slew rate is determined by the maximum value of i_{O1}.

Let v_{ID} in Fig. 21.1 be positive enough to cut Q_2 off. In this case we have $i_{C1} = +I_Q$, $i_{C2} = 0$, and $i_{O1} = I_Q$. Now let v_{ID} be negative enough to cut Q_1 off. In this case we have $i_{C1} = 0$, $i_{C2} = I_Q$, and $i_{O1} = -I_Q$. It follows that $-I_Q \leq i_{O1} \leq +I_Q$ so that

$$\frac{-I_Q}{C_c} \leq \frac{dv_O}{dt} \leq \frac{+I_Q}{C_c} \qquad (21.13)$$

Thus the slew rate is given by

$$SR = \frac{I_Q}{C_c} \qquad (21.14)$$

Example 2 *Calculate the slew rate of the op amp of Example 1.*

Solution. $SR = I_Q/C_c = 1.63 \, \text{V}/\mu\text{s}$.

21.7 Relations between Slew Rate and Gain-Bandwidth Product

If C_c is eliminated between Eqs. (21.7) and (21.14), we obtain the relations

$$SR = \omega_x I_Q \left(R_E + \frac{2V_T}{I_Q} \right) = 4\pi f_x V_T \left(1 + \frac{I_Q R_E}{2V_T} \right) \qquad (21.15)$$

This equation clearly shows that the slew rate is fixed by the gain-bandwidth product if $R_E = 0$. If $R_E > 0$, the slew rate and gain bandwidth product can be specified independently.

Example 3 *Emitter resistors with the value $R_E = 300\,\Omega$ are added to the input diff amp in the op amp of Example 1. If ω_x is to be held constant, calculate the new value of the slew rate and the new value of C_c.*

Solution. $SR = 2\pi f_x I_Q (r_e + R_E) = 6.34 \, \text{V}/\mu\text{s}$. $C_c = I_Q/SR = 78.9 \, \text{pF}$. The slew rate is greater by about a factor of 4 and C_c is smaller by the same factor.

The above example illustrates how the slew rate of an op amp can be increased without changing its gain-bandwidth product. When R_E is added, ω_x decreases. To make it equal to its original value, C_c must be decreased, and this increases the slew rate. It can be seen from Eq. (21.14) that the slew rate can also be increased by increasing I_Q. However, this causes ω_x to increase. To make it equal to its original value, R_E must be increased. Therefore, the general rule for increasing the slew rate is to either decrease C_c, increase I_Q, or both. Then R_E is increased to bring ω_x back down to its original value. The change in R_E does not affect the slew rate.

21.8 Closed-Loop Transfer Function

Figure 21.3 shows the op amp with a two resistor voltage divider connected as a feedback network. The output voltage can be written

$$V_o = G(s)(V_i - V_f) = G(s)(V_i - bV_o) \qquad (21.16)$$

21.9. TRANSIENT RESPONSE

where b is the gain of a voltage divider given by

$$b = \frac{R_1}{R_1 + R_F} \quad (21.17)$$

Note that $-1 \leq b \leq +1$. Eq. (21.16) can be solved for V_o/V_i to obtain

$$\frac{V_o}{V_i} = \frac{G(s)}{1+bG(s)} = \frac{\omega_x/s}{1+b\omega_x/s} = \frac{1}{b} \times \frac{1}{1+s/b\omega_x} \quad (21.18)$$

where Eq. (21.8) is used for $G(s)$. This is a low pass transfer function having a dc gain and a pole frequency, respectively, given by

$$A_{0f} = \frac{1}{b} \quad (21.19)$$

$$\omega_{0f} = 2\pi f_{0f} = b\omega_x = \frac{b}{(R_E + r_e)C_c} \quad (21.20)$$

where the f in the subscript implies "with feedback."

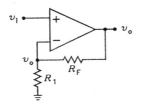

Figure 21.3: Op amp with feedback.

Figure 21.2(b) shows the Bode plot for $|V_o/V_i|$ for two values of b. As b is increased, the gain A_{0f} decreases and the bandwidth ω_{0f} increases. For any gain, the product of the gain and the bandwidth is equal to the gain bandwidth product ω_x. That is $A_{0f}\omega_{0f}$, or alternately $A_{0f}f_{0f}$, is a constant.

Example 4 *An op amp has the gain bandwidth product $f_x = 2\,\text{MHz}$. Calculate the upper $-3\,dB$ frequency f_u if the op amp is operated at a voltage gain of 20.*

Solution. The upper -3 dB frequency is equal to the pole frequency of the closed-loop transfer function. Thus $f_u = f_{0f} = f_x/20 = 100\,\text{kHz}$.

21.9 Transient Response

Let the input voltage to the non-inverting amplifier in Fig. 21.3 be a step of amplitude V_{I1}. We can write $v_I(t) = V_{I1}u(t)$, where $u(t)$ is the unit step function. The Laplace transform of $v_I(t)$ is $V_i(s) = V_{I1}/s$. The Laplace transform of the output voltage is given by

$$V_o(s) = \frac{V_{I1}}{s} \times \frac{1}{b} \times \frac{1}{1+s/b\omega_x} = \frac{V_{I1}}{b} \times \frac{1}{s(1+s/b\omega_x)} \quad (21.21)$$

The time domain output voltage is obtained by taking the inverse Laplace transform to obtain

$$v_O(t) = \frac{V_{I1}}{b}\left(1 - e^{-b\omega_x t}\right)u(t) \quad (21.22)$$

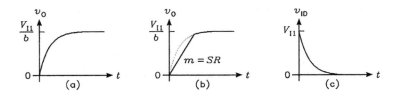

Figure 21.4: (a) Transient response. (b) Transient response with slewing. (c) Differential input voltage.

A plot of $v_O(t)$ is shown in Fig. 21.4(a).

The maximum time derivative of $v_O(t)$ occurs at $t=0$ and is given by

$$\left.\frac{dv_O}{dt}\right|_{\max} = \frac{V_{I1}}{b}\frac{d}{dt}\left[\left(1-e^{-b\omega_x t}\right)u(t)\right]\bigg|_{t=0} = \omega_x V_{I1} = 2\pi f_x V_{I1} \qquad (21.23)$$

If the derivative exceeds the slew rate of the op amp, the output voltage will be distorted as shown in Fig. 21.4(b). The maximum value of V_{I1} before the op amp slews is given by

$$V_{I1\max} = \frac{SR}{\omega_x} = \frac{SR}{2\pi f_x} \qquad (21.24)$$

Example 5 *Calculate the maximum value of V_{I1} for the op amps of Examples 1 and 3.*

Solution. For Example 1, $V_{I1} = SR/\omega_x = 51.9\,\text{mV}$. For Example 3, $V_{I1} = 202\,\text{mV}$. This is greater by about a factor of 4.

21.10 Input Stage Overload

For the step input signal to the op amp with feedback in Fig. 21.3, the differential input voltage is given by

$$v_{ID}(t) = v_I(t) - bv_O(t) = V_{I1}e^{-b\omega_x t}u(t) \qquad (21.25)$$

A plot of $v_{ID}(t)$ is shown in Fig. 21.4(c). The peak voltage occurs at $t=0$ and is $v_{ID\text{peak}} = V_{I1}$. If the op amp is not to slew, the diff amp input stage must not overload with this voltage.

Figure 21.5 shows example plots of the collector currents in Q_1 and Q_2 of Fig. 21.1 as a function of v_{ID} for $I_Q = 1\,\text{mA}$ and two values of the emitter resistors R_E. The plots assume that base currents can be neglected. Points on each curve are labeled where the collector currents are equal to $0.1 I_Q$ and $0.9 I_Q$. We will take the region between these points on any curve to be the active range for the diff amp. If v_{ID} is constrained to lie in this region, the op amp cannot slew. The plots labeled $R_E = 300\,\Omega$ show an active range that is much wider than those labeled $R_E = 0$. The value of $|v_{ID}|$ where the diff amp leaves the active region is given by

$$v_{ID\max} = V_T \ln 9 + 0.8 I_Q R_E \qquad (21.26)$$

With $V_T = 0.0259\,\text{V}$ and $I_Q = 1\,\text{mA}$, this equation gives $v_{ID\max} = 57\,\text{mV}$ for $R_E = 0$ and $v_{ID\max} = 300\,\text{mV}$ for $R_E = 300\,\Omega$.

Example 6 *Calculate v_{IDmax} for the op amps of Examples 1 and 3.*

Solution. For Example 1, $v_{ID\max} = V_T \ln 9 = 56.9\,\text{mV}$. For Example 3, $v_{ID\max} = V_T \ln 9 + 0.8 I_Q R_E = 177\,\text{mV}$. These values are less than the values for V_{I1} calculated in Example 5 because neither transistor in the diff amp is completely cut off.

21.11. THE BIFET OP AMP

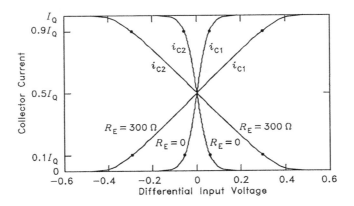

Figure 21.5: Collector currents versus differential input voltage.

21.11 The BiFet Op Amp

We have seen above that the addition of emitter resistors to the diff amp transistors reduces the gain bandwidth product of the op amp. If the compensation capacitor is then reduced to bring the gain bandwidth product back up to its original value, the slew rate is increased. Another method of accomplishing this is to replace the BJTs with JFETs. The circuit is shown in Fig. 21.6 and is called a bifet op amp. For a specified bias current, the JFET has a much lower transconductance than the BJT. In effect, this makes it look like a BJT with emitter resistors. For this reason, resistors in series with the JFET sources are omitted in Fig. 21.6.

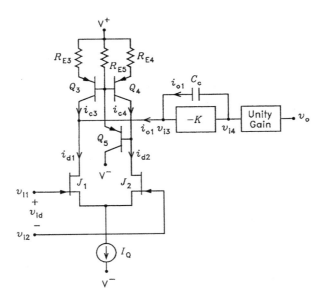

Figure 21.6: Model of the bifet op amp.

The JFET drain current can be written

$$i_D = I_{DSS}\left(1 - \frac{v_{GS}}{V_{TO}}\right)^2 \tag{21.27}$$

where I_{DSS} is the drain-to-source saturation current, V_{TO} is the threshold voltage (which is negative), v_{GS} is the gate to source voltage, and $V_{TO} \leq v_{GS} \leq 0$. For the drain current in either JFET in the diff amp to be in the range of $0.1I_Q$ to $0.9I_Q$, the maximum differential input voltage is given by

$$v_{ID\max} = |V_{TO}|\sqrt{\frac{0.4I_Q}{I_{DSS}}} \tag{21.28}$$

The JFET transconductance is given by

$$g_m = \frac{\partial I_D}{\partial V_{GS}} = \frac{2I_{DSS}}{-V_{TO}}\left(1 - \frac{V_{GS}}{V_{TO}}\right) = \frac{2}{-V_{TO}}\sqrt{I_D I_{DSS}} \tag{21.29}$$

To convert a formula derived for the op amp with a BJT diff amp into a corresponding formula for the JFET diff amp, the BJT intrinsic emitter resistance r_e is replaced with $1/g_m$ for the JFET. Thus the gain bandwidth product of the bifet op amp is given by

$$\omega_x = 2\pi f_x = \frac{g_m}{C_c} \tag{21.30}$$

Figure 21.7 shows a plot of the drain currents in the two FETs in Fig. 21.6 versus differential input voltage. For this plot, the values $I_Q = 1\,\text{mA}$, $I_{DSS} = 3\,\text{mA}$, and $V_{TO} = -2\,\text{V}$ are assumed. Although these JFET parameters are typical, they can vary widely from device to device. The maximum differential input voltage for these values is $v_{ID\max} = 0.73\,\text{V}$.

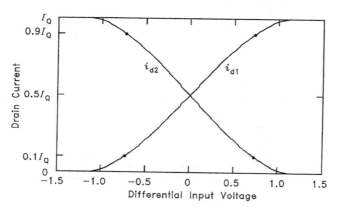

Figure 21.7: Drain current versus differential input voltage.

Example 7 *A bifet op amp is to be designed for $f_x = 5$ MHz and $I_Q = 0.5\,\text{mA}$. The diff amp transistors have the parameters $I_{DSS} = 3\,\text{mA}$ and $V_{TO} = -2\,\text{V}$. Calculate the required value for C_c and calculate the slew rate of the op amp.*

Solution. $g_m = (2/2)\sqrt{0.00025 \times 0.003} = 8.66 \times 10^{-4}\,\text{S}$. $C_c = g_m/(2\pi f_x) = 27.6\,\text{pF}$. $SR = I_Q/C_c = 18\,\text{V}/\mu\text{s}$.

Example 8 *Calculate v_{IDmax} for the diff amp of Example 7.*

21.12. SINE-WAVE RESPONSE

Solution. $v_{IDmax} = 2\sqrt{0.4 \times 0.5/3} = 0.516\,\text{V}$.

Example 9 *If the JFETs in Example 7 are replaced with BJTs, what value of R_E is required to make the circuit have the same gain bandwidth product? Assume I_Q splits equally between the two diff amp transistors.*

Solution. We set $r_e + R_E$ for the BJT equal to $1/g_m$ for the JFET and solve for R_E. For the BJT, $r_e = V_T/I_E = 51.8\,\Omega$. For the JFET, $1/g_m = 817\,\Omega$. It follows that $R_E = 817 - 51.8 = 765\,\Omega$.

21.12 Sine-Wave Response

Let the input voltage to the op amp with feedback be a sine wave. If the op amp does not slew and is not driven into peak clipping, the output voltage can be written $v_O(t) = V_{O1}\sin\omega t$. The time derivative is given by $dv_O/dt = \omega V_{O1}\cos\omega t$. The maximum value of $|dv_O/dt|$ occurs at $\omega t = n\pi$, where n is an integer, and is given by $|dv_O/dt|_{\max} = \omega V_{O1}$. For a physical op amp, this cannot exceed the slew rate, i.e. $\omega V_{O1} < SR$. It follows that the maximum frequency that the op amp can put out the sine wave without slewing is given by

$$f_{\max} = \frac{SR}{2\pi V_{O1}} \qquad (21.31)$$

Conversely, the peak output voltage without slewing is given by

$$V_{O1\max} = \frac{SR}{2\pi f} \qquad (21.32)$$

The full power bandwidth frequency $FPBW$ is defined as the highest frequency at which the op amp can put out a sine wave with a peak voltage equal to the op amp clipping or saturation voltage V_{SAT}. It is given by

$$FPBW = \frac{SR}{2\pi V_{SAT}} \qquad (21.33)$$

Example 10 *The op amps of Examples 1 and 3 have a peak sine-wave output voltage of 13 V. Calculate the full power bandwidth frequency if the op amps are not to slew at this voltage level.*

Solution. For the op amp of Example 1, $f_{\max} = 1.63 \times 10^6/(2\pi 13) = 20.0\,\text{kHz}$. For the op amp of Example 3, $f_{\max} = 6.34 \times 10^6/(2\pi 13) = 77.6\,\text{kHz}$.

21.13 Full Slewing Response

Figure 21.8(a) shows the output voltage of an op amp with a sine wave input for two cases, one where the op amp is not slewing and the other where the op amp is driven into full slewing. The full slewing waveform is a triangle wave. The slew-limited peak voltage is given by the slope multiplied by one-fourth the period, i.e.

$$V_P = SR \times \frac{T}{4} = \frac{SR}{4f} \qquad (21.34)$$

where $T = 1/f$. When the op amp is driven into full slewing, an increase in the amplitude of the input signal causes no change in the amplitude of the output signal. If the frequency is doubled, the amplitude of the output signal is halved.

Figure 21.8(b) shows the peak output voltage versus frequency for a sine wave input signal. At low frequencies, the peak voltage is limited to the op amp clipping or saturation voltage V_{SAT}. As frequency is increased, the peak voltage becomes inversely proportional to frequency when the op amp is driven into full slewing and is given by $SR/(4f)$. The figure also shows the peak voltage below which the op amp does not slew. It is given by $SR/(2\pi f)$.

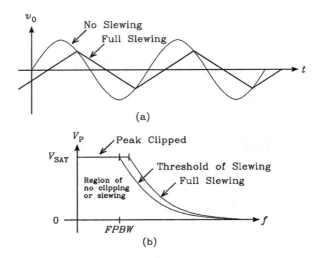

Figure 21.8: (a) Output waveform for no slewing and full slewing. (b) Peak output voltage versus frequency.

21.14 Intermediate Circuits

Figure 21.9 shows an intermediate op amp circuit consisting of an input differential amplifier driving a second gain stage. Resistive loads are used on both the diff amp and the second stage. Without the current mirror load on the diff amp, it can be shown that the gain-bandwidth product and the slew rate have values equal to 1/2 those derived above for the current mirror load. In the following, a current mirror load will be added to the diff amp as we develop the circuit. The feedback resistors R_{F1} and R_{F2} set the closed-loop gain. Resistor R_1 sets the input resistance. The resistors labeled R_B help prevent parasitic oscillation problems that can occur. The diff amp tail supply is a JFET current source. For a desired drain current I_D in J_1, the series source resistor R_S is given by

$$R_S = \frac{|V_{TO}|}{I_D}\left(1 - \sqrt{\frac{I_D}{I_{DSS}}}\right) \tag{21.35}$$

where I_{DSS} is the drain-to-source saturation current and V_{TO} is the threshold or pinch-off voltage, which is negative.

Example 11 *Typical parameters for the 2N5457 JFET are $I_{DSS} = 3$ mA and $V_{TO} = -2$ V. Calculate the required value of R_S for a drain current of 1.5 mA.*

Solution. From Eq. (21.35), we have $R_S = (2/0.0015)\left(1 - \sqrt{1.5/3}\right) = 390\,\Omega$.

To solve for the closed-loop voltage gain of the circuit in Fig. 21.9, we can write

$$v_o = A(v_i - v_f) \qquad v_f = \frac{R_{F1}}{R_{F1} + R_{F2}}v_o = bv_o \tag{21.36}$$

where A is the open-loop gain, i.e. the gain with feedback removed. These equations can be solved to obtain

$$\frac{v_o}{v_i} = \frac{A}{1 + bA} \simeq \frac{1}{b} = 1 + \frac{R_{F2}}{R_{F1}} \tag{21.37}$$

where the approximation holds for $bA \gg 1$. The quantity $(1 + bA)$ is called the amount of feedback. It is the ratio of the open-loop gain to the closed-loop gain.

21.14. INTERMEDIATE CIRCUITS

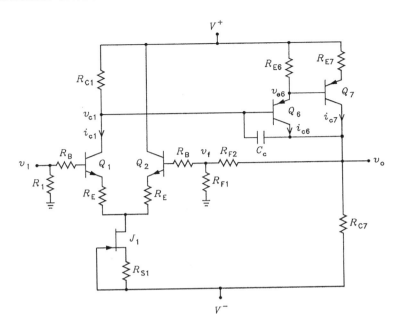

Figure 21.9: Intermediate op amp circuit.

To calculate the open-loop gain A, the feedback must be removed. To do this, R_{F2} is first removed from the circuit. Then a resistor equal to R_{F2} is connected from the v_f node to ground and a resistor equal to $R_{F2} + R_{F1}$ is connected from the v_o node to ground. The open-loop gain A can then be written as the product of terms

$$A = \frac{v_o}{v_i} = \frac{i_{c1}}{v_i} \times \frac{v_{c1}}{i_{c1}} \times \frac{i_{c6} + i_{c7}}{v_{c1}} \times \frac{v_o}{i_{c6} + i_{c7}} \tag{21.38}$$

To evaluate the terms in this equation, we will neglect base currents, i.e. assume that $\beta \to \infty$ for each transistor. Because there is no feedback, the differential input voltage is equal to v_i, i.e. $v_{id} = v_i$. Thus we can use Eq. (21.1) to write $i_{c1}/v_i = 1/[2(r_e + R_E)]$ For the second term in the gain expression, we have $v_{c1}/i_{c1} = -R_{C1}$. To solve for the third term, we must solve for i_{c6} and i_{c7} separately. The small-signal Thévenin equivalent circuit seen looking into the emitter of Q_6 is the resistor r_{e6} in series with the voltage v_{c1}. Thus we can write $i_{c6} = i_{e6} = -v_{c1}/(r_{e6} + R_{E6})$. The small-signal Thévenin equivalent circuit seen looking into the emitter of Q_7 is the resistor r_{e7} in series with v_{e6}, where voltage division can be used to write $v_{e6} = v_{c1}R_{E6}/(r_{e6} + R_{E6})$. Thus we can write $i_{c7} = i_{e7} = -v_{e6}/(r_{e7} + R_{E7}) = -[v_{c1}R_{E6}/(r_{e6} + R_{E6})]/(r_{e7} + R_{E7})$. For the final term, we have $v_o = (i_{c6} + i_{c7})[R_{C7} \| (R_{F2} + R_{F1})]$. When these equations are combined, we obtain

$$A = \frac{1}{2(r_e + R_E)} \times (-R_{C1}) \times \frac{-R_{E6}}{r_{e6} + R_{E6}} \left(\frac{1}{R_{E6}} + \frac{1}{r_{e7} + R_{E7}} \right) \times [R_{C7} \| (R_{F2} + R_{F1})] \tag{21.39}$$

A good approximation to this equation can be made by assuming that $r_{e6} \ll R_{E6}$ and that $R_{E6} \gg r_{e7} + R_{E7}$. In this case, we can write

$$A \simeq \frac{R_{C1}}{2(r_e + R_E)} \times \frac{R_{C7} \| (R_{F2} + R_{F1})}{r_{e7} + R_{E7}} \tag{21.40}$$

This equation is useful for design purposes. After it is used to calculate target values for the resistors, the actual gain can be calculated with Eq. (21.39).

CHAPTER 21. THE OPERATIONAL AMPLIFIER

Example 12 *The circuit of Fig. 21.9 is to be designed for a closed loop gain of 10, a slew rate of $5\,\text{V}/\mu\text{s}$, and a gain-bandwidth product of $5\,\text{MHz}$. The target gain of the $Q_6 - Q_7$ stage is to be -100. For the specified currents $I_Q = 1.5\,\text{mA}$, $I_{C6} = 0.2\,\text{mA}$, and $I_{C7} = 1.8\,\text{mA}$, calculate values for C_c, the resistors in the circuit, and the open-loop gain A. Assume that I_Q divides equally between Q_1 and Q_2, that $V_T = 0.0259\,\text{V}$, and $V^+ = -V^- = 15\,\text{V}$. Assume base-emitter junction voltages of $0.65\,\text{V}$. Neglect all base currents.*

Solution. Without the current-mirror load, the slew rate and gain-bandwidth product are given by $SR = I_Q/2C_c$ and $f_x = 1/\left[4\pi\left(R_E + r_e\right)C_c\right]$, where $r_e = 2V_T/I_Q = 34.5\,\Omega$. Thus we have $C_c = 0.0015/\left(2 \times 5 \times 10^6\right) = 150\,\text{pF}$ and $R_E = 1/\left(4\pi \times 5 \times 10^6 \times 150 \times 10^{-12}\right) - r_e = 71.6\,\Omega$. For a quiescent output voltage of zero, we must have $0 - V^- = (I_{C6} + I_{C7})R_{C7}$. This equation yields $R_{C7} = 7.5\,\text{k}\Omega$. For negligible loading by the feedback circuit, $R_{F2} + R_{F1}$ should be much larger than R_{C7}. We will choose $R_{F2} + R_{F1} = 10R_{C7} = 75\,\text{k}\Omega$. For a gain of 10, we have $10 = 1 + R_{F2}/R_{F1}$. These equations can be solved to obtain $R_{F1} = 7.5\,\text{k}\Omega$ and $R_{F2} = 67.5\,\text{k}\Omega$. The approximate gain of the second stage is given by $A \simeq -\left[R_{C7}\|(R_{F2} + R_{F1})\right]/(r_{e7} + R_{E7})$, where $r_{e7} = 0.0259/0.0018 = 14.4\,\Omega$. For this to be -100, it follows that $R_{E7} = 53.8\,\Omega$. The dc voltage across R_{E6} is $I_{E7}R_{E7} + V_{EB7} = I_{E6}R_{E6}$. It follows that $R_{E6} = 3.73\,\text{k}\Omega$. The dc voltage across R_{C1} is $I_{E7}R_{E7} + V_{EB7} + V_{EB6} = I_{C1}R_{C1}$. Solution for R_{C1} yields $R_{C1} = 1.86\,\text{k}\Omega$. The value of R_B is not important as long as it is not too small or too large. A convenient value is $R_B = 100\,\Omega$.

To calculate the open-loop gain, we must first calculate r_e for Q_1 and Q_2, r_{e6}, and r_{e7}. These are $r_e = 2V_T/I_Q = 34.5\,\Omega$, $r_{e6} = V_T/I_{E6} = 130\,\Omega$, $r_{e7} = V_T/I_{E7} = 14.4\,\Omega$. Thus A is given by

$$\begin{aligned} A &= \frac{1}{2(34.5 + 71.6)} \times (-1860) \times \frac{3730}{130 + 3730}\left(\frac{1}{3730} + \frac{1}{14.4 + 53.8}\right) \times [7500\|75000] \\ &= 862.3 \end{aligned} \tag{21.41}$$

The output resistance of the circuit in Fig. 21.9 is given by the output resistance of the circuit with feedback removed divided by the amount of feedback. The output resistance of the circuit with feedback removed is $R_{C7}\|(R_{F1} + R_{F2})$, where we assume that the collector output resistance for Q_6 and Q_7 is $r_{ic} = \infty$. Thus the output resistance is given by

$$r_{\text{out}} = \frac{R_{C7}\|(R_{F1} + R_{F2})}{1 + bA} \tag{21.42}$$

We see from this expression that the larger the open-loop gain, the smaller the output resistance. The circuit clips on the positive cycle when Q_6 and Q_7 saturate. It clips on the negative cycle when Q_6 and Q_7 cut off. The clipping voltages are approximately given by

$$\begin{aligned} V_{+\text{clip}} &= V^+ \frac{R_{C7}\|R_L}{R_{E7} + R_{C7}\|R_L} + V^- \frac{R_{E7}\|R_L}{R_{C7} + R_{E7}\|R_L} \\ V_{-\text{clip}} &= V^- \frac{R_L}{R_L + R_{C7}} \end{aligned} \tag{21.43}$$

where R_L is the load resistance at the output.

Example 13 *Calculate the output resistance and clipping voltages for the op amp of Example 12 for $R_L = \infty$ and $R_L = 1\,\text{k}\Omega$.*

Solution. $r_{\text{out}} = [7500\|75000]/(1 + 862.3/10) = 78.2\,\Omega$. For $R_L = \infty$, $V_{+\text{clip}} = 15 \times 7500/(7553.8) - 15 \times 53.8/(7553.8) = 14.8\,\text{V}$ and $V_{-\text{clip}} = -15\,\text{V}$. For $R_L = 1\,\text{k}\Omega$, $V_{+\text{clip}} = 15 \times 882/936 - 15 \times 53.4/7550 = 14\,\text{V}$ and $V_{-\text{clip}} = -15 \times 1000/8500 = -1.76\,\text{V}$.

Example 14 *Calculate the closed-loop gain of the amplifier of Example 12 for $R_L = \infty$ and $R_L = 1\,\text{k}\Omega$.*

Solution. For $R_L = \infty$, the closed-loop gain is $v_o/v_i = A/(1 + bA) = 862.3/(1 + 86.23) = 9.89$. For $R_L = 1\,\text{k}\Omega$, the gain is reduced by the factor $R_L/(r_{\text{out}} + R_L) = 1000/1078.2 = 0.928$. Thus the gain is $9.89 \times 0.928 = 9.17$.

21.14. INTERMEDIATE CIRCUITS

The above examples illustrate how the performance of the circuit can change when a load resistor is added. To reduce this dependance on R_L, a complementary common-collector stage can be added between the output and the load as shown in Fig. 21.10. The CC stage provides current gain to isolate R_L from the output of the second stage, thus causing the gain and clipping voltages to be almost independent of R_L. Positive output current is supplied by Q_8 and negative output current is supplied by Q_9. The circuit has one problem, however. For $-V_\gamma \leq v'_O \leq +V_\gamma$, where V_γ is the turn-on or threshold voltage of Q_8 and Q_9, the output voltage is zero. Fig. 21.11(a) shows a plot of v_O versus v'_O for the circuit. There is a deadband region in the plot in which $v_O = 0$. Fig. 21.11(b) shows the plot of v_O versus time for v'_O a sine wave. It can be seen that the center portion of the waveform is clipped out. This is called center clipping or crossover distortion.

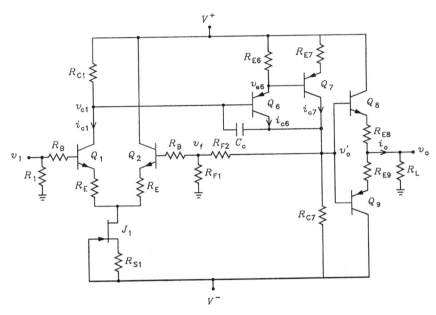

Figure 21.10: Op amp circuit with a complementary CC stage added at the output.

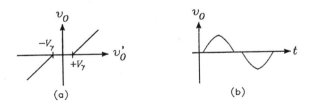

Figure 21.11: (a) Plot of v_O versus v'_O. (b) Plot of v_O versus time for v'_O a sine wave.

The crossover distortion can be reduced by connecting R_{F2} between the v_f and v_o nodes in Fig. 21.10. When v'_O is in the deadband region, i.e. $-V_\gamma \leq v'_O \leq +V_\gamma$, the circuit loses feedback and the gain increases from the closed-loop value to the open-loop value. This causes v'_O to change rapidly until either Q_8 or Q_9 cuts on. The circuit then has feedback and the gain decreases back to its closed-loop value. The crossover distortion is not completely eliminated by the feedback. To reduce it further, a bias voltage can be applied between the bases of Q_8 and Q_9 to cause these transistors to be quiescently on. Fig. 21.12 shows the circuit with a V_{BE} multiplier bias circuit connected between the bases of Q_8 and Q_9. If the base current in Q_{10} is neglected, its collector-emitter voltage is given by

$$V_{CE10} = \frac{V_{BE10}}{R_3} \times R_2 + V_{BE10} = V_{BE10}\left(1 + \frac{R_2}{R_3}\right) \tag{21.44}$$

Let I_E be the desired emitter bias current in Q_8 and Q_9. It follows that the required value of V_{CE10} is

$$V_{CE10} = V_{BE8} + I_E(R_{E8} + R_{E9}) + V_{EB9} \tag{21.45}$$

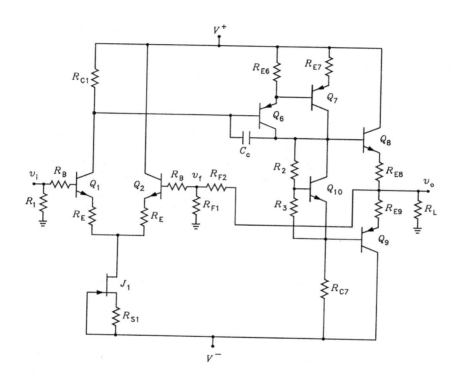

Figure 21.12: Circuit with the feedback connection changed and a V_{BE} multiplier added.

Example 15 *The quiescent current in Q_8 and Q_9 in the op amp of Fig. 21.12 is specified to be 2 mA. For $R_{E8} = R_{E9} = 100\,\Omega$, calculate the required voltage across Q_{10}. Calculate the values for R_2 and R_3 if the current through R_2 is 0.2 mA and $I_{C10} = 1.8$ mA. Neglect base currents and assume base-emitter junction voltages of 0.65 V.*

Solution. $V_{CE10} = 0.65 + 0.002(100 + 100) + 0.65 = 1.7\,\text{V}$, $R_3 = 0.65/0.0002 = 3.25\,\text{k}\Omega$, $R_2 = (1.7 - 0.65)/0.0002 = 5.25\,\text{k}\Omega$.

21.15 Completed Op-Amp Circuit

The completed circuit of the op amp is shown in Fig. 21.13. Compared to the circuit in Fig. 21.9, the resistive load on the diff amp has been replaced with a current mirror and the resistive load on the second

21.15. COMPLETED OP-AMP CIRCUIT

stage has been replaced with a current source. These changes cause the open loop gain to increase to a very high value. For all practical purposes, the open-loop gain is too high to measure with the feedback removed. It is for this reason that the resistive loads are used in the intermediate circuits. For the intermediate circuits, the second stage could have been realized by omitting the Q_6 common-collector stage with little change in performance. However, Q_6 is necessary in the circuit of Fig. 21.13 to provide an extra emitter-base junction voltage drop across Q_3. Without Q_6, Q_3 would be saturated.

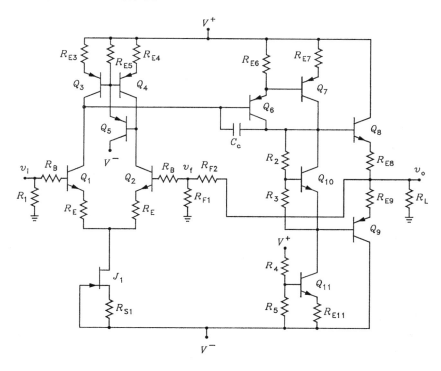

Figure 21.13: Completed op-amp circuit.

The current source load on the second stage is realized by Q_{11}. The collector current in Q_{11} is given by

$$I_{C11} = \frac{(V^+ - V^-)R_5/(R_4 + R_5) - V_{BE9}}{(R_4 \parallel R_5)/\beta + R_{E9}/\alpha} \simeq \frac{(V^+ - V^-)R_5/(R_4 + R_5) - V_{BE9}}{R_{E9}} \quad (21.46)$$

where the approximation assumes that $\beta \to \infty$. If the base currents in Q_8 and Q_9 are neglected, it follows from the circuit that $I_{C11} = I_{C6} + I_{C7}$. Symmetrical clipping of the op amp output voltage requires that $R_{E7} = R_{E11}$. In the design of the current source, the current through R_4 should be chosen to be at least 10 times the maximum expected base current in Q_{11}. For example, if $\beta_{11} \geq 100$, the current through R_4 might be chosen to be equal to $I_{C11}/10$. This makes it possible to neglect the base current in Q_9 in calculating R_4 and R_5. In this case we have

$$R_4 = \frac{V^+ - (V_{BE11} + I_{C11}R_{E11} + V^-)}{I_{R4}} \qquad R_5 = \frac{V_{BE11} + I_{C11}R_{E11}}{I_{R4}} \quad (21.47)$$

where I_{R4} is the current through R_4.

Example 16 *For $R_{E11} = 56\,\Omega$, use Eq. (21.47) to calculate the values of R_4 and R_5 for $I_{C11} = 2\,\text{mA}$ in the op amp of Example 12. Assume $\beta_{11} \geq 100$ and $V_{BE11} = 0.65\,\text{V}$.*

Solution. Let $I_{R4} = 0.2\,\text{mA}$. Thus $R_4 = [15 - (0.65 + 0.002 \times 56 - 15)]/0.0002 = 146.2\,\text{k}\Omega$ and $R_5 = (0.65 + 0.002 \times 56)/0.0002 = 3.81\,\text{k}\Omega$.

21.16 Preliminary Derivations

1. Derive an expression for the current in the resistor R_{E5} in Fig. 21.1. Assume that all of the base currents are zero and that the tail current of the diff-amp has been specified as I_Q.

2. Derive the relationship $f_x = A_o f_o$ from Fig. 21.2a. Hint: Both the vertical and horizontal axes are on a log scale.

21.17 Preliminary Calculations

The calculations are integrated throughout the procedure. The procedure is in three parts.

21.18 Preliminary SPICE Simulations

The SPICE simulations are integrated throughout the procedure. The procedure is in three parts.

21.19 Procedure Part 1, A Discrete Op Amp

The circuit shown in Fig. 21.9 will be designed, simulated, assembled, and analyzed in this procedure. Future procedures will examine refinements or elaborations of this shell discrete op amp circuit. This circuit features an input differential amplifier driving a second gain stage. Resistive loads are used on both the diff-amp and the second stage. A JFET current source is used to bias the diff-amp.

21.19.1 Determination of Emitter Resistor and Compensation Capacitor

To begin the design process some overall characteristics for this circuit must be specified; these are the slew rate and the gain-bandwidth product of the amplifier and the quiescent or dc bias current for the current source. These are selected as:

$$SR = 20 \quad \text{V}/\mu\text{s}$$
$$f_x = 5 \quad \text{MHz}$$
$$I_Q = 2 \quad \text{mA}$$

from which Eq. 21.14 can be used to determine the value of the compensating capacitor, C_c, and Eq. 21.7 can be used to determine the value of the emitter degeneration resistor, R_E.

21.19.2 Design of JFET Current Source

Now that the value of the bias current has been determined the JFET current source must be designed. Namely the resistor R_S in Fig. 21.9 must be determined. The current source is designed and assembled separately prior to the insertion into the discrete op amp circuit.

The circuit that is used to design the current source is shown in Fig. 21.14. The drain current will be measured with a digital multimeter configured as a digital ammeter.

The drain current for an N Channel JFET is given by

$$I_D = I_{DSS}\left[1 - \frac{V_{GS}}{V_P}\right]^2 \tag{21.48}$$

where I_D is the drain current, V_{GS} is the gate-to-source voltage, I_{DSS} is the drain-to-source saturation current and V_P is the pinchoff voltage. Both the pinchoff voltage and gate-to-source voltage, V_{GS}, are negative for the N Channel JFET.

21.19. PROCEDURE PART 1, A DISCRETE OP AMP

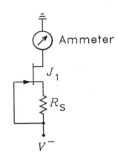

Figure 21.14: Current source circuit.

An alternative formulation for the drain current is

$$I_D = \beta (V_{GS} - V_{TO})^2 \qquad (21.49)$$

where $\beta = I_{DSS}/V_{TO}^2$ is the transconductance parameter and $V_{TO} = V_P$ is the threshold voltage. The parameter β (BETA) and V_{TO} (VTO) are needed for the SPICE simulation of the circuit.

Assemble the circuit shown in Fig. 21.14 using a potentiometer (aka pot) connected as a variable resistor for R_S. A 1 kΩ pot is preferable but a 10 kΩ pot will suffice. With $R_S = 0$ measure I_D and record this as I_{DSS}. Adjust R_S until $I_D = I_{DSS}/4$. Measure R_S and calculate V_{TO} from

$$V_{TO} = \frac{-I_{DSS} R_S}{2} \qquad (21.50)$$

Now use the values of I_{DSS} and V_{TO} to calculate the value of R_S to produce a drain current of $I_D = 2\,\mathrm{mA}$ using the formula

$$R_S = \frac{-V_{TO}}{I_D} \left[1 - \sqrt{\frac{I_D}{I_{DSS}}} \right] \qquad (21.51)$$

Vary the pot until the drain current is equal to the desired value of 2 mA and measure the value of R_S. Compare this with the value obtained in Eq. 21.51. Remove the pot and use the nearest standard resistor for R_S. Measure the value of the current again to assure that it is close to the design value of 2 mA. This current source will be used for the remainder of this and subsequent experiments to bias the diff-amp.

21.19.3 BJT Transistor Parameter Measurement

Matched Transistors

Use a transistor curve tracer to display the output characteristics (collector current versus collector-to-emitter voltage for stepped values of base current) of a number of NPN BJTs. Try to obtain two with reasonably matched characteristics to use as Q_1 and Q_2 in the diff-amp. Determine the current gain $\beta = I_C/I_B$ of the transistors to be used in the diff-amp.

The SPICE parameters for the BJT transistors are required for the simulations. These are the Early voltage, V_A (VA), forward current gain β_0 (BF), saturation current I_{S0} (IS), base-collector zero-bias depletion capacitance, c_{jco} (CJC), base-emitter zero-bias depletion capacitance, c_{jeo} (CJE), and forward transit time, τ_F (TF). This will require the output and transfer characteristics for each BJT transistor that is used. The procedure described is for NPN transistors; obvious interchanges of subscripts are required for PNP transistors.

Early Voltage, V_A

Use the output characteristic to determine the Early voltage as

$$V_A = \frac{I_C}{m} - V_{CB} = \frac{I_C}{m} - (V_{CE} - V_{BE}) \tag{21.52}$$

where m is the slope of one of the lines of constant base current is the region where it is flat

$$m = \left.\frac{\Delta I_C}{\Delta V_{CE}}\right|_{I_B=constnat} \tag{21.53}$$

The slope should be reasonably small and the Early voltage rather large. For PNP transistors the subscripts need to be reversed in Eq. 21.52. The Early voltage is the SPICE parameter VA. Normally the Early voltage for PNP is smaller than that for NPN transistors.

Forward Current Gain, β_0

Measure I_C/I_B for some convenient value of I_C and V_{CE}. The forward current gain β_0 is given by

$$\beta_0 = \frac{I_C/I_B}{1 + V_{CB}/V_A} \tag{21.54}$$

where V_{CB} is the quiescent value of the collector-to-base voltage. If this bias value is known the exact value for β_0 can be determined. An approximate value can be obtained by assuming that $V_A \gg V_{CB}$ which makes $\beta_0 \approx I_C/I_B$. The forward current gain is the SPICE parameter BF.

Saturation Current I_{S0}

Use a transistor curve tracer to display the transfer characteristic of the transistor. This is a plot of the collector current as a function of the base-to-emitter voltage. From the plot a convenient value of I_C and V_{BE} can be used to obtain I_{S0} from

$$I_{S0} = \frac{I_C}{1 + V_{CB}/V_A} \exp\left(-\frac{V_{BE}}{V_T}\right) \tag{21.55}$$

which is the SPICE parameter IS.

c_{jco}

The base-collector zero-bias depletion capacitance, c_{jco} is given by

$$c_{jco} = c_\mu \left[1 + \frac{V_{CB}}{\phi_c}\right]^{m_c} \tag{21.56}$$

where $\phi_c = 0.7\,\text{V}$, $m_c = 0.5$, $c_\mu = c_{cb}$. Although the parameter c_{cb} could be measured it would be difficult and, therefore, a value from transistor data sheets will be used. For 2N4401 NPN BJT $c_{cb} = 6.5\,\text{pF}$ @ $V_{CB} = 5\,\text{V}$, $I_E = 0$, and $f = 100\,\text{kHz}$. For 2N4403 PNP BJT $c_{cb} = 8.5\,\text{pF}$ @ $V_{BC} = 10\,\text{V}$, $I_E = 0$, and $f = 140\,\text{kHz}$. The capacitance c_{jco} is the SPICE parameter CJC.

τ_F

The forward transit time τ_F is given by

$$\tau_F = \frac{1}{2\pi f_T} \tag{21.57}$$

where f_T is the unity gain frequency $\approx$ gain bandwidth product of the transistor. From transistor data sheets $f_T = 250\,\text{MHz}$ for the 2N4401 NPN BJT and 200 MHz for the 2N4403 PNP BJT. The forward transit time τ_F is the SPICE parameter TF.

21.19. PROCEDURE PART 1, A DISCRETE OP AMP

c_{jeo}

The base-emitter zero-bias depletion capacitance c_{jeo} is given by

$$c_{jeo} = c_\pi \left[1 - \frac{V_{BE}}{\phi_e}\right]^{m_e} \tag{21.58}$$

where $V_{BE} = -\phi_e$, $m_e = 0.3$, and $c_\pi = c_{eb}$. The parameter c_{eb} could be measured but will be obtained from transistor data sheets. For 2N4401 NPN BJT $c_{eb} = 30\,\text{pF}$ @ $V_{BE} = 0.5\,\text{V}$, $I_C = 0$, and $f = 100\,\text{kHz}$. For 2N4403 PNP BJT $c_{eb} = 30\,\text{pF}$ @ $V_{EB} = 0.5\,\text{V}$, $I_C = 0$, and $f = 140\,\text{kHz}$. The parameter c_{jeo} is the SPICE parameter CJE.

21.19.4 Differential Amplifier

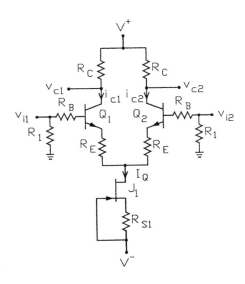

Figure 21.15: Differential amplifier.

- Assemble the differential amplifier shown in Fig. 21.15. Use $R_B = 1\,\text{k}\Omega$, $R_1 = 10\,\text{k}\Omega$, and $R_C = 3\,\text{k}\Omega$. Some of these resistors are temporary values which will change as circuits are added to the diff-amp.

- Ground v_{i1} and v_{i2} and measure the dc voltages in the circuit to verify that the bias currents are correct.

- Ground v_{i2} and apply a 1 kHz sine wave to v_{i1}. Use the oscilloscope to measure the ac output voltage at each collector. Calculate the ac collector current in each transistor and the ratio of this current to v_{i1} which is G_m. Compare the value to the theoretical value

$$G_m = \frac{1}{2(R_E + r_e)} \tag{21.59}$$

- Increase the amplitude of the ac input voltage until the outputs are just clipped on both the positive and negative peaks. The theoretical peak-to-peak clipping voltage at each output is $I_Q R_C$. Take measurements to verify this formula.